SYMBOLS FOR INSTRUCTOR'S CORRECTIONS

C — Show construction
D — Show dimensions; sh
I — Improve form or spac
H — Too heavy
NH — Not heavy enough
ND — Not dark enough
SL — Sharpen pencil or compass lead
GL — Use guide lines
A — Improve arrowheads

— Error in encircled area

SYMBOLS FOR USE ON DRAWINGS

TL — True Length
EV — Edge View
TS — True Size
LI — Line of Intersection

— Parallel

— Perpendicular

— Piercing Point of line and surface

Descriptive Geometry

Metric

MACMILLAN BOOKS BY THESE AUTHORS

Computer Graphics Projects for Design and Descriptive Geometry by E. G. Paré and M. Shook, 1985.

Descriptive Geometry: Metric, 7th ed., by E. G. Paré, R. O. Loving, I. L. Hill, and R. C. Paré, 1987

Descriptive Geometry Worksheets, Series A, 6th ed., by E. G. Paré, R. O. Loving, I. L. Hill, and R. C. Paré, 1986

Descriptive Geometry Worksheets, Series B, 5th ed., by E. G. Paré, R. O. Loving, I. L. Hill, and R. C. Paré, 1984

Engineering Graphics, 4th ed., by F. E. Giesecke, A. Mitchell, H. C. Spencer, I. L. Hill, R. O. Loving, and J. T. Dygdon, 1987

Engineering Graphics Problems, Series 1, 4th ed., by H. C. Spencer, I. L. Hill, R. O. Loving, and J. T. Dygdon, 1987

Technical Drawing, 8th ed., by F. E. Giesecke, A. Mitchell, H. C. Spencer, I. L. Hill, and J. T. Dygdon, 1986

Technical Drawing Problems, Series 1, 6th ed., by F. E. Giesecke, A. Mitchell, H. C. Spencer, I. L. Hill, and J. T. Dygdon, 1981

Technical Drawing Problems, Series 2, 4th ed., by H. C. Spencer, I. L. Hill, and J. T. Dygdon, 1980

Technical Drawing Problems, Series 3, 3rd ed., by H. C. Spencer, I. L. Hill, and J. T. Dygdon, 1980

World financial center architectural model. (© Wolfgang Hoyt/Esto.)

Descriptive Geometry *Metric*

7TH EDITION

E. G. Paré

Professor of Mechanical Engineering
Washington State University

R. O. Loving

Professor Emeritus of Engineering Graphics
and Former Chairman of the Department
Illinois Institute of Technology

I. L. Hill

Professor Emeritus of Engineering Graphics
and Former Chairman of the Department
Illinois Institute of Technology

R. C. Paré

Associate Professor of
Mechanical Engineering Technology
University of Houston

MACMILLAN PUBLISHING COMPANY
New York

COLLIER MACMILLAN PUBLISHERS
London

Macmillan Publishing Company
866 Third Avenue, New York, New York 10022

Collier Macmillan Canada, Inc.

Library of Congress Cataloging-in-Publication Data

Descriptive geometry.

    Includes index.
    1. Geometry, Descriptive.   I. Paré, E. G. (Eugene George)
QA501.D387   1987   516.6   86–18852
ISBN 0-02-391320-7 (Hardcover Edition)
ISBN 0-02-946530-3 (International Edition)

Printing: 1 2 3 4 5 6 7 8   Year: 7 8 9 0 1 2 3 4 5 6

ISBN 0-02-391320-7

Preface

In this text the authors have endeavored to fulfill the need for a descriptive geometry textbook in which the fundamentals of this graphic science are presented in the same pedagogically sound units of work as they are usually introduced in daily presentations. By division of the text material into relatively short, homogeneous chapters, convenient textbook reference is available to students. This simplified organization follows that of the authors' *Descriptive Geometry Worksheets*, Series A and B, in which new principles are introduced in order of need as well as difficulty.

Solution illustrations throughout the text have been broken into the necessary steps to make the construction easy to follow. Considerable care has been exercised to provide solutions in pictorial form whenever they can be used to aid visualization. Emphasis is focused on those applications which serve to illuminate fundamentals and to introduce new engineering experiences.

Conveniently located at the end of each chapter are abstract and laboratory problems based on the text material of the chapter unit. Chapter 24 contains condensed review material and many problems, the solutions of which entail combinations of principles. For the convenience of students and instructors, problems are given in layout form exactly as they are to be reproduced on the drawing paper. *All quantitative information is given in metric units.*

Continued in this seventh edition are full-size self-testing problems that permit the student to evaluate systematically his or her comprehension of pertinent fundamentals. These projects are presented at the conclusion of the chapters, and for maximum convenience and time economy they may be solved directly in the text. Carefully delineated solutions are shown in Appendix D.

It is anticipated that the numerous pictorial projects will accomplish dual objectives, for they serve not only to reinforce the related orthographic principles but, more importantly, stimulate the development of the student's depth perception and visualization ability, which are so essential to creative design.

This latest edition provides, in a new Chapter 25, a selection of Computer Graphics Projects that have been developed and tested at Washington State University and the University of Houston. Layout projects at the end of each chapter have been altered, and a few illustrations have been updated to reflect latest drafting standards.

The authors continue to acknowledge the valuable suggestions and assistance received from many colleagues and students. Additional suggestions and comments for the improvement of this text are always welcome.

E. G. P.
R. O. L.
I. L. H.
R. C. P.

Contents

Descriptive Geometry
Metric

1 Orthographic Projection

From the early beginning, the human race has endeavored to record and communicate ideas in the mind through the medium of visual illustrations. Today these illustrations may take the form of an artist's painting, a photographer's print, an engineer's pictorial or multiview sketch, or a designer's carefully made technical drawing. For a rapid and superficial representation of an idea, the pictorial type of drawing is often used. To provide the completeness of detail necessary for the design and construction of a machine, structure, device, or research apparatus, the technical drawing based on the principles of *orthographic projection* is employed. (See §1.5 for a definition of orthographic projection.)

1.1 Graphic Solutions

In addition to the communication of ideas, the principles of orthographic projection may be used for the solution of many engineering problems. Graphic solutions may be effectively employed when the required accuracy of the results falls within the limits of accuracy of the graphic method or of the original data, especially since a large proportion of the empirical data used in engineering calculations is considerably less precise than graphic methods. (See Appendix A for a discussion of graphical accuracy.)

The student should consider descriptive geometry problem solutions as if they were engineering or scientific reports. In this light, graphic solutions should be always as accurate as possible, absolutely clear and understandable without supplemental oral explanation. To accomplish this, the minimum amount of lettering and notation should include the following:

1. Identification in all views of isolated or unconnected points and at least *one* of the endpoints of isolated lines.
2. Identification of important construction points.
3. Identification in all views of at least *one* prominent point on planes or solid objects.

1

4. A clear indication by dimensions or notes of the uses of all given data and of the sources of all required data.
5. Indication of all *folding lines* employed in the problem solution (§1.10).

1.2   Descriptive Geometry

Descriptive geometry is the science of graphic representation and solution of space problems. The fundamentals of descriptive geometry are based on the principles of orthographic projection—the same principles employed in a basic course in graphics. In descriptive geometry the theory of orthographic projection is applied to the drawing-board representation and solution of engineering problems more advanced than those usually encountered in an elementary course in engineering graphics.

The majority of the problems investigated in this course will require solutions that are not obtainable in the usual three principal views of orthographic drawing: the front, top, and side views. For this reason many of the solutions will employ one or more auxiliary or revolved views. Although most problems presented in engineering and technical drawing courses are derived from the field of machine drawing, the problems included in a descriptive geometry course are selected from several technologies and engineering areas: civil, aeronautical, mechanical, chemical, electrical, architectural, and others.

Descriptive geometry is a versatile tool that the prospective engineer and technology student should learn to use broadly and skillfully. It is a means by which a great savings of time and effort can be made. The problems presented in this text have a careful balance of theoretical and practical applications so that the student not only will be fully trained in the fundamentals but also will be introduced to the extensive practical scope of this graphic method.

1.3   Projections

In order to represent an object by a line drawing on a plane, imaginary *projectors* emanating from various points on the object are extended until they pierce a *picture plane* or *projection plane*. The projectors may be thought of as *visual rays* extending from the object to an observer. The various piercing points are then connected with lines to form a *view* of the object.

Two common types of projection are *perspective* projection, Figure 1.1, and *orthographic* projection, Figure 1.2.

1.4 Perspective Projection

The basic theory of perspective projection is illustrated in Figure 1.1. The projectors emanate from points on the object and converge to an observer whose position is designated as the *station point*. The intersections of the projectors with the projection plane provide the framework for the pictorial of the house, closely resembling that which would be seen by an observer actually looking at the house from the same station point. The size of the pictorial will vary as the relative positions of the eye, the projection plane, and the object are altered.

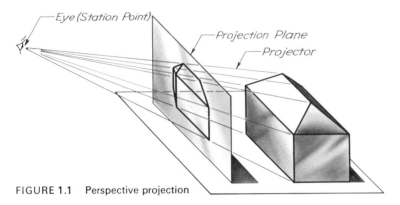

FIGURE 1.1   Perspective projection

Although perspective projection produces a realistic pictorial view of an object, its distortion of angles and distances prevents its meeting the exacting requirements demanded of a technical drawing. Perspective projection is used primarily by architects and commercial artists to describe in a general way the physical external appearance of a structure or a product. (See Chapter 20 for a more thorough treatment of perspective projection, including methods of construction.)

1.5 Orthographic Projection

Orthographic projection is a method of representing an object by a line drawing on a projection plane that is perpendicular to parallel projectors, Figure 1.2.

In contrast with perspective projection, it is important to note that in orthographic projection the size of the view of the object will not vary with the distance between the object and the projection plane.

The object in space may be turned and tilted, a procedure that results in an *axonometric* pictorial having foreshortened edges. This text is, however, primarily concerned with that class of orthographic projections called *multiview drawing*, in which the main purpose is to obtain views of an object on which true measurements can be made. Therefore, in

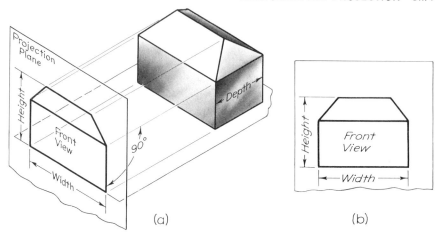

FIGURE 1.2   Orthographic projection

Figure 1.2 the front face is oriented parallel to the projection plane so that the established view shows the true *width* and *height* of the house. It will be noted, however, that the *depth* dimension is not shown in this *front* view. Thus a single orthographic view in itself cannot fully describe an object. An additional view on a projection plane perpendicular to the

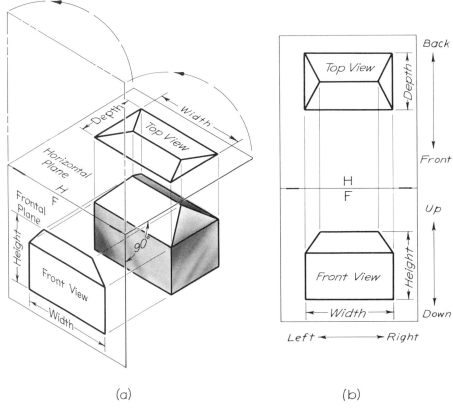

FIGURE 1.3   Multiview drawing

first is needed to give the depth of the house. Figure 1.3(a) illustrates the projection of the house on two perpendicular planes. It will be noted that the width dimension appears in both views while the height appears only in the front view and the depth only in the top view.

Since it would be indeed inconvenient to make or carry about a drawing on planes at right angles to each other, a conventional means of showing several views on a single plane has been developed.

Figure 1.3(a) suggests the manner in which the *horizontal* projection plane H is rotated into the same plane as the front view (the *frontal* plane, F). Or it may be assumed that the frontal projection plane is rotated into the plane of the top view. Figure 1.3(b) shows these two views in the resulting conventional arrangement as they would be drawn on paper. Two such directly related views of an object are called *adjacent* views.

1.6 Multiview Drawing

A multiview drawing is a systematic arrangement of orthographic views on a single plane (the drawing paper). The relationship of the views in the arrangement is based on the principle that *any two adjacent views lie on perpendicular planes of projection.* The two standard arrangements of views in general use are called *first-angle projection* and *third-angle projection.*

1.7 First-Angle Projection

In first-angle projection, Figure 1.4(a), the object is placed in quadrant I, formed by the intersection of a horizontal and a frontal plane. The top view of the object is obtained by projecting from the object *downward* to the horizontal plane, and the front view is obtained by projecting from the object *backward* to the frontal plane. The horizontal plane is then rotated downward into the frontal plane, resulting in the arrangement of views in Figure 1.4(b). Note that in first-angle projection the observer is always assumed to be looking *through the object* toward the projection plane. The resulting arrangement of six views is shown in Figure 1.5.

First-angle projection is used by many foreign countries for all types of engineering drawings, and this arrangement of views is sometimes employed in the United States for architectural and structural drawings.

1.8 Third-Angle Projection

In third-angle projection, Figures 1.3 and 1.6(a), the object is placed in quadrant III, formed by the intersection of a horizontal and a frontal

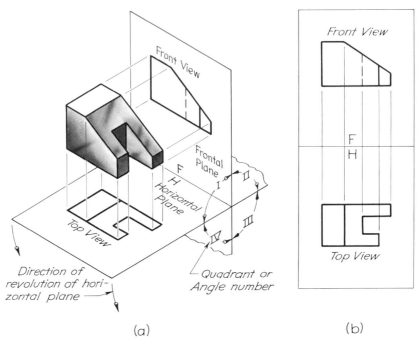

(a)   (b)

FIGURE 1.4   First-angle projection—two views

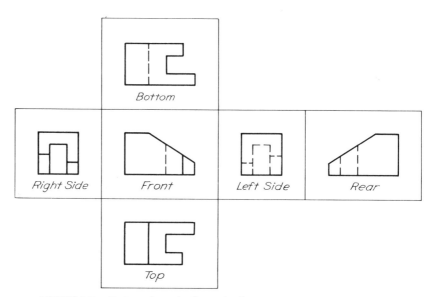

FIGURE 1.5   First-angle projection—six views

plane. The top view of the object is obtained by projection from the object *upward* to the horizontal plane, and the front view is obtained by projecting from the object *forward* to the frontal plane. The horizontal plane is then rotated upward into the frontal plane, resulting in the relative positions of views shown in Figure 1.6(b). Note that the observer is

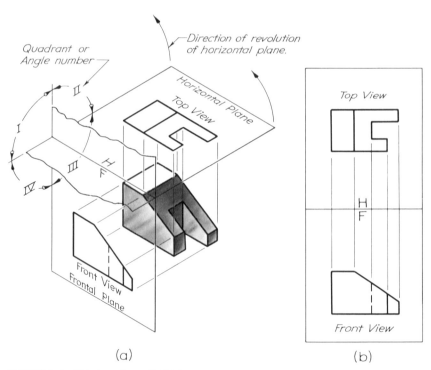

FIGURE 1.6   Third-angle projection

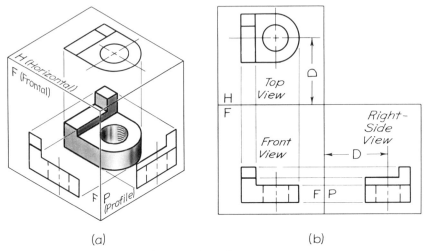

FIGURE 1.7   Third-angle projection—three views

always assumed to be looking *through the projection plane* toward the object. Compare this concept with that of first-angle projection.

Third-angle projection is used in the United States for practically all types of technical drawings.

Of course a great many technical drawings require more details than can be clearly indicated in only two views. A right-side (profile) view may be added by projecting from the object to a profile plane P as illustrated in Figure 1.7(a). The relative locations of the views on the drawing paper resulting from the rotation of the profile plane into the frontal plane are shown in Figure 1.7(b). Another acceptable but less frequently used arrangement of these views may be obtained by rotating the profile plane into the horizontal plane, Figure 1.8.

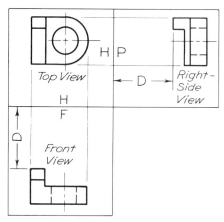

FIGURE 1.8   Alternative position for right-side view

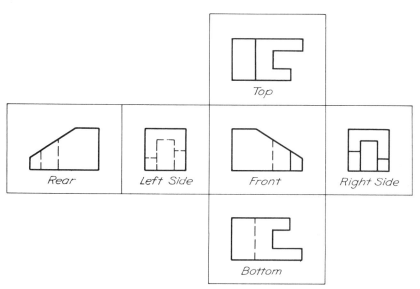

FIGURE 1.9   Third-angle projection—six basic views

Occasionally it becomes desirable on a technical drawing to use such views as a *left-side* view, a *bottom* view, or a *rear* view. These views, together with the top, front, and right-side views, constitute the six basic views, Figure 1.9. These are the views that may be obtained by projecting from an object onto the faces of an enclosing rectangular "projection box." Note that the rear view is adjacent to the left-side view.

1.9 Visualization

Traditionally, a view has been defined as the result of projecting from an object onto a projection plane. This is the historically basic theory of descriptive geometry resulting in the internationally understood arrangements of views.

For purposes of visualization, however, most engineers and technicians find that a more direct approach is to consider that each view is an actual picture of the object as seen with a line of sight perpendicular to the corresponding projection plane.

To obtain a front view of an object, the observer views the object from a location in front of the object, shifting position for each point so that the line of sight is always perpendicular to an imaginary frontal projection plane F, Figure 1.10(a).

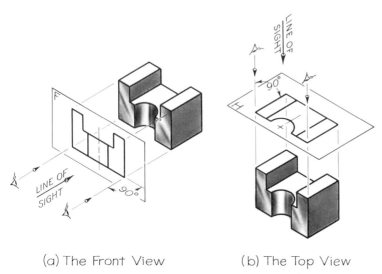

(a) The Front View   (b) The Top View

FIGURE 1.10   Obtaining views by the line-of-sight method

To procure a top view, the object is considered stationary, and the observer views the object looking down from a "bird's-eye view," Figure 1.10(b).

Other views may be obtained in a similar manner, with the observer changing position successively until a sufficient number of views is secured to describe adequately all the features of the object. For ease in construction and interpretation, the relative positions of these views on the drawing paper must always conform to the arrangement of views secured by third-angle projection.

To read a technical drawing, or let us say to *visualize* an object from its given orthographic views, the observer should always consider the views to be the object itself and should always be acutely aware of the position occupied by the observer relative to the object observed. Thus, when reading the front view, the observer must face the front of the object with the line of sight directed front to back. A glance at the front view discloses the height and width of the object, and in addition the location for the top, bottom, left-side, and right-side surfaces of the object.

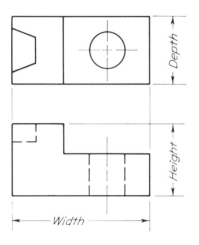

(a) Given: Front and Top Views

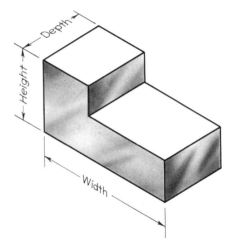

(b) Mental Picture of General Shape

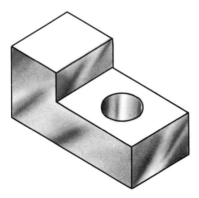

(c) Hole Added to Mental Picture

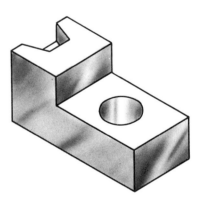

(d) Notch Added to Complete the Mental Picture

FIGURE 1.11   Visualization

The observer must never forget that the line of sight for the top view is directed *downward* in space toward the object so that depth and width dimensions appear in the top view while height dimensions are not available in this view. These principles enable the observer to build up a mental three-dimensional outline, Figure 1.11(a) and (b).

In Figure 1.11(a), a circle is observed in the top view. A study of the corresponding lines that represent this feature in the front view shows that the circle represents a hole rather than a protruding cylinder or boss. This feature is then added to the mental picture, Figure 1.11(c). Other features are similarly checked in adjacent views until a complete mental picture of the object is developed, Figure 1.11(d).

1.10 Folding Lines

Most technical drawings in industry do not include those lines between the views which represent the intersections of the projection planes (the lines marked H/F or F/P). However, since descriptive geometry constructions often include abstract forms, such as points, lines, and planes, the inclusion of these lines will be useful as shown later.

The lines of intersection of the mutually perpendicular projection planes are referred to as *folding lines*. The folding line between the front and top views is labeled H/F as shown in Figure 1.12(a) and (b), the F indicating the frontal plane and the H the horizontal plane. *It should be noted that when the front view is examined, this folding line represents the edge view of the horizontal plane, Figure 1.12(b). When the top view is studied, the same folding line represents the edge view of the frontal plane.*

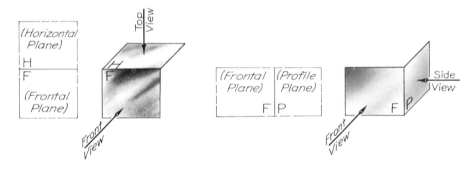

FIGURE 1.12   The H/F folding line   FIGURE 1.13   The F/P folding line

The folding line between the front and side views is labeled F/P as shown in Figure 1.13(a) and (b), the P indicating the profile plane. When the front view is examined, this folding line represents the edge view of the profile plane; when the side view is examined, the folding line represents the edge view of the frontal plane, Figure 1.13(b).

The use of folding lines serves as a reminder that adjacent views lie in projection planes that were at right angles to each other in space before rotation into the plane of the paper.

1.11   Views of a Point

Fundamentally a mental picture of an object is created or read from given views by a point-by-point analysis: two points locating a line, lines defining surfaces, and finally surfaces combining to form objects. Consequently, a logical starting place for a thorough investigation of the theory of descriptive geometry is an analysis of the nature of and relationship between views of a single isolated point.

Theoretically a *point* has location only and no dimensions. For accuracy an isolated point is best indicated on the drawing by fine inter-secting dashes, as shown in Figure 1.14, rather than by a dot.

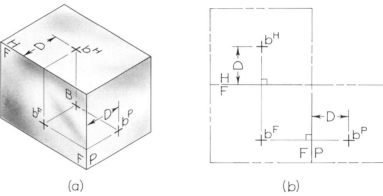

(a)   (b)

FIGURE 1.14   Three views of a point

In Figure 1.14 three views of a point B are shown. Since the top and profile views of any object lie in planes that are perpendicular to the frontal plane before rotation into the frontal plane, the top and profile projections of point B are the same distance D from the frontal plane, Figure 1.14(a). It follows that the projections of that point in the top and profile views must be the same distance D from the folding lines representing edge views of the frontal plane, Figure 1.14(b).

Note that the point in *space* is designated by a capital letter B while each of its views is identified with the lowercase letter plus the appropriate superscript: b^F, b^H, and b^P. An illustration showing typical lettering and line weights is printed on the inside of the front cover. Note that the superscripts which identify the view need be shown only once per view.

1.12   Views of a Line

In geometry a *line* theoretically has no width. In practice, of course, lines are drawn with various widths according to established conventions. A line may be considered to consist of an infinite number of collinear points. The position of a *straight* line is established by locating any two noncoincident points on the line. For drafting accuracy, points should be selected that are separated by an appreciable distance (Appendix A.2).

To obtain a view of a line, the views of the endpoints of the line may be established as illustrated in the following problem.

Problem

Add a profile view of a line AB, given the top and front views, Figure 1.15(a).

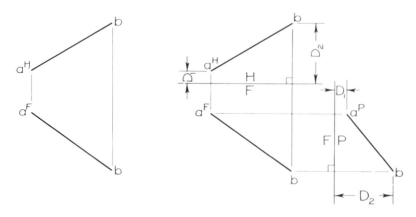

(a) Two Views Given   (b) Right-Side View Constructed

FIGURE 1.15   Construction of a third view of a line

Analysis

A straight-line segment is established by locating its endpoints. The projections of the endpoints in the front and profile views will be at the same elevation. The projections of these points in the top and profile views will be the same distance behind the frontal plane.

Graphic Solution

The folding line F/H is inserted in any convenient position between the top and front views and perpendicular to the projection lines connecting them, Figure 1.15(b). The folding line F/P is added at any convenient distance to the right of the front view and perpendicular to the projection lines from the front view to the side view. The elevations of the points for the profile projection are obtained by extending horizontal projection lines from the front view to the profile view. The measurements D_1 and D_2, which represent the distances the points lie behind the frontal plane, are then transferred with dividers from the top view to the profile view. The line joining the projections of the two points is the required profile view of the line AB.

A similar procedure may be followed in a problem that requires the addition of a top view of a line having given the front and profile views. Drawing an additional view of a surface or solid object entails successive repetition of this procedure to locate all points and lines in turn.

1.13   Visibility

An essential step in drawing an orthographic view is the correct determination of the visibility of the lines that make up the view. The *outline*
of a view will always be visible, but the lines within the outline may be
visible or hidden, depending on the relative positions of those lines with
respect to the line of sight.

Frequently the visibility of lines may be determined by inspection,
such as those in Figure 1.16(a). The screened lines in the top view will be
visible or hidden, depending upon the visibility of point O from which
all these lines emanate.

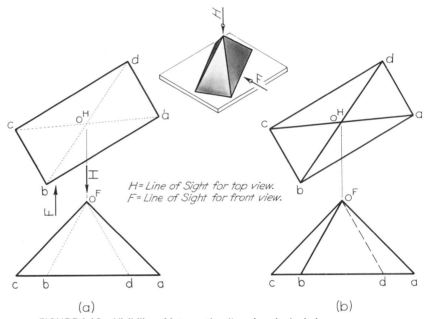

H = Line of Sight for top view.
F = Line of Sight for front view.

FIGURE 1.16   Visibility of intersecting lines in principal views

The line of sight for a top view is directed toward the object from
above. Thus, on the drawing, if the front view is regarded as the object
itself, the line of sight for the top view may be represented by an arrow
pointing downward, as indicated by arrow H. Since o^F is the point
nearest arrow H, point O must be visible in the top view. Therefore, the
four edges o^Ha, o^Hb, o^Hc, and o^Hd are visible in the top view, Figure
1.16(b).

The line of sight for a front view is directed toward the front of the
object. Thus, on the drawing, if the top view is now regarded as the object, the line of sight for the front view is represented by an arrow pointing toward the top view as indicated by arrow F, Figure 1.16(a). Since
edge o^Hb is nearest arrow F, o^Fb is visible, Figure 1.16(b). Since edge
o^Hd is behind the body of the object, o^Fd is hidden.

The determination of the visibility of nonintersecting lines, such as

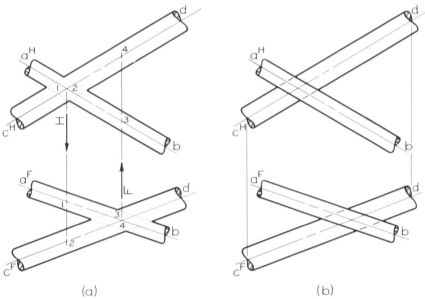

FIGURE 1.17 Visibility of nonintersecting rods in principal views

those illustrated by two rods AB and CD in Figure 1.17(a), requires a more detailed study. Inspection reveals that the apparent points of crossing of the rods in the top view and in the front view are not in vertical alignment. Therefore, the rods do not intersect in space. Since the rods are nonintersecting, one of the rods must be above the other at the apparent point of crossing in the top view. Thus the apparent point of intersection of center lines a^Hb and c^Hd in the top view actually represents *two* points, one point on rod AB and the other point on rod CD. These are located on the respective lines in the front view, where it can be observed that point 1 on a^Fb is at a higher elevation than point 2 on c^Fd; that is, point 1 is nearer to line of sight H than is point 2. Since rod AB is thus higher than rod CD at this location, it is nearer to an observer looking downward at the object, as indicated by line of sight H. Thus rod AB is completely visible in the top view, and rod CD is hidden where it passes below AB, Figure 1.17(b).

The apparent point of crossing in the front view, Figure 1.17(a), also represents two points, which are labeled 3 and 4 for convenience. These points are then located in the top view, where it can be seen that point 3 on a^Hb is nearer an observer represented by the line of sight F than is point 4 on c^Hd. Thus rod AB is completely visible in the front view and rod CD is hidden where it passes behind AB, Figure 1.17(b).

The preceding discussion of visibility has been in terms of the top and front views. However, the intersection and visibility in any two adjacent views may be determined by using the identical approach, because any two adjacent views can be reoriented by turning the drawing paper or by shifting the view point so that the two views assume the relative positions of front and top views.

　　In the two successive views of Figure 1.18, it is established that the rods are nonintersecting. Consequently, the apparent point of crossing of center lines e^1g and j^1k in view 1 is labeled 3 and 4, and then these two points are located on the center lines in view 2. In third-angle projection *the line of sight for a view is always directed from that view toward the adjacent view.* Therefore, the line of sight for view 1 in Figure 1.18 is toward view 2, as indicated by arrow 1. Thus point 3 in view 2 is nearer the observer looking along arrow 1, and rod EG is completely visible in view 1. The visibility in view 2 may be determined similarly by observing the positions of points 5 and 6 in view 1 in relation to the line of sight 2.

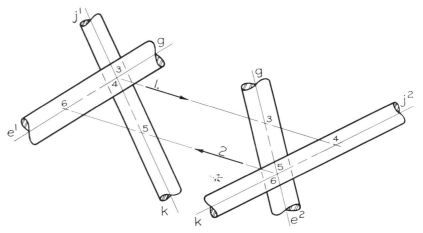

FIGURE 1.18　Visibility of nonintersecting rods in any two adjacent views

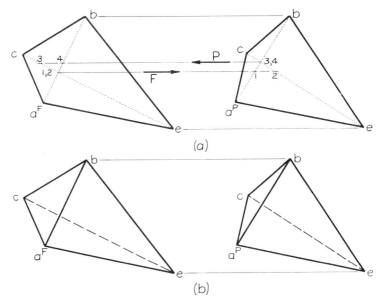

(a)

(b)

FIGURE 1.19　Visibility of nonintersecting lines

The foregoing principles may also be applied to the determination of visibility of the interior lines in the views of the *tetrahedron* (see Appendix C.1) in Figure 1.19(a). The apparent point of crossing of $a^F b$ and ce is labeled 1 and 2, and these points are located in view P. Since point 1 on $a^P b$ is nearer the observer (line of sight F), line AB is visible in the front view, Figure 1.19(b). In similar fashion, observation of the relative positions of points 3 and 4 and line of sight P reveals that line AB is visible in the profile view.

In comparing Figures 1.17, 1.18, and 1.19, it should be observed that the visibility must be determined independently for each view; that is, the visibility of one view is not necessarily a clue to the visibility in the other view.

1.14 Problem Layouts

Although this text contains some theoretical descriptive geometry problems, the emphasis is focused on those practical problems which serve both to teach the basic fundamentals and to introduce a variety of new engineering experiences.

At the end of each chapter, problem material will be found that is based on principles previously introduced. Chapter 24, "Review," contains problems whose solutions entail a combination of principles.

Each worksheet problem presented in this text is accompanied by a layout illustration. The layout dimensions are so stated and the space requirements of the solutions so planned that any conventional border

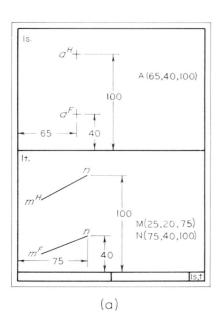
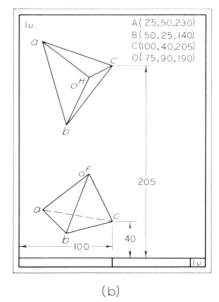

(a)   (b)

FIGURE 1.20   The coordinate system

on an A4 (210 mm × 297 mm), 8.5″ × 11.0″, 9.0″ × 12.0″, or similar size sheet will prove satisfactory.

Some data are given exactly by location dimensions or by a coordinate system; others are purposely presented so that the student is required to plan the spacing. *Coordinate dimensions are given in millimeters and are always to full scale, even if the scale of the problem is otherwise specified.* It will be noted that the "origin" of all coordinate dimensions is the lower left corner of the working space. The first coordinate locates the views of the point from the left border. The second and third coordinates establish the respective positions of the front and top views of the point from the bottom border line of the working space. Figure 1.20(a) demonstrates the location of a point A and of point N of a line MN on a divided sheet; Figure 1.20(b) shows the location of the views of point C of a solid on a full sheet.

The amount of lettering the student should include in a solution is, of course, at the discretion of the instructor (or see §1.1). The notation in the illustrations throughout this book is shown in an extensive form for ease of correlation with the text material. A representative illustration is printed for reference on the inside of the front cover.

PROBLEMS

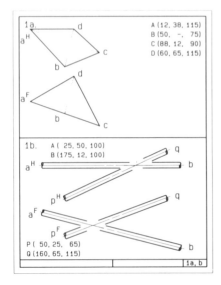

1a.
A (12, 38, 115)
B (50, –, 75)
C (88, 12, 90)
D (60, 65, 115)

1b.
A (25, 50, 100)
B (175, 12, 100)
P (50, 25, 65)
Q (160, 65, 115)

1a, b

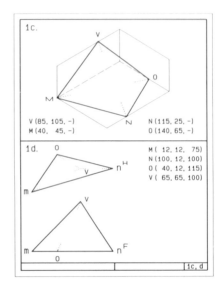

1c.
V (85, 105, –)
M (40, 45, –)
N (115, 25, –)
O (140, 65, –)

1d.
M (12, 12, 75)
N (100, 12, 100)
O (40, 12, 115)
V (65, 65, 100)

1c, d

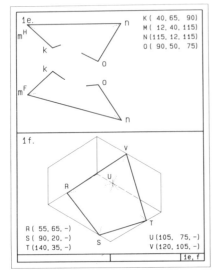

1e.
K (40, 65, 90)
M (12, 40, 115)
N (115, 12, 115)
O (90, 50, 75)

1f.
R (55, 65, –)
S (90, 20, –)
T (140, 35, –)
U (105, 75, –)
V (120, 105, –)

1e, f

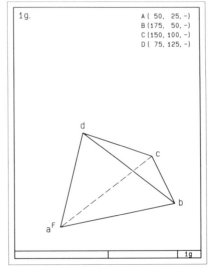

1g.
A (50, 25, –)
B (175, 50, –)
C (150, 100, –)
D (75, 125, –)

1g

1a. Complete the visibility of the pyramid and add a right-side view.

1b. Show the appropriate visibility of the two nonintersecting 6 mm DIA rods.

1c. Complete the isometric drawing of the tetrahedron VMNO including the visibility of the missing edges.

1d. Complete the visibility of the pyramid and add a right-side view.

1e. Complete the visibility of the two planes that have the line MN in common. Add a right-side view.

1f. Complete the isometric drawing of the pyramid' VRSTU including the visibility of the missing edges.

1g. Design a top view of the tetrahedron consistent with the given front view.

19

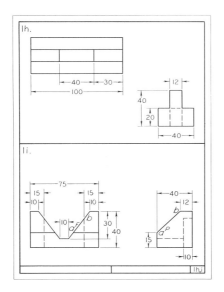

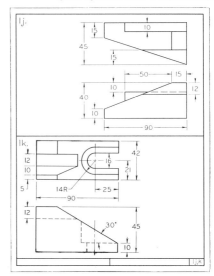

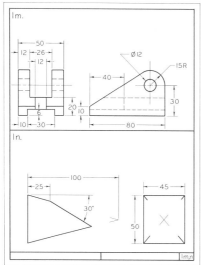

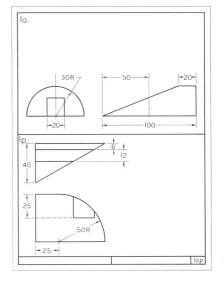

1h. Duplicate the given views and add a front view. Omit dimensions unless assigned.

1i. Duplicate the given views and add a top view. Label A and B in each view.

1j. Duplicate the given views and add a left-side view.

1k. Duplicate the given views and add a right-side view.

1m. Duplicate the given views and add a top view.

1n. Complete the given views of the truncated pyramid and add a top view.

1o. Duplicate the given views and add a top view.

1p. Duplicate the given views and add a right-side view.

1q. Indicate whether the following statements are true or false. If assigned, provide written statements or sketches to justify your answers.

(1) A horizontal plane appears edgewise in a top view.

(2) A front view shows the height and depth of an object.

(3) First-angle projection is obsolete everywhere.

(4) Third-angle projection pertains to the fact that three views of an object are used on the drawing.

(5) In a front view a profile plane appears edgewise.

(6) A folding line is the line of intersection of two adjacent projection planes.

(7) The visibility of the lines in a view is independent of the relative orientation of the lines in any other view.

(8) If a line is visible in a top view, it must also be visible in a profile view.

SELF-TESTING PROBLEMS

The following full-size problems and those found at the ends of subsequent chapters are intended to be solved directly in the text as a self-testing measure in order that students may have an immediate evaluation of their comprehension of the pertinent fundamentals. Solutions for these problems are given in Appendix D.

1A. In the given views change the screened lines to visible or hidden lines to indicate correctly the visibility of the lines of the tetrahedron. Then add a right-side view.

1B. In the isometric pictorial of an intersecting plane ABCE and pyramid VMNK, establish the visibility by changing the screened lines to visible or hidden lines. Disregard the visibility of the isometric "box."

I A.

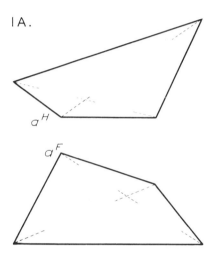

I B.

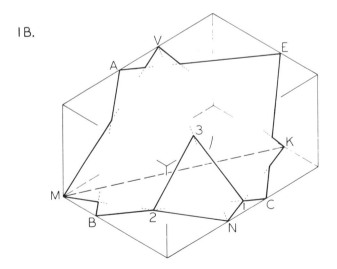

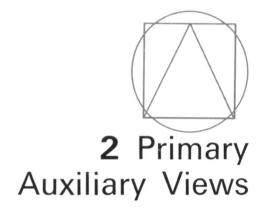

2 Primary Auxiliary Views

Although the essential details of most objects can be shown in front, top, and profile views, many of the more complex problems with which this text is concerned require for their solutions the use of *auxiliary* views. *An auxiliary view is a view projected on any plane other than one of the three principal planes of projection* (frontal, horizontal, or profile).

A *primary* auxiliary view is obtained by projection on a plane that is perpendicular to one of the three principal planes of projection and is inclined to the remaining two.

2.1 Views Projected from the Top View

An *auxiliary elevation* view, illustrated pictorially in Figure 2.1(a), is obtained by projection on a plane that is perpendicular to the horizontal plane of projection and inclined to the profile and frontal planes of projection. The procedure for constructing an auxiliary view is essentially the same as for drawing any additional basic view. This similarity should be particularly noted in the following example.

With the top and front views of the house drawn in conventional multiview arrangement as in Figure 2.1(b), the steps to obtain a profile view and an auxiliary elevation view follow:

Step 1. Establish the line of sight.

> To draw any additional view, it is first necessary to determine that line of sight which will establish a view showing the desired information. In this case let it be assumed that the desired lines of sight are P (for a right-side view) and 1 (for an auxiliary elevation view), as indicated in Figure 2.1(b). The projection lines for the new views are drawn parallel to the respective lines of sight, Figure 2.1(c).

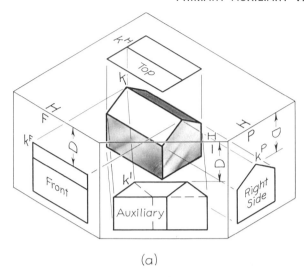

(a)

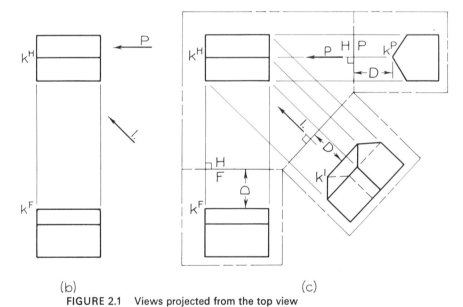

(b)   (c)

FIGURE 2.1   Views projected from the top view

Step 2.   Introduce the necessary folding lines.

One folding line H/F is drawn between the given adjacent views and perpendicular to the projection lines joining them.

Folding lines represent the edge views of projection planes (§1.10). It is a fundamental principle of multiview projection that the plane of projection for a view must be perpendicular to the line of sight for that view. Consequently, the remaining folding lines H/P and H/1 are drawn perpendicular to the established lines of sight P and 1, respectively, at a convenient distance from the top view.

Step 3.   Transfer distances to the new view.

> Since in this case the front view, view P, and view 1 lie on three separate planes perpendicular to the horizontal projection plane, the corresponding points on these three views of the house lie the same distances below the horizontal plane. Therefore, the measurements to be transferred are obtained from the front view. For example, see dimension D, which is the distance point K lies below the horizontal plane. These measurements are transferred as indicated to the corresponding projection lines in each of the new views.

Step 4.   Determine the visibility and complete the views.

These four steps may be used to draw any number of additional auxiliary elevation views. For example, see those obtained by the lines of sight 2 and 3 in Figure 2.2. It should be noted as a general principle that *all views projected from a top view contain identical height dimensions.*

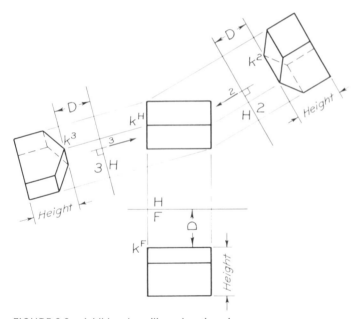

FIGURE 2.2   Additional auxiliary elevation views

2.2   Views Projected from the Front View

Another related group of primary auxiliary views may be obtained by projection on planes perpendicular to the frontal plane and inclined to the horizontal and profile planes, Figure 2.3(a).

The same four steps used in the previous example are again employed

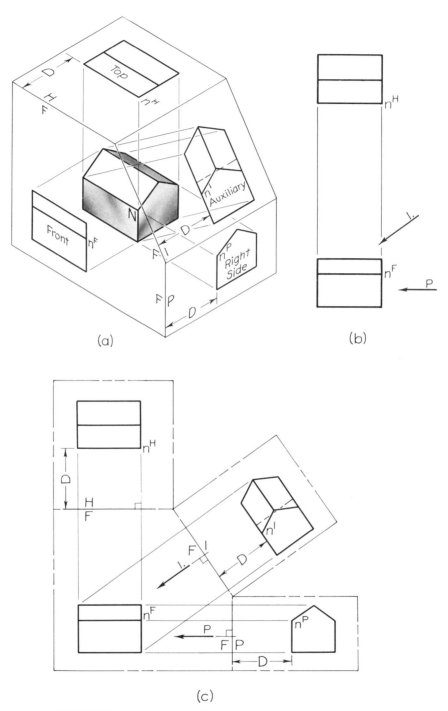

(a)

(b)

(c)

FIGURE 2.3   Views projected from the front view

to obtain the right-side view P and the auxiliary view 1 of the house, given the top and front views, Figure 2.3(b).

Step 1. Establish the line of sight.

In this case the lines of sight P and 1 are given, establishing the directions of the projection lines.

Step 2. Introduce the necessary folding lines.

Folding line H/F is drawn between the given views and perpendicular to the projection lines joining them, Figure 2.3(c). The other folding lines F/P and F/1 are drawn perpendicular to the established lines of sight P and 1, respectively, at a convenient distance from the front view.

Step 3. Transfer distances to the new view.

Since in this case the top view, view P, and view 1 lie on three separate planes perpendicular to the frontal projection plane, the corresponding points on these three views of the house lie the same distances behind the frontal plane. Therefore, the measurements to be transferred are obtained from the top view. For example, see dimension D, which is the distance point N lies behind the frontal plane. These measurements are transferred as indicated to the corresponding projection lines in each of the new views.

Step 4. Determine the visibility and complete the views.

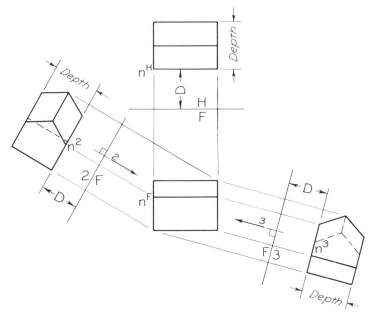

FIGURE 2.4 Additional auxiliary views projected from the front view

These four steps may be used to draw any number of additional auxiliary views projected from the front view. For example, see those obtained by the lines of sight 2 and 3, Figure 2.4. *All views projected from a front view contain common depth dimensions.*

2.3   Views Projected from the Side View

Still another related group of primary auxiliary views may be obtained by projection on planes perpendicular to the profile plane and inclined to the horizontal and frontal planes, Figure 2.5(a).

The following steps used in the two previous examples are again employed to obtain the auxiliary view 1 of the house, given the front and side views, Figure 2.5(b).

Step 1.   Establish the line of sight.

In this case the line of sight 1 is given, establishing the direction of the projection lines.

Step 2.   Introduce the necessary folding lines.

One folding line F/P is drawn between the given views and perpendicular to the projection lines joining them. The other folding line P/1 is drawn perpendicular to the established line of sight 1 at a convenient distance from the side view.

Step 3.   Transfer distances to the new view.

Since in this case the front view and view 1 both lie on separate planes perpendicular to the profile projection plane, the corresponding points on these two views of the house lie the same distances to the left of the profile plane. Therefore, the measurements to be transferred are obtained from the front view. For example, see dimension D, which is the distance point M lies to the left of the profile plane. The measurements are transferred as indicated to the corresponding projection lines in the new view.

Step 4.   Determine the visibility and complete the view.

These four steps may be used to draw any number of additional auxiliary views projected from the profile view. For example, see those obtained by the lines of sight 2 and 3 in Figure 2.6. *All views projected from a side view contain common width dimensions.*

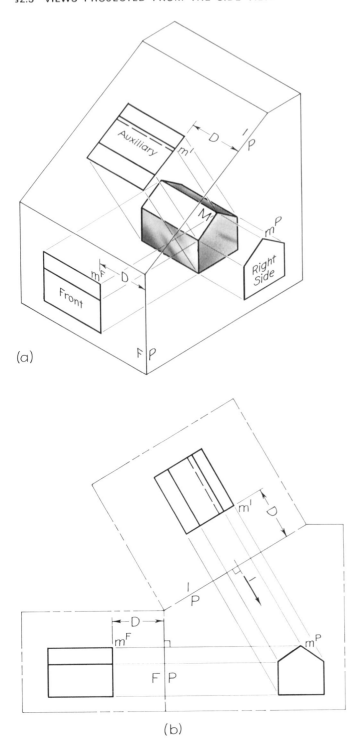

FIGURE 2.5   Views projected from the side view

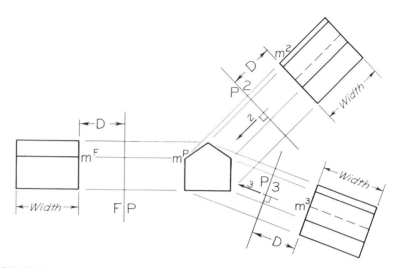

FIGURE 2.6   Additional auxiliary views projected from the side view

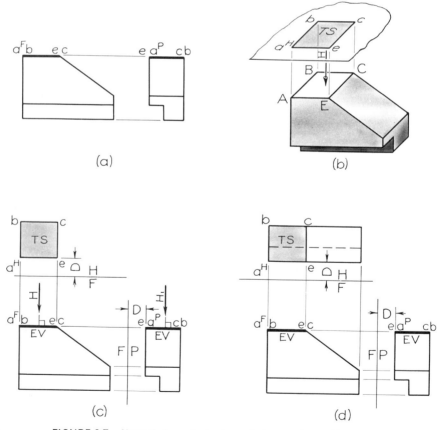

(a)

(b)

(c)

(d)

FIGURE 2.7   Normal view of a horizontal plane surface with edge view given

2.4 Normal View of a Plane with Edge View Given

The preceding discussion in this chapter has dealt with the basic mechanics of drawing primary auxiliary views. The following material is concerned with one of the most frequent applications of such views.

A plane will be shown in its true size and shape, TS, *in a view for which the line of sight is normal* (perpendicular) *to that plane.* For example, let it be assumed that in Figure 2.7(a) a view is required showing the true size and shape of surface ABCE. The line of sight H for the required view must then be perpendicular to the surface ABCE as shown pictorially in Figure 2.7(b). Since the *edge view*, EV, of surface ABCE shows as the horizontal line a^Fc in the front view, the line of sight H appears perpendicular to a^Fc, Figure 2.7(c). Folding lines F/H and F/P and transfer distances such as D are then used to construct the required true size and shape view of surface ABCE. The resulting top view a^Hbce is called a *partial* top view, since only the surface ABCE is shown. The complete top view is drawn in Figure 2.7(d).

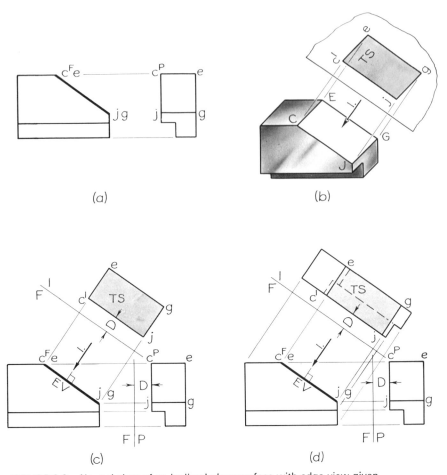

(a)   (b)

(c)   (d)

FIGURE 2.8   Normal view of an inclined plane surface with edge view given

It should be noted in Figure 2.7(c) that a second line of sight H_1 is also normal to an edge view, eb, of the surface ABCE. The required normal view could therefore be positioned above arrow H_1 instead of above arrow H, although this is not common practice.

A normal view of surface CEGJ, Figure 2.8(a), may be obtained by following the same basic procedure as in the preceding example. Line of sight 1 is established perpendicular to the edge view of surface CEGJ, represented by line $c^F g$, Figure 2.8(c). The resulting view in Figure 2.8(c) is a partial auxiliary view showing the true size and shape of the surface.

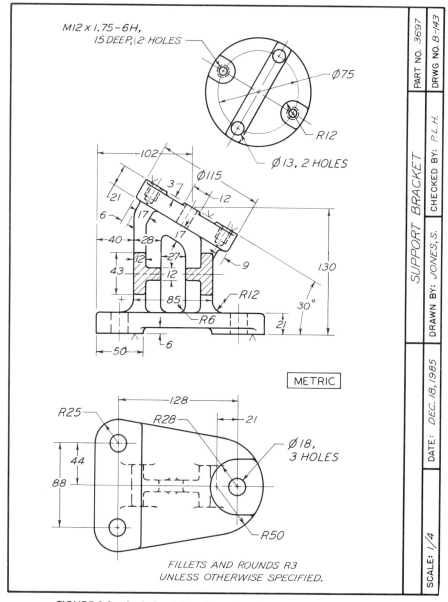

FIGURE 2.9   An industrial working drawing employing a primary auxiliary view

The remaining points on the object may be projected to this view to obtain the complete auxiliary view, Figure 2.8(d).

In practice only the essential features of each view are included unless ambiguity would result from the omission of the remainder of the view. Note in Figure 2.9, which is a working drawing from industry, that both the auxiliary view and the bottom view are partial views. Elliptical curves and a great many hidden lines are omitted in these views, since their inclusion would merely create confusion and increase the drafting time required to make the drawing without clarifying the shape description.

2.5 Plotting a Curve by Projection

Since circles and cylinders are among the most common geometrical shapes in industrial design, it frequently occurs that a circular feature is viewed obliquely, appearing elliptical in the resulting view. This is likely to occur in the construction of auxiliary views. It may also happen that a cylindrical or other curved shape is cut, or *truncated*, in an inclined direction, resulting in an elliptical or other noncircular line of intersection. The construction of such a curve entails the location of a series of appropriately spaced points along the curve, Figure 2.10. It should be noted that the principles involved in locating the points are in no way different from those used in constructing any other portions of the views. Thus the points may be located by transferring distances from the folding line F/P to the corresponding projection lines in the auxiliary view 1.

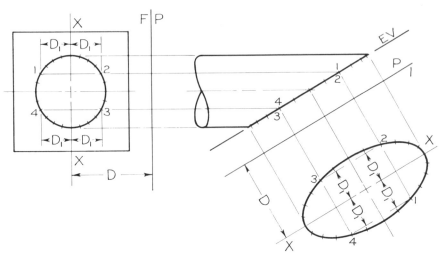

FIGURE 2.10 Plotting a curve in a primary auxiliary view

If the object is symmetrical, as in Figure 2.10, it may be more convenient to establish first the axis of symmetry, center line X-X. A single distance D_1 may then be used to locate four points in turn, as indicated for points 1, 2, 3, and 4. Repetition of this process rapidly establishes enough points for the drawing of a smooth curve.

PROBLEMS

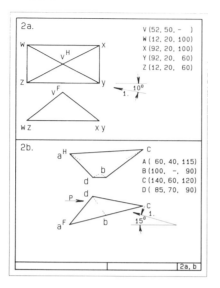

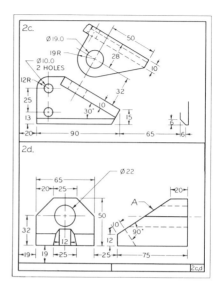

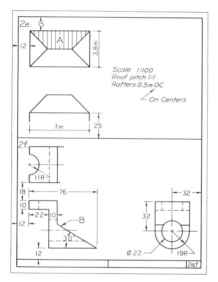

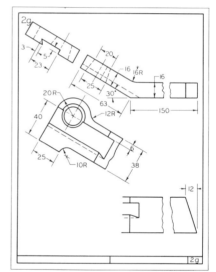

2a. Add the auxiliary view of the pyramid as indicated by arrow 1.

2b. Complete the visibility of the tetrahedron and add views P and 1.

2c. Complete the right-side view of the *angle bracket*.

2d. Draw a partial auxiliary view that shows the true size of surface A of the *locating slide*.

2e. Add the views necessary to show the true size of roof plane A and the true lengths of the rafters.

2f. Draw a partial auxiliary view showing the true shape of surface B and complete the top view.

2g. Complete the front view of the *tool holder*.

34

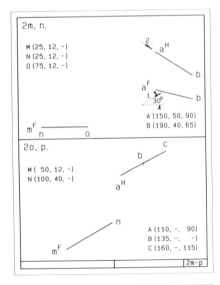

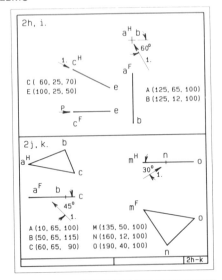

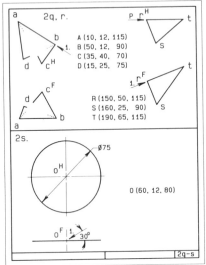

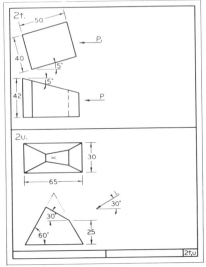

2h. Add view 1 of the vertical line AB.

2i. Add views P and 1 of the horizontal line CE.

2j. Add view 1 of the frontal triangle MNO.

2k. Add auxiliary view 1 of the horizontal triangle ABC.

2m. Add views 1 and 2 of the oblique line AB.

2n. Add a top, true-size view of the 30–60° triangle MNO.

2o. Add a true-size view of the 45° triangle ABC.

2p. Add a true-size view of an equilateral triangle MNO that appears edgewise in the given front view. Locate point O in the front view.

2q. Draw the views P and 1.

2r. Complete the visibility of the given views of the intersecting planes and then add view 1.

2s. Add the indicated auxiliary view 1.

2t. Add the views P and P_1.

2u. Add the auxiliary view 1.

36

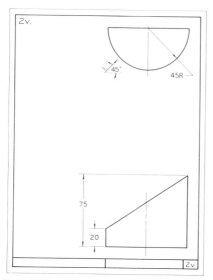

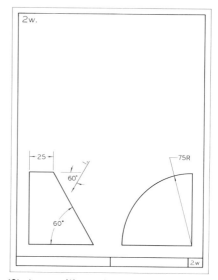

2v. Add the complete auxiliary view 1.

2w. Add the complete auxiliary view 1.

2x. Indicate whether the following statements are true or false. If assigned, provide a written statement or sketch to justify the answers.

(1) All views projected from a top view contain common height dimensions.

(2) An auxiliary view cannot show a line as a point.

(3) An auxiliary view may be used to show the true size of a plane surface.

(4) Projectors for an auxiliary view are always drawn at 30°, 60°, or 45° with the horizontal.

(5) A circle always appears true shape in an auxiliary view.

(6) The folding line for a new view is always perpendicular to the projection lines for the new view.

SELF-TESTING PROBLEMS

2A. Add the auxiliary views 1 and 2. Label views that show the plane either in edge view or true size.

2B. Add a partial auxiliary view that shows the true size of the base plane of the pyramid. Then add a complete auxiliary view that includes the true size of surface N.

2A.

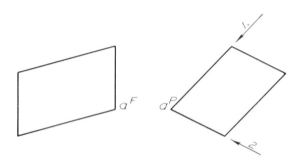

· 2B.

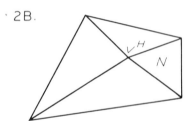

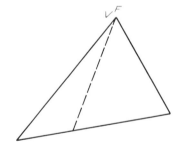

2C. The curve from point A to point C lies in the inclined plane ABCE of the isometric pictorial. Provide appropriate construction and the orthographic views needed to show the true representation of this curve.

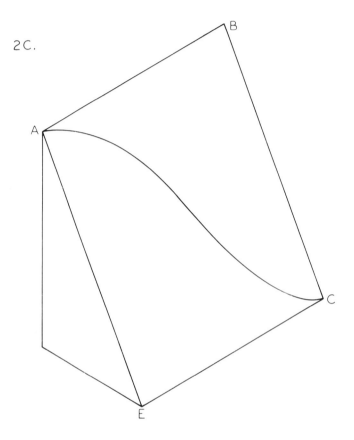

3 Lines

The term *line* is generally used to designate a *straight* line unless otherwise specified. Theoretically a line is of indefinite length, but frequently when the conditions of the problem make it obvious, the term implies a definite line *segment*.

3.1 True Length of a Line

A line is shown in true length, TL, when the line of sight is normal (perpendicular) to the line. For example, in Figure 3.1 the lines of sight F and H for the front and top views are each perpendicular to edge AB of the rectangular block. The front and top views a^Fb and a^Hb are therefore each true length.

In orthographic projection a view of a line cannot be longer than the line itself. Any line of sight not normal to a line results in a view that is shorter than true length. For example, in Figure 3.1(b) line of sight 1, which is not normal to edge AB of the solid, results in the *foreshortened* view a^1b. Similarly, in Figure 3.1(c), which shows edge AB as an individual line, view a^1b is foreshortened.

3.2 Principal Lines

Lines that lie in or parallel to a principal plane of projection (frontal, horizontal, or profile) are called *principal lines*. An *inclined* line is a principal line that is parallel to one principal plane of projection but is inclined to the other two. Such lines are widely used in descriptive geometry and are designated according to the planes to which they are parallel.

A *frontal line* lies in or parallel to a frontal plane. Figure 3.2(a) and (b) illustrates an inclined frontal line AB as an edge of a solid object. Figure

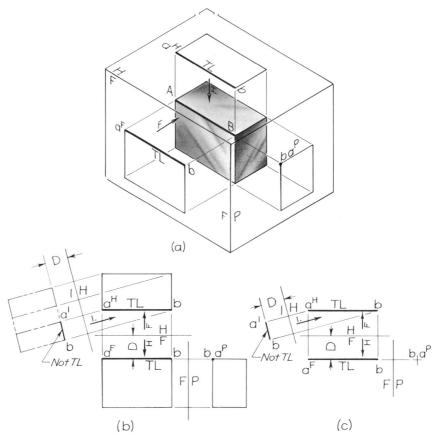

FIGURE 3.1   True length of a line

3.2(c) shows the same line AB independently. Note that the top view is parallel to folding line H/F and is therefore a horizontal line on the drawing paper. This is the characteristic by which any frontal line is easily recognized. Since the line of sight for the front view is perpendicular to line AB, the front view a^Fb is true length.

It will be noted that the front view also shows the horizontal and profile projection planes in edge view (see §1.9). *The true angle between any line and any plane appears in any view that shows simultaneously the line in true length and the plane in edge view.* Hence, as indicated in Figure 3.2, the true angles θ_H and θ_P between line AB and the horizontal and profile planes, respectively, may be measured in the front view.

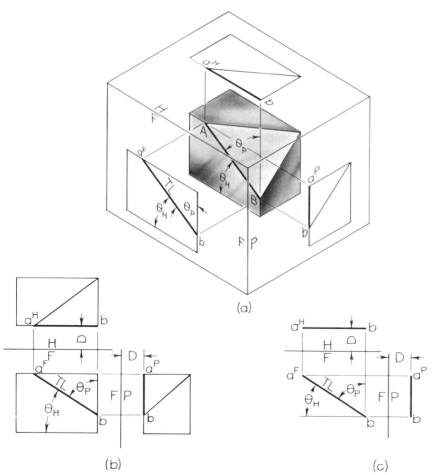

FIGURE 3.2   Principal lines—the frontal line

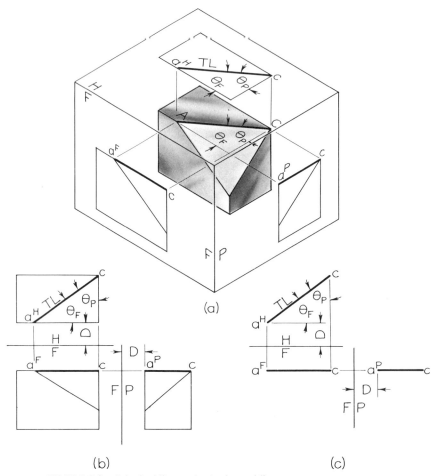

FIGURE 3.3   Principal lines—the horizontal line

A *horizontal line* lies in or parallel to a horizontal plane. Figure 3.3 illustrates an inclined horizontal line AC as an edge of a solid object, (a) and (b), and as an independent line, (c). Horizontal lines are also called *level lines*.

The distinguishing characteristic of a horizontal line is that its front view is parallel to the H/F folding line and therefore appears horizontal on the drawing paper. Since the line of sight for the top view is normal to a horizontal line, the top view a^Hc is true length. The top view also shows the frontal and profile projection planes in edge view. Hence the true angles, θ_F and θ_P, respectively, between line AC and these planes are shown in the top view as indicated.

A *profile line* lies in or parallel to a profile plane. Figure 3.4 illustrates a profile line BC as an edge of a solid object and as an independent line.

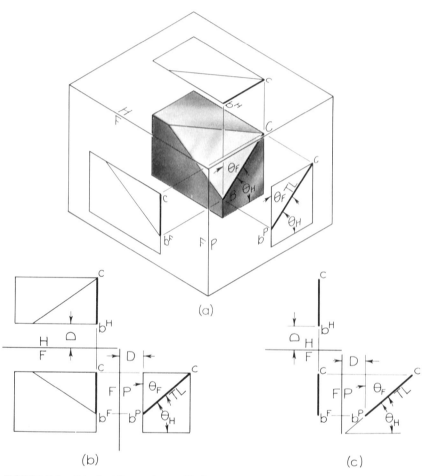

(a)

(b)   (c)

FIGURE 3.4   Principal lines—the profile line

The distinguishing characteristic of a profile line is that its front view is parallel to the F/P folding line and appears vertical on the drawing paper. Since the line of sight for the profile view is normal to a profile line, the profile view b^{P}c is true length. The profile view also shows the fontal and horizontal projection planes in edge view. Therefore, the true angles, θ_F and θ_H, respectively, between the line BC and these planes are shown in the profile view.

3.3   True Length of an Oblique Line

An *oblique* line is one that is not parallel to any of the principal planes of projection: horizontal, frontal, or profile. This being the case, none of the lines of sight for the principal views is perpendicular to an oblique line, and therefore these views of the line are foreshortened. For example, note in Figure 3.5 the foreshortened front and top views of a pencil placed in an oblique position.

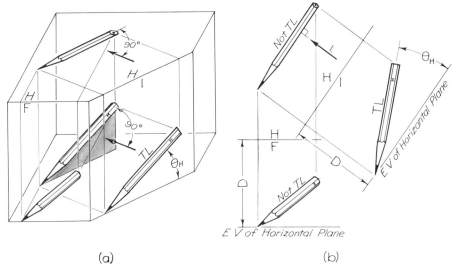

(a)   (b)

FIGURE 3.5   True length of an oblique line and its angle with a horizontal plane

In order to find the true length of an oblique line—the pencil—an additional view must be constructed with the line of sight perpendicular to the oblique line. For example, if a horizontal line of sight is assumed perpendicular to the pencil, as shown by the pictorial arrow in Figure 3.5(a), the top view of this arrow appears perpendicular to the top view of the pencil. The resulting auxiliary elevation view shows the true length of the pencil, Figure 3.5(a) and (b). It may be noted that *a line is shown true length in a particular view when the adjacent view of the line is parallel to the folding line between the two views.*

In this same auxiliary elevation view, the horizontal plane upon which the pencil point is resting shows in edge view and parallel to folding line H/1. The student should take particular note that *any auxiliary elevation shows all horizontal planes in edge view and parallel to the folding line between that auxiliary view and the top view.* Since in the auxiliary elevation view of Figure 3.5 the line is shown true length and the horizontal plane in edge view, the true angle θ_H between the two is shown as indicated.

The true length of an oblique line may be found just as readily by assuming a frontal line of sight perpendicular to the line as indicated pictorially in Figure 3.6(a). In this case the line of sight appears perpen-

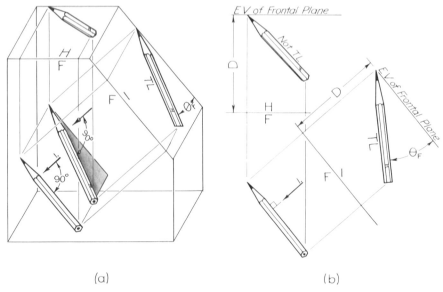

(a) (b)

FIGURE 3.6   True length of an oblique line and its angle with a frontal plane

dicular to the front view of the pencil, as shown by arrow 1. The resulting auxiliary view, projected from the front view, shows the true length of the pencil, Figure 3.6(b).

In this auxiliary view the frontal plane against which the pencil point is resting shows in edge view and parallel to folding line F/1. It should be observed that *any auxiliary view projected from the front view shows all frontal planes in edge view and parallel to the folding line between the auxiliary view and the front view.* With the pencil now in true length and the frontal plane in edge view, the true angle θ_F between them is shown.

The true length of an oblique line may be shown by still another type

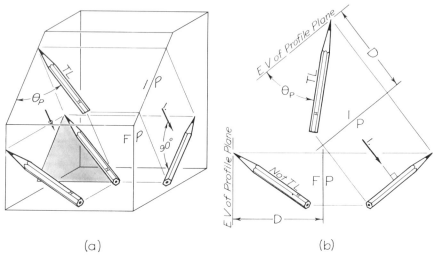

(a) (b)

FIGURE 3.7   True length of an oblique line and its angle with a profile plane

of primary auxiliary view. If a line of sight parallel to the profile plane is assumed perpendicular to an oblique pencil, as shown pictorially in Figure 3.7(a), the line of sight appears in the side view as arrow 1 perpendicular to the side view of the pencil. The profile plane touching the pencil point shows in the resulting auxiliary view as a line parallel to the folding line P/1. *Any auxiliary view projected from a side view shows all profile planes in edge view and parallel to the folding line between the auxiliary view and the side view.* With the pencil now in true length and the profile plane in edge view, the true angle θ_P between them is shown.

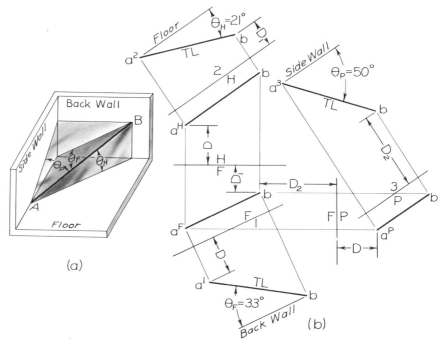

FIGURE 3.8 Comparison of the separate constructions for θ_H, θ_F, and θ_P

In summary of the material just discussed, Figure 3.8 shows the true length of an oblique line AB obtained by each of the three types of primary auxiliary views. Although each auxiliary view shows the line in true length, the *true angle* between the line and *one* and *only one* of the principal planes of projection appears in any particular view. The student should specifically note, Figure 3.8(b), that no pair of these angles is either complementary or supplementary. Hence *a separate auxiliary view must be drawn to obtain each of the angles* θ_H, θ_F, and θ_P.

3.4 Bearing, Slope, and Grade

On the drawing paper, a line is represented by its projections on at least two planes of projection. In engineering practice, however, the position of a line in space is often described by its *bearing* and *slope*, or by its *bearing* and *grade*.

Bearing is a term used to describe the direction or course of a line on the earth's surface. For practical purposes small portions of the earth's surface are usually regarded as planes; a map of such a portion is then merely a top view. Hence the bearing of a line is the angular relationship of the top view of the line with respect to due north or south, expressed in degrees. *North is assumed to be directed toward the top of the drawing unless specifically given otherwise.*

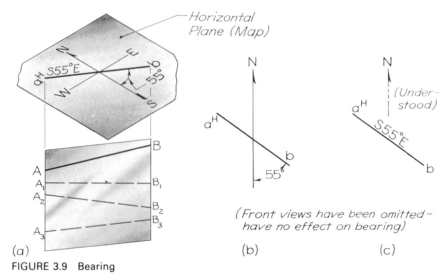

FIGURE 3.9   Bearing

In Figure 3.9 line AB, implying A toward B, has a bearing of S55°E; or it could be stated that line BA, B toward A, has a bearing of N55°W. In this traditional system it is customary to make the specified angle less than 90°, selecting north or south accordingly as the base direction. It should be noted that lines AB, A_1B_1, A_2B_2, and A_3B_3 all have the same bearing since their top views coincide. In other words, the bearing of a line specifies the direction of its top view and is in no way affected by the angle between the line and the horizontal plane. Figure 3.9(b) shows how the bearing angle is measured, while Figure 3.9(c) indicates how it is usually noted on a map.

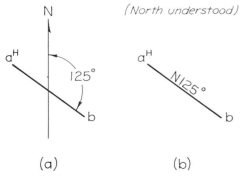

FIGURE 3.10   Azimuth bearing

Another method of specifying the direction of the top view of a line is *azimuth bearing*, which finds its greatest use in navigation, civil engineering, and when bearing data are computerized. The azimuth bearing of a line is the *clockwise departure* in degrees of the top view of the line from a base direction, usually north. Figure 3.10(a) shows line AB again, A toward B, with an azimuth bearing of N125° and Figure 3.10(b) shows how it is indicated on a map. It could also be said that line BA, B toward A, has an azimuth bearing of N305°.

The *slope* of a line may be defined as the angle in degrees that the line makes with a horizontal plane. This is the same angle as that designated as θ_H in §§3.2 and 3.3, in which it was pointed out that the angle must be measured in that particular view which shows both the horizontal plane in edge view and the line in true length.

EXAMPLE: BEARING, SLOPE, AND TRUE LENGTH

Problem

A 160 m segment AB of a power line has a bearing of N60° and a downward slope of 20° from the given point A, Figure 3.11(a). Complete the front and top views.

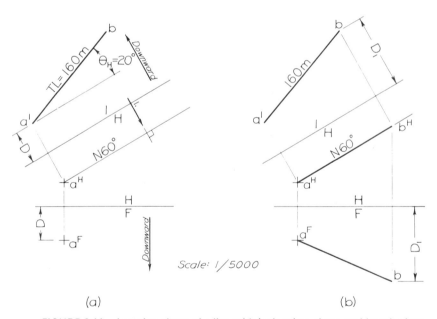

(a)   (b)

FIGURE 3.11   Locating views of a line with its bearing, slope, and length given

Analysis

The bearing of a line establishes the direction of its top view. The slope of a line, θ_H, and the true length will be seen in an auxiliary view having a line of sight normal to the top view of the line.

Graphic Solution

A construction line of indefinite length is drawn from a^H at the given bearing N60°, Figure 3.11(a). The line of sight 1 is positioned normal to this construction line. Point a^1 in the auxiliary view is established by the conventional use of folding lines and transfer distance D. Since this view will show the true slope of the line, the 20° angle may be set off in the auxiliary view as indicated. The direction "downward" is away from the folding line H/1, as it can be seen in the front view that any point lower than point A will be at a greater distance than a^F from the H/F folding line. With a construction line having a 20° downward slope now established, the true length of 160 m is set off to scale from a^1 to b. Point b is then located in the top view, Figure 3.11(b), by returning to the top view along a projection line perpendicular to the folding line H/1. The transfer distance D_1 is set off along a vertical projector to establish b in the front view.

The *grade* of a line is another means of describing the inclination of a line with respect to a horizontal plane. Grade is given by the following expression:

$$\text{Percent grade} = \frac{\text{Vertical rise}}{\text{Horizontal run}} \times 100$$

These terms are illustrated pictorially in Figure 3.12. It will be noted that the grade is the tangent of θ_H, multiplied by 100.

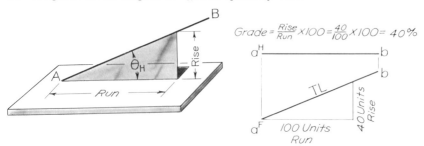

FIGURE 3.12   Percent grade   FIGURE 3.13   Grade of a frontal line

For the special case of a frontal line, the true slope is seen in its front view, and hence the rise and run may be measured as shown in Figure 3.13. For ease in calculation the run is set off as 100 units of an appropriate size. A metric scale is convenient for such measurements. In this case the 1/40 scale was used. With the corresponding rise measured in the same units, the percent grade is obtained directly without calculations.

For the general case of an oblique line, Figure 3.14, the true slope is not shown in the front view. If an auxiliary elevation view showing the true length of the line is drawn as indicated, the true slope will be seen and the grade calculation may be made. The horizontal run, 100 units, must be set off parallel to folding line H/1 in order to be horizontal *in*

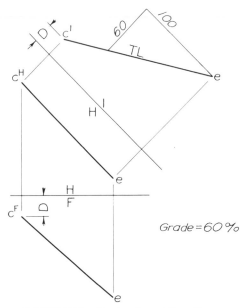

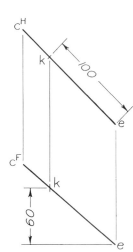

FIGURE 3.14   Grade of an oblique line

FIGURE 3.15   Grade measured in the front and top views

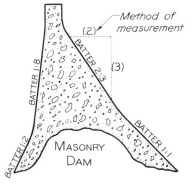

(a) Batter
(Civil Engineering)

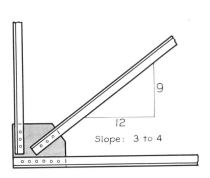

(b) Slope of Beam
(Structural Engineering)

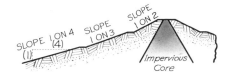

(c) Slopes on Earth Dam
(Civil Engineering)

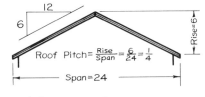

(d) Pitch of Roof
(Architectural Drafting)

FIGURE 3.16   Other engineering methods of describing inclination

space. The corresponding rise, 60 units, must then be measured in a direction perpendicular to H/1 in order to be vertical in space.

Actually, the grade of an oblique line may be obtained from the front and top views without additional views. Horizontal distances such as the run may always be measured on a map or top view. Hence the run of 100 units may be set off along the top view as shown in Figure 3.15, thus establishing point k. Vertical distances such as the rise appear in any elevation view. Therefore, the rise appears in the front view as the difference in elevation, 60 units, between points e and k.

In some engineering fields inclinations are specified in other ways peculiar to the particular field. In Figure 3.16 some of these are illustrated: (a) *batter*, (b) *slope* of a beam, (c) *slope* on an earth dam, and (d) *pitch* of a roof.

3.5 Points on Lines

If a point is on a line (in space), the views of the point appear on the corresponding views of the line. Any two successive views of a point must lie on a projection line perpendicular to the folding line between the two views. Consequently, if a point is known to be on a line, it may usually be located in successive views of the line by simple projection as illustrated for point C in Figure 3.17. The exception occurs when the views of the line are perpendicular to the folding line, Figure 3.18(a).

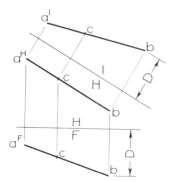

FIGURE 3.17   Point on a line

In this case, the top view of the given point X cannot be determined by direct projection. An additional view is drawn for which the folding line is not perpendicular to the given views of the line, Figure 3.18(b). Point X is projected to this new view, and the transfer distance D_1 is used to locate the required top view of X, Figure 3.18(c).

If given views of a line are nearly perpendicular to the folding line between them, projection from one line to the other may be quite inaccurate, in which case the preceding procedure is recommended (see Appendix A.2).

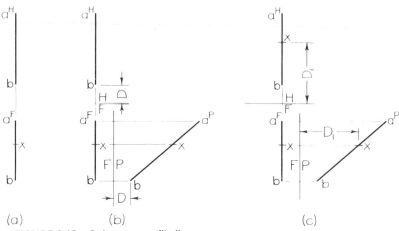

FIGURE 3.18   Point on a profile line

Points on lines may be determined by spatial relations. A point may be said to be above or below, in front of or behind, or to the right or to the left of another point. These directions are oriented as directed in Figure 3.19(a). A specific illustration appears in Figure 3.19(b), in which

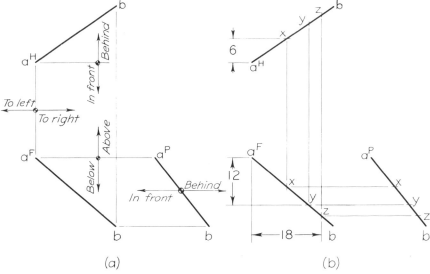

FIGURE 3.19   Points on line by spatial description

point X is on line AB and 6 mm behind point A. Note that this description does not necessarily imply that point X is *directly* behind A nor even that point X is 6 mm from A (true distance along line AB). In this same problem point Y is on line AB and 12 mm below point A. Point Z is on line AB and 18 mm to the right of A.

Points dividing a line segment in a given ratio will divide any view of the line in the same ratio. See Figure 3.20, in which the true length auxiliary view of the line GE is divided into three equal segments. When the points

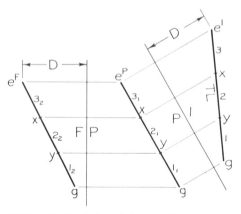

FIGURE 3.20 Points dividing a line into parts

X and Y are projected back to the profile and front views, it becomes evident that these views of the line are divided in this same ratio, and thus the division could have been made without the use of the auxiliary view. (If the actual lengths of the segments are desired, the true-length view of line AB is necessary.)

Intersecting lines are lines that contain a common point (parallel lines and perpendicular lines are discussed in Chapters 9 and 10, respectively). In order for a point to be common to two lines, the views of the point must lie on a single projection line perpendicular to the folding line between two adjacent views. In Figure 3.21 points x in the front and top views lie on a common vertical projector. These are then views of a single point X which is on both lines, and therefore the lines intersect. In any additional view, such as the side view in Figure 3.21, the corresponding view of the common point X must still fall on the proper projection line.

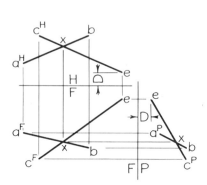

FIGURE 3.21 Intersecting lines

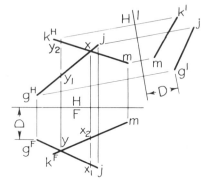

FIGURE 3.22 Nonintersecting lines

In contrast, note that in Figure 3.22, projection to the top view of the apparent point of crossing at y in the front view results in *two* separate points in the top view, y_1 and y_2. Similarly, the projection of x to the front view results in two points, x_1 and x_2. From this it can be seen that the two lines do not have a single point in common and therefore are nonintersecting. This is further substantiated by the additional view, in which the line segments do not cross within the limits of the view as drawn.

EXAMPLE: INTERSECTING LINES

Problem

Complete the top view of the hoist frame, Figure 3.23(a), given the information that the braces CE and GE intersect the beam AB at point E.

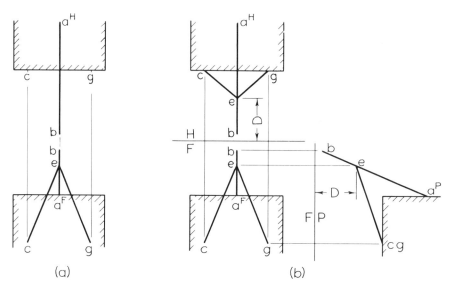

FIGURE 3.23   Intersecting structural members—hoist frame

Graphic Solution

Since the top view of point E cannot be obtained directly by projecting from the front view to the top view, the side view of the hoist frame is drawn, Figure 3.23(b). Point e in the side view is located by extending a horizontal projection line from point e in the front view to intersect aᴾb. Transfer distance D then establishes point e in the top view, and the view is completed as shown.

PROBLEMS

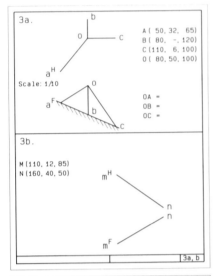

3a.

A (50, 32, 65)
B (80, -, 120)
C (110, 6, 100)
O (80, 50, 100)

Scale: 1/10

OA =
OB =
OC =

3b.

M (110, 12, 85)
N (160, 40, 50)

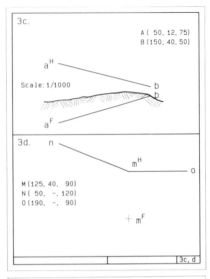

3c.

A (50, 12, 75)
B (150, 40, 50)

Scale: 1/1000

3d.

M (125, 40, 90)
N (50, -, 120)
O (190, -, 90)

| 3a, b |
| 3c, d |

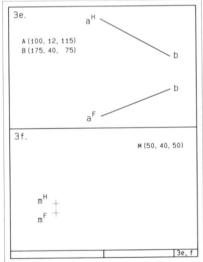

3e.

A (100, 12, 115)
B (175, 40, 75)

3f.

M (50, 40, 50)

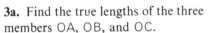

3g.

A(50, -, 40)

| 3e, f |
| 3g |

3a. Find the true lengths of the three members OA, OB, and OC.

3b. Determine the true length of line MN and its true angles with horizontal and profile planes.

3c. Determine the bearing, length, and grade of shaft AB.

3d. Line MN has a downgrade of 30 percent, and line MO has a downgrade of 40 percent. Complete the front views of these lines.

3e. Determine the bearing, length, and grade of line AB.

3f. Complete the views of a 65 mm line MN that has a bearing of N60°

and a downgrade of 30 percent from point M to point N.

3g. An observation aircraft at A, elevation 1000 m is flying at 400 knots air speed on a bearing of N30° and is gaining altitude at a rate of 200 m in 1000 m. A speedboat is sighted from A at N50° on a 20° angle of declination. Twenty seconds later, the boat is sighted at N105° on a 60° angle of declination. Determine the course of the speedboat and its speed in knots. Scale 1/40 000. One knot = 1852 m per hour.

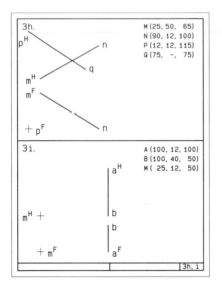

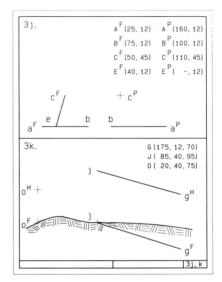

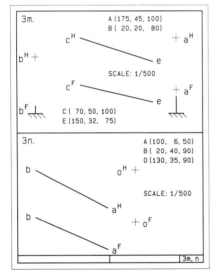

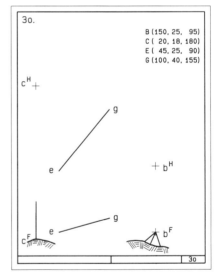

3h. Complete the front view of the two intersecting lines and add a right-side view.

3i. Complete the views of line MN that intersects line AB at a point 12 m from point B. Determine the grade of MN. Scale 1/400.

3j. Complete the views of the intersecting lines AB and CE.

3k. Find the bearing, length, and grade of a shaft to be driven from point O to join tunnel GJ at a point 20 m from point J.

3m. A television antenna is at point A and the base of a receiving antenna is at point B. The maximum height of an obstruction is represented by line CE. What minimum height must be exceeded by the receiving antenna to provide a "line-of-sight" condition?

3n. Point S lies on line AB and 12 m behind point A in space. Determine the bearing, length, and grade of line OS.

3o. Determine the *vertical angle*, θ_H, for a transit telescope at point B to sight a stadia rod at point C, with the line of sight just clearing an obstruction whose maximum height is represented by line EG.

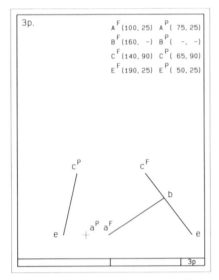

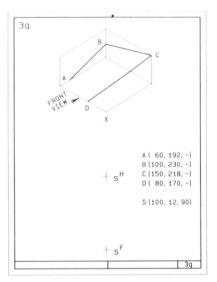

3p. Complete the side view of the intersecting lines CE and AB. Determine the bearing and grade of line AB.

3q. Broken line ABCD is shown in isometric pictorial form. Placing corner X at the starting point S, construct the necessary multiview projections of ABCD, labeling and measuring the true-length views of the three line segments. If assigned, determine the bearing and/or grade of segment DC.

3r. Indicate whether the following statements are true or false. If assigned, provide written statements or sketches to justify the answers.

(1) In an orthographic view a line can project longer or shorter than the line itself.

(2) A principal line appears true length in a front, top, or side view.

(3) The true slope of a line is always observed in a front view.

(4) A line may lie in only two principal planes at the same time.

(5) Two successive auxiliary views are needed to obtain the true length of an oblique line.

(6) Only one auxiliary view exists that produces the true-length view of a given line.

(7) A line must be true length before its bearing can be measured.

(8) The grade and angular slope of a line have the same numerical value.

(9) The grade of a line can be greater than 200 percent.

(10) *Rise* is always measured perpendicular to a horizontal plane.

SELF-TESTING PROBLEMS

3A. For the line AB determine the slope angle and θ_p.

3B. Determine the bearing and grade of the line CE. From the midpoint M of line CE add the principal views of a 25 mm frontal line MN that has a 120 percent downgrade from point M to point N.

3A.

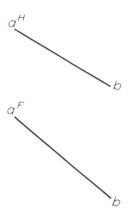

3B.

3C. Supply the necessary orthographic views to represent all lines of the pyramid in true length.

3C.

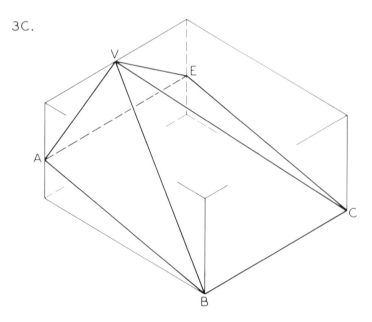

4 Planes

A plane is a surface such that a straight line connecting any two points in that surface lies wholly within the surface. The general term "plane" implies a plane indefinite in extent unless other specified. Any two lines in a plane must either intersect or be parallel.

4.1 Representation of Planes

A plane may be uniquely represented by two intersecting lines, Figure 4.1(a). A plane may also be represented by two parallel lines, by

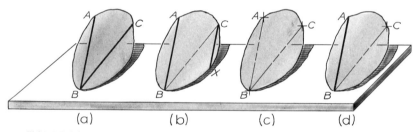

FIGURE 4.1 Representation of a plane in space

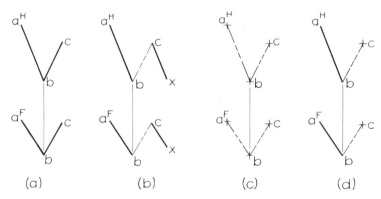

FIGURE 4.2 Representation of a plane in multiview projection

three points not in a straight line, or by a point and a line, Figure 4.1(b), (c), and (d). Multiview drawings of these representations are shown in Figure 4.2(a), (b), (c), and (d), respectively. In graphical constructions involving planes, representations (b), (c), and (d) are usually converted to (a), as suggested by the phantom lines.

4.2 Points and Lines in Planes

If a line is known to be in a plane, then any point on that line is in the plane. A line may be drawn in a plane by keeping it in contact with (intersecting) any two given lines in the plane. A line may also be located in a plane by drawing the line through a known point in the plane and parallel to a line in the plane, §9.1.

EXAMPLE: LOCATING A POINT IN A PLANE

Problem
Given the front and side views of a plane MON and the front view of a point A in the plane, determine the side view of the point.

Analysis
An infinite number of lines that contain the point may be drawn in the plane. Any convenient one may be selected, such as XY, Figure 4.3(a). The views of the point must lie on the corresponding views of the line.

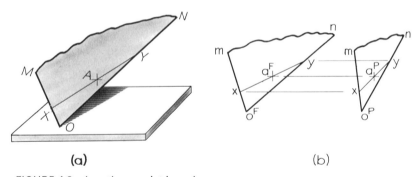

(a)   (b)

FIGURE 4.3   Locating a point in a plane

Graphic Solution
A line xy is drawn through a^F, representing the front view of line XY on the plane, Figure 4.3(b). The side views of the contact points X and Y with the given lines of the plane are located by projection and line xy is drawn in the side view. The intersection with xy of a horizontal projection line from a^F locates a^P, the required side view of point A.

4.3   Lines in Planes

The preceding ideas may also be applied to the location of the views of given lines lying in planes. For example, in Figure 4.4(a) a five-sided plane figure ABCEG is given with the front view incomplete. In the top view, Figure 4.4(b), a^{H}g is extended to intersect bc at x. The front view of point X is located by projection from the top view of point X to bc, thus establishing a^{F}x. Projection from g locates g on a^{F}x, and the front view is then completed as shown.

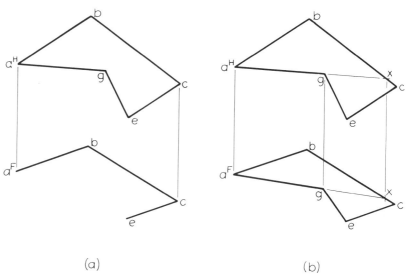

(a)   (b)

FIGURE 4.4   Oblique lines in plane

4.4   Principal Lines in Planes

Many problems in descriptive geometry require the addition of frontal, horizontal, or profile lines in a plane.

In Figure 4.5(a) the frontal line AF has been located in the plane ABC by first drawing its top view a^{H}f parallel to an imaginary folding line H/F. (In this particular case it is not actually necessary to draw or show folding line H/F, because it is not used in the subsequent construction.) Since point F is the intersection of lines AF and BC, its front view may be located by projecting to bc, thus establishing a^{F}f. Other frontal lines, such as A_1F_1, may be drawn on the given plane. Note that any such additional frontal line is parallel to line AF. As a general principle, *all frontal lines in the same plane are parallel* unless the plane itself is frontal, in which case all lines in the plane are frontal but are not necessarily parallel.

In a similar fashion a horizontal line, Figure 4.5(b), may be located by drawing its front view in a horizontal position and determining its top

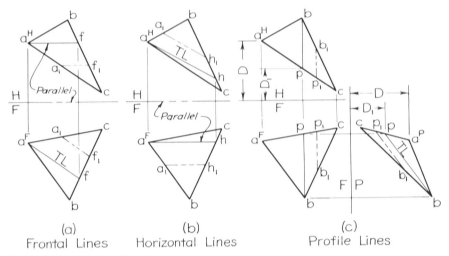

FIGURE 4.5   Principal lines in a plane

view by projection. *All horizontal lines in the same plane are parallel*, unless the plane itself is horizontal, in which case all lines in the plane are horizontal but not necessarily parallel.

In Figure 4.5(c) profile line BP has been located in the plane ABC by first drawing bp either in the front view or in the top view, each of which appears as a vertical line on the drawing paper, and then locating bp in the side view by projection. *All profile lines in the same plane are parallel*, with the exception that profile lines in a profile plane are not necessarily parallel.

4.5   Locus

A *locus* is the path of a point, line, or curve moving in some specified manner. Or a locus can be thought of as the assemblage of all possible positions of a moving point, line, or curve. Thus the locus of points in a plane and at a specified distance from a given point is a circle, while the locus of points in space at a specified distance from a given point is a sphere. The concept of a locus appears in the solution of the example in §4.6.

4.6   Space Analysis

The student may have noted in several previous examples that the explanation was divided into two general parts, designated as *analysis* and *graphical solution*. The analysis is usually a *space* analysis in which a logical procedure is formulated for the solution of the problem *in space* —as if dealing with a three-dimensional model. The graphical solution then consists of the representation of these steps on the drawing board in terms of multiview projection.

In many cases it will be found desirable to outline the space analysis in written form before translating it to the drawing-board solution. Such an analysis will require the student to consider carefully and to visualize in their logical order the fundamental steps necessary to reach the desired solution. A written analysis will also provide the student with a constant reminder of the successive constructions to apply to the solution of the problem. It will help prevent the student from rushing headlong into a graphic solution without knowing where to go or how to get there.

EXAMPLE: APPLICATION OF SPACE ANALYSIS

Problem

In a given plane ABC locate a point K that lies 6 mm above horizontal line AB and 5 mm in front of frontal line AC. Scale: full size.

(a) *Establish a horizontal line HH in plane ABC that is 6 mm above line AB, Figure 4.6(a).*

This line is the locus of all points in plane ABC which are 6 mm above line AB in space. See Figure 3.19(a).

(b) *Establish a frontal line FF in plane ABC that is 5 mm in front of line AC, Figure 4.6(b).*

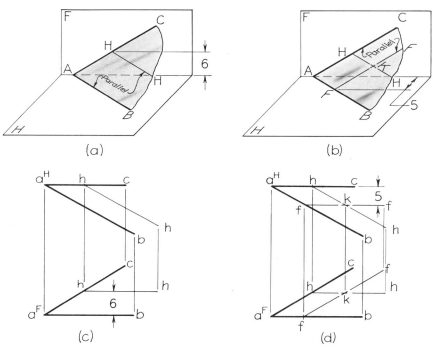

FIGURE 4.6   Locus problem—point in a plane

This line is the locus of all points in plane ABC and 5 mm in front of line AC.

(c) *The intersection of lines* HH *and* FF *is the required point* K.

Point K is the only point that fulfills all the specifications.

Graphic Solution

In Figure 4.6(c) the drawing-board representation of analysis step (a) is shown. The remaining steps that complete the solution are shown in Figure 4.6(d). The vertical projector between the views of point K checks the accuracy of the solution.

4.7 Pictorial Intersections

Let us consider the boundaries of a plane surface MNK that results from the intersection of this plane with the surfaces of the prism of the isometric pictorial, Figure 4.7(a). In the following solution, two basic

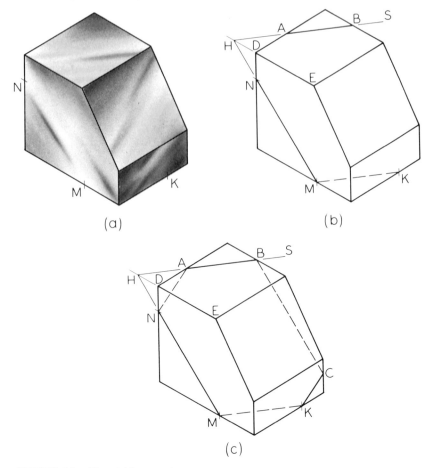

(a)

(b)

(c)

FIGURE 4.7 Pictorial intersection

principles previously introduced in this chapter must be emphasized. The first relates to the fact that *lines in a single plane must either intersect or be parallel.* The second principle deals with the axiom that parallel horizontal lines in a plane may be created by a series of horizontal cutting planes; or more generally stated, *parallel lines in a plane are established by a series of parallel cutting planes.* These principles were previously related to orthographic views, but they pertain in equal validity to both oblique and isometric pictorials.

In Figure 4.7(b) the direct introduction of visible line MN and hidden line MK provides the initial required lines of intersection of the plane MNK and the surfaces of the pictorial. Since lines MN and DE both lie in the same frontal plane of the pictorial, they must either be parallel or intersect. When these two lines are extended, their intersection point H is established. It should now be recognized that point H also lies in the upper horizontal plane of the pictorial, and that horizontal line HS in the plane MNK may be added parallel to horizontal line MK. By this procedure visible line AB of the intersection boundary is obtained.

Frontal line BC may then be introduced parallel to frontal line MN, Figure 4.7(c). The introduction of parallel lines KC and NA completes the solution.

PROBLEMS

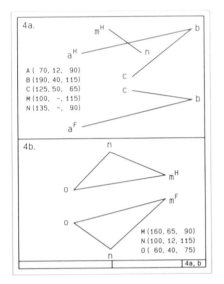

4a.
A (70, 12, 90)
B (190, 40, 115)
C (125, 50, 65)
M (100, -, 115)
N (135, -, 90)

4b.
M (160, 65, 90)
N (100, 12, 115)
O (60, 40, 75)

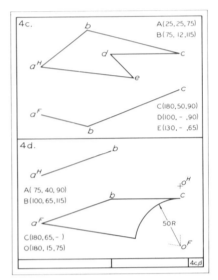

4c.
A (25, 25, 75)
B (75, 12, 115)
C (180, 50, 90)
D (100, -, 90)
E (130, -, 65)

4d.
A (75, 40, 90)
B (100, 65, 115)
50R
C (180, 65, -)
O (180, 15, 75)

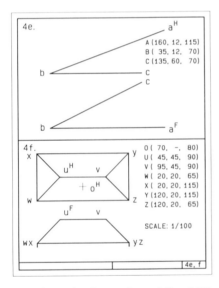

4e.
A (160, 12, 115)
B (35, 12, 70)
C (135, 60, 70)

4f.
O (70, -, 80)
U (45, 45, 90)
V (95, 45, 90)
W (20, 20, 65)
X (20, 20, 115)
Y (120, 20, 115)
Z (120, 20, 65)

SCALE: 1/100

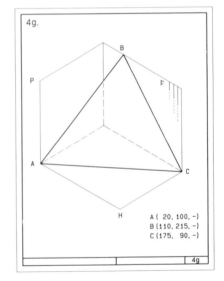

4g.
A (20, 100, -)
B (110, 215, -)
C (175, 90, -)

4a. Show the front view of line MN that lies in the plane ABC.

4b. Through point M add the views of a frontal line that lies in plane MNO. Through point O add the views of a horizontal line that lies in the plane.

4c. Complete the front view of the plane figure ABCDE.

4d. Complete the top view of the plane figure.

4e. Locate a point P in plane ABC which lies 25 mm above line AB and 20 mm behind line CB.

4f. Show the views of a hole in the roof plane cut for a 1.2 m × 1.2 m chimney centered at point O.

4g. In the isometric, locate a point X in plane ABC that lies 40 mm above line AC and 30 mm in front of plane F. Then locate a point Y in plane ABC that lies 25 mm to the right of line AB and 12 mm in front of plane F. Draw *heavy* that portion of line XY that lies within triangle ABC.

67

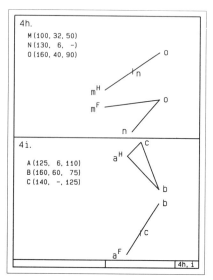

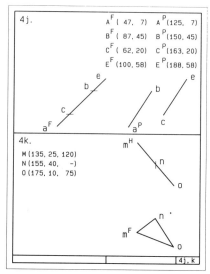

4h. Locate the views of a point in plane MON that is equidistant from the sides ON and OM and 50 mm from point O.

4i. Locate the views of a point in plane ABC that is equidistant from the three points.

4j. Locate the views of a point in plane ABCE that is equidistant from parallel lines AB and CE and that is equidistant from points A and B.

4k. Locate the center and plot the front view of a circle that passes through the points M, N, and O.

4m. Indicate whether the following statements are true or false. If assigned, provide written statements or sketches to justify the answers.

(1) An infinite number of planes can be passed through a single line.

(2) Any two horizontal lines determine a single plane.

(3) Only one plane can be passed through three points not in a straight line.

(4) Four points cannot lie in a single plane.

(5) Frontal lines in an oblique plane appear parallel in any principal view.

(6) Two lines in a plane must either intersect or be parallel.

(7) The lateral surface of a cylinder may be considered to be a plane surface, since the parallel elements of the cylinder lie wholly within the surface.

(8) The surface of a cone may be considered to be a plane surface, since the elements intersect at the vertex.

SELF-TESTING PROBLEMS

4A. Complete the front view of the plane figure ABCDE.

4B. To the given views of the plane MNO add a 65 mm horizontal line that lies 25 mm above horizontal line MO. Then in the plane add a 50 mm frontal line that lies 18 mm behind point N.

4 A.

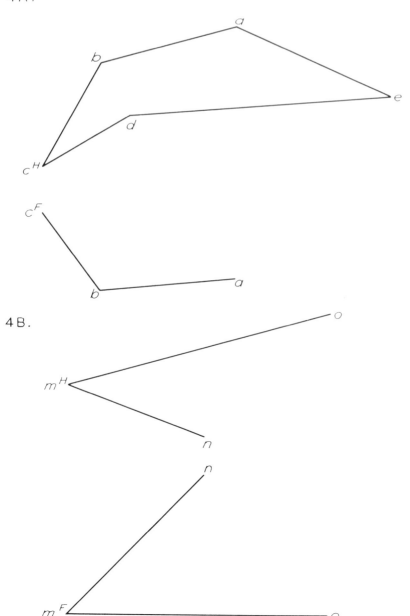

4 B.

4C. In the isometric pictorial, locate the point in plane ABC that lies 10 mm above the horizontal plane H and 22 mm from plane P.

4D. In the isometric pictorial, locate the point in plane ABC that lies 15 mm from plane F and 25 mm from plane P.

4C.

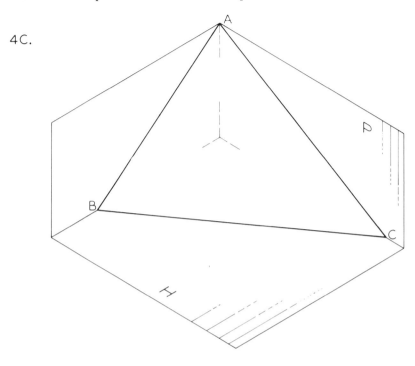

4D.

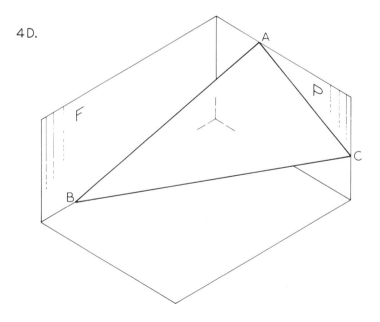

5 Successive Auxiliary Views

A primary auxiliary view is an auxiliary view obtained by projection from one of the six basic views. As was stated in Chapter 2, a primary auxiliary plane of projection is one that is perpendicular to one of three principal projection planes and inclined to the remaining two.

A *secondary* auxiliary view is one projected from a primary auxiliary view. Since any two adjacent views must lie on mutually perpendicular projection planes, the plane of projection for the secondary auxiliary view must be perpendicular to that of the primary auxiliary view. A secondary auxiliary view is used to obtain information available only in a view projected on a plane that is oblique to the principal planes.

In some instances it may be necessary to construct a third auxiliary view projected from a secondary auxiliary view. Theoretically, it is possible to continue this chain of additional auxiliary views indefinitely, adding a fourth, a fifth, and so on. Auxiliary views in such a series are called *successive* auxiliary views.

TABLE 5.1 *Uses of Auxiliary and Additional Views*

Use	Position of Line of Sight	
	In space	*On multiview drawing*
1. True length of line (TL)	Perpendicular to line	Perpendicular to any view of the line or directed toward a point view of the line
2. Point view of line	Parallel to line	Parallel to true-length view of line
3. Edge view of plane (EV)	Parallel to plane	Parallel to true-length view of line in plane or directed toward a true-size view of the plane
4. Normal or true-size view of plane (TS)	Perpendicular to plane	Perpendicular to edge view of plane

5.1   Construction of Successive Auxiliary Views

As an illustration of the mechanics of drawing successive auxiliary views, Figure 5.1 shows the construction of two additional views of a point A, given the top and front views and successive lines of sight 1 and 2.

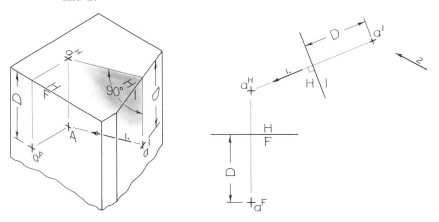

(a) The Primary Auxiliary View

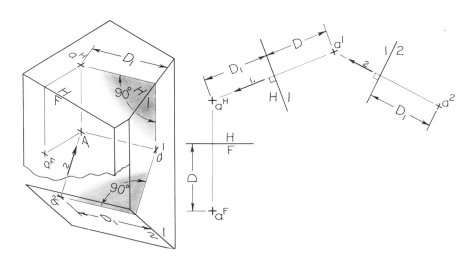

(b) The Secondary Auxiliary View

FIGURE 5.1   Successive auxiliary views of a point

In Figure 5.1(a) the auxiliary elevation indicated by line of sight 1 has been drawn. Since the auxiliary plane and the frontal plane are perpendicular to the horizontal plane, views a^1 and a^F lie the same distance D below the horizontal plane as shown in the pictorial, Figure 5.1(a). In the multiview drawing, the view a^1 is therefore located by transferring the distance D from the front view as indicated.

With the primary auxiliary view a¹ completed, the secondary auxiliary view indicated by line of sight 2, Figure 5.1(a) and (b), may be drawn as follows, using the conventional steps for a new view:

Step 1. Establish the line of sight.

> In this case the line of sight is given as arrow 2, Figure 5.1(b). In practical applications the line of sight is established according to the information desired.

Step 2. Introduce the necessary folding lines.

> One of the necessary folding lines is H/1. The other folding line 1/2 is drawn perpendicular to the line of sight 2 at a convenient distance from view 1.

Step 3. Transfer distance(s) to the new view.

> In this case, the top view and the new view 2 both lie on planes perpendicular to the projection plane of view 1. Consequently, views a² and aᴴ lie the same distance from the plane of view 1. The distance to be transferred is then D_1, which is the distance point A lies from the plane of view 1.

Step 4. Complete the view.

Additional successive auxiliary views may be obtained by repeated application of the preceding steps, since in any chain of three successive views the two "outside" views both lie on projection planes that are perpendicular to the plane of the central view. In Figure 5.2 a third auxiliary view a³ has been constructed for a given line of sight 3. Theoretically a fourth auxiliary view a⁴ could then be added, then a fifth view a⁵, and so on in an endless chain. Few descriptive geometry problems require more than two successive auxiliary views in their solutions.

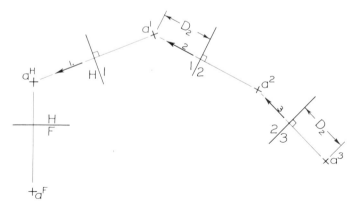

FIGURE 5.2   Additional successive auxiliary views of a point

5.2   Point View of a Line

A fundamental operation is the construction of a view showing a given line as a point.

A line will appear as a point in any view for which the line of sight is parallel to the line in space.

In Figure 5.3 line AB is shown in true length in the front view. Therefore, line of sight 1, being parallel to a^Fb, is parallel to line AB in space. The resulting auxiliary view a^1b is a point of view of line AB. In contrast, line of sight P, while parallel to a^Hb, is not parallel to line AB in space, since a^Hb is not true length. Consequently, view P cannot show line AB in point view, as is proved by the resulting view a^Pb.

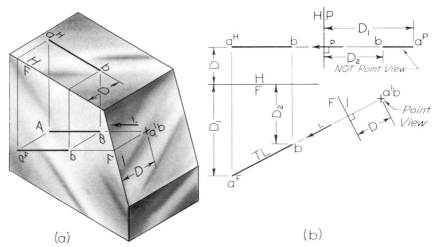

(a)   (b)

FIGURE 5.3   Point view of a frontal line

In order to produce a point view of a line, the line of sight must be parallel to a true-length view of the line. If no given view of the line is true length, it is necessary to construct a true-length view before a point view can be obtained.

In Figure 5.4(a) let it be assumed that the front and top views of an oblique line EG are given, with a point view required. Neither e^Fg nor e^Hg is true length. Therefore, the first step is to construct a true-length auxiliary view. In this case the line of sight is selected perpendicular to the top view of line EG, resulting in the desired true-length view e^1g.

Line of sight 2 is then established parallel to the view e^1g, Figure 5.4(b). The resulting view e^2g is the required point view of line EG. Note that the distance D_1 is the same for any point on e^Hg.

The point view of a line is used as an intermediate step in a number of descriptive geometry constructions such as in §5.3 (see also §§8.1, 10.4, and 13.2). It may also be used to find the true distance from the line to some point in space as in the following "clearance" problem.

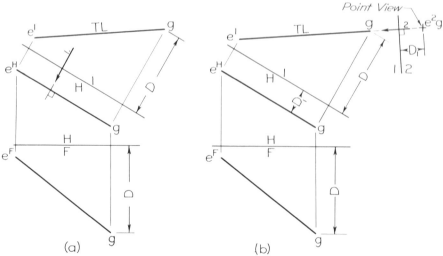

FIGURE 5.4   Point view of an oblique line

EXAMPLE: TRUE DISTANCE FROM A POINT TO A LINE (POINT-VIEW METHOD)

Problem

Find the true clearance between the spherical tank with its center at point O and the cylindrical pipe with center line XY, Figure 5.5.

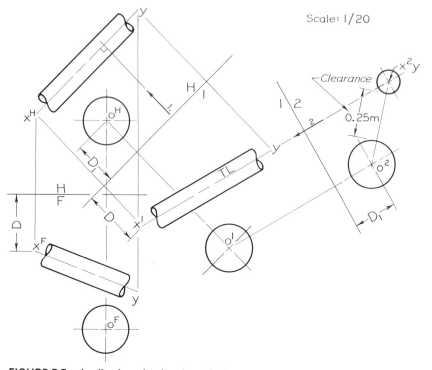

FIGURE 5.5   Application of point view of a line to a clearance problem

Analysis

If center line XY is viewed as a point, the true distance from XY to point O is observed and the desired clearance may be measured.

Graphic Solution

Line of sight 1 is assumed perpendicular to x^Hy in order to obtain the true-length auxiliary view x^1y. Point O is also projected to this view, appearing as o^1. Line of sight 2 is then introduced parallel to x^1y, resulting in the point view x^2y and the view o^2. This view also shows the true circular shape of the pipe. Since any view of the sphere is circular, the clearance between the two may be measured in view 2 as indicated.

5.3   Edge View of a Plane

Graphical representations of many engineering situations require the construction of views showing plane surfaces in *edge view* (EV).

A plane will appear in edge view in any view for which the line of sight is parallel to the plane. If the line of sight is to be parallel to the plane, it must be parallel to a line in the plane, thus producing a point view of that line. Consequently, it may be said that a plane will appear in edge view in any view in which a line in the plane appears as a point; that is, *a plane will appear in edge view in any view for which the line of sight is parallel to a true-length line in the adjacent view.*

Since the line of sight F in Figure 5.6(b) is parallel to the true-length view a^Hb, the front view shows line AB as a point and surface ABCE in edge view. In this instance, line EC is also shown in point view since it appears true length in the top view.

In the case of the oblique plane ABEC of Figure 5.7(b), line CE is horizontal and shows true length in the top view. Line of sight 1 is therefore assumed parallel to c^He, resulting in the edge view bc^1.

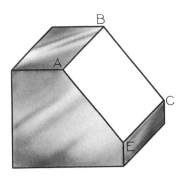
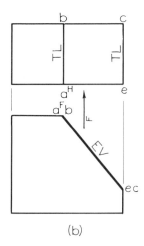

(b)

FIGURE 5.6   Edge view of a plane surface

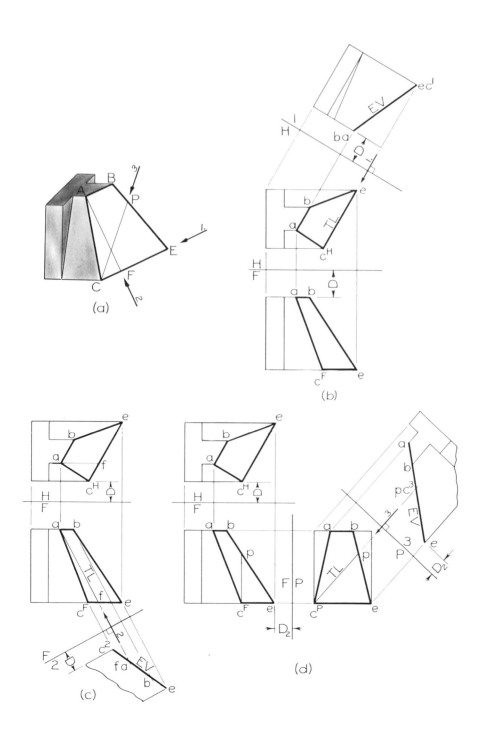

FIGURE 5.7 Edge views of an oblique plane surface

It is also possible to construct an edge view of the plane ABEC by projection from the front view, Figure 5.7(c). To do this, it is necessary to have a line in the plane appearing true length in the front view. Since none of the four edges of the surface meets this requirement, a frontal line AF is introduced in the plane. Line of sight 2 is then assumed parallel to the true-length front view af, resulting in the edge view c^2e.

In similar fashion, Figure 5.7(d), a profile line CP may be added to the plane. After the side view is drawn, line of sight 3 is assumed parallel to the true-length view c^Pp. The auxiliary view thus obtained, ae, is again an edge view of the same surface ABEC.

5.4   Normal View of a Plane

A *normal* or *true-size-and-shape* (TS) view of a plane is obtained in any view for which the line of sight is perpendicular to the plane. On the drawing paper this line of sight will appear perpendicular to the edge view of the plane.

Therefore, in obtaining a normal view of an oblique plane such as plane ABCE in Figure 5.8(a), the first step is to construct an edge view of this plane. A horizontal line EH is added to the plane and the line of sight 1 is established parallel to the true-length view e^Hh. The primary auxiliary view ac is an edge view as shown.

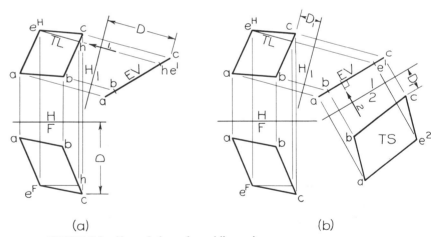

(a)   (b)

FIGURE 5.8   Normal view of an oblique plane

A second line of sight 2 is then introduced perpendicular to ac, Figure 5.8(b). Since this line of sight is perpendicular to the plane ABCE, the secondary auxiliary view of the plane, abce², is the required normal view.

In practice, normal views of plane surfaces are used to provide true size-and-shape views so that they may be properly dimensioned and clearly described to the shop technician or other user of the drawing, Figure 5.9.

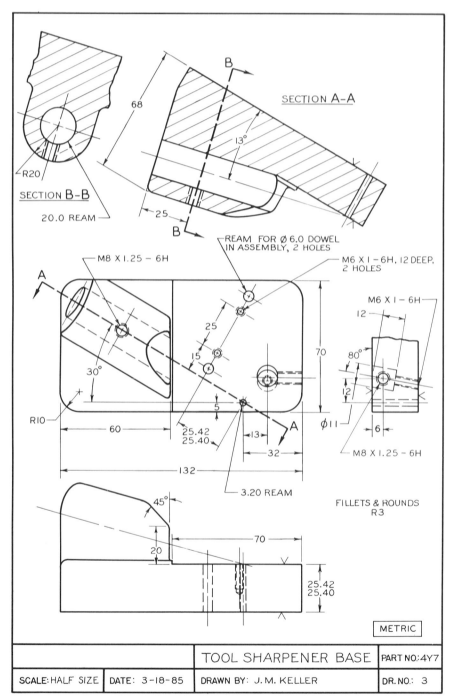

FIGURE 5.9   An industrial drawing employing successive auxiliary views

Normal views are also used in the solution of space problems involving plane geometry constructions, such as

1. Finding the shortest distance from a point to a line (plane method),
2. Measuring the angle between two intersecting lines,
3. Constructing the bisector of an angle,
4. Inscribing a circle in a triangle.

EXAMPLE: A PLANE GEOMETRY CONSTRUCTION IN AN OBLIQUE PLANE

Problem

Find the front and top views of a 2.5 m radius bend joining two pipes represented by their intersecting center lines BA and BC, Figure 5.10(a).

Analysis

The arc of the required bend must lie in the plane ABC. Its true circular shape will therefore show in a normal view of this plane.

Graphic Solution

To the given front and top views of ABC, a frontal line BF is added as shown, Figure 5.10(a). (Note the addition of the line AC to secure point F.) Since $b^F f$ is true length, auxiliary view 1 showing plane ABC in edge view may now be drawn as indicated. Line of sight 2 is established perpendicular to the edge view of the plane, and the normal view of the plane, ab^2c, is constructed. In this view the center of the 2.5 m radius arc is located such that the arc is tangent to lines b^2a and b^2c, Appendix B.10. The points of tangency are indicated as t and t_1.

Conveniently spaced intermediate points 1, 2, and 3 are now assumed on the arc in view 2, Figure 5.10(b). These points plus t and t_1 are projected to the edge view of the plane in view 1. They are then projected to the front view and located by means of transfer distances such as D_2.

The points are now established in the top view by projecting upward from the front view and transferring distances such as D_3 from view 1. The front and top views are completed by fairing smooth curves through the points. If required, the center O may be projected to these views as shown.

Frequently elliptical views of a circle in an oblique plane may be more conveniently drawn by establishing the major and minor axes and then employing the *trammel* method (Appendix B.6), an *ellipse template*, or a special device for drawing ellipses such as the *Ellipsograph*.

Figures 5.11 and 5.12 illustrate two methods by which the major and minor axes of such elliptical views may be found. Both methods are based on the facts that the major axis in any view must always be that diameter of the circle which appears true length in that view, and that the minor axis is perpendicular to the major axis. In Figure 5.11, a frontal line FF is drawn through the center point O, establishing the

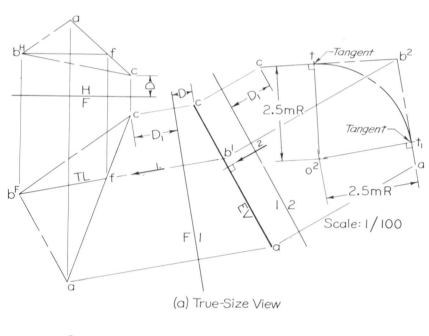

(a) True-Size View

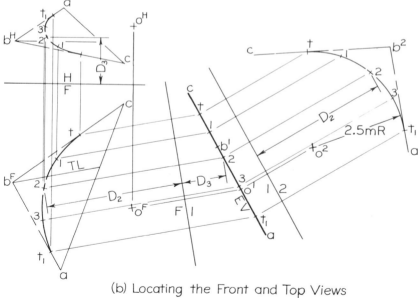

(b) Locating the Front and Top Views

FIGURE 5.10 Drawing a circular arc in an oblique plane

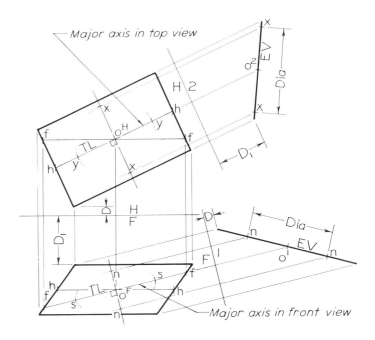

FIGURE 5.11   Finding the axes of the elliptical views of a circle in an oblique plane

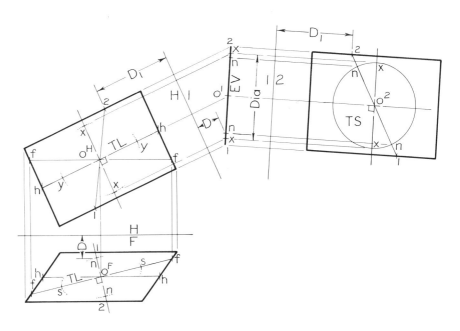

FIGURE 5.12   Finding the axes of the elliptical views of a circle in an oblique plane (alternative method)

direction of true-length lines in the front view. The major axis ss is then
set off as shown, equal to the diameter of the circle as given. Auxiliary
view 1, showing the edge view of the plane of the circle, is then constructed. Along this edge view the diameter of the circle is set off, establishing
points n which when projected to the front view determine the endpoints
of the minor axis in the front view.

In similar fashion the major axis yy in the top view lies on a true-
length line hh through o^H. Auxiliary view 2 is then employed to secure
the endpoints x and x of the minor axis in the top view. It should be
noted that the axes in the front view and in the top view are projections
of two entirely different pairs of diameters of the circle.

In Figure 5.12 the major axis yy and minor axis xx in the top view
and major axis ss in the front view have all been established as in Figure
5.11. As an alternative method, minor axis nn in the front view is
obtained as follows: Its direction is known to be at right angles to major
axis ss, and when extended it intersects lines of the given plane at
points 1 and 2. These points are projected to the top view, to view 1,
and finally to normal view 2, establishing line 1-2 in these views. In
view 2, line 1-2 intersects the circle at points n. These points are
returned to view 1, from which transfer distances such as D are used
to establish points n on line 1-2 in the front view.

PROBLEMS

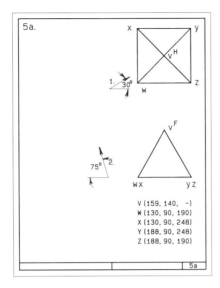

5a.

V (159, 140, -)
W (130, 90, 190)
X (130, 90, 248)
Y (188, 90, 248)
Z (188, 90, 190)

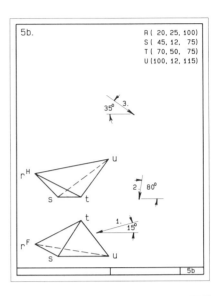

5b.

R (20, 25, 100)
S (45, 12, 75)
T (70, 50, 75)
U (100, 12, 115)

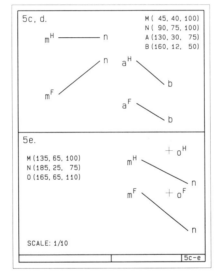

5c, d.

M (45, 40, 100)
N (90, 75, 100)
A (130, 30, 75)
B (160, 12, 50)

5e.

M (135, 65, 100)
N (185, 25, 75)
O (165, 65, 110)

SCALE: 1/10

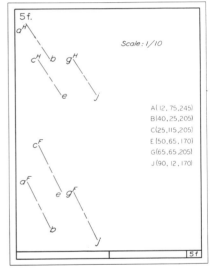

5f.

Scale : 1/10

A(12, 75,245)
B(40, 25,205)
C(25,115,205)
E(50,65,170)
G(65,65,205)
J(90, 12,170)

5a. Draw the successive auxiliary views 1 and 2 of the right square pyramid.

5b. Draw the successive auxiliary views 1, 2, and 3 of the tetrahedron.

5c. Obtain a point view of line MN.

5d. Obtain a point view of line AB.

5e. Determine the clearance between

a 100 mm diameter cylinder established by its center line MN and a 150 mm diameter sphere having its center at point O.

5f. Determine the minimum clearance between three 100 mm diameter pipes having the given center lines.

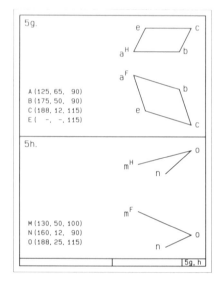

5g.

A (125, 65, 90)
B (175, 50, 90)
C (188, 12, 115)
E (-, -, 115)

5h.

M (130, 50, 100)
N (160, 12, 90)
O (188, 25, 115)

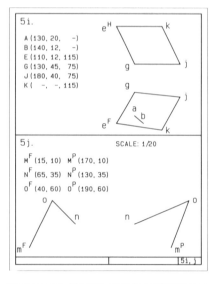

5i.

A (130, 20, -)
B (140, 12, -)
E (110, 12, 115)
G (130, 45, 75)
J (180, 40, 75)
K (-, -, 115)

5j.   SCALE: 1/20

M^F (15, 10) M^P (170, 10)
N^F (65, 35) N^P (130, 35)
O^F (40, 60) O^P (190, 60)

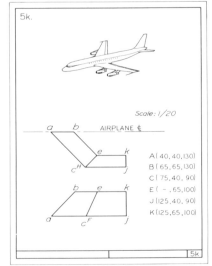

5k.

Scale: 1/20

AIRPLANE ₵

A (40, 40, 130)
B (65, 65, 130)
C (75, 40, 90)
E (-, 65, 100)
J (125, 40, 90)
K (125, 65, 100)

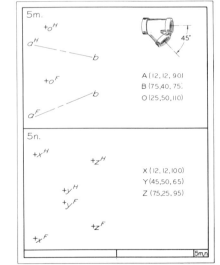

5m.

45°

A (12, 12, 90)
B (75, 40, 75)
O (25, 50, 110)

5n.

X (12, 12, 100)
Y (45, 50, 65)
Z (75, 25, 95)

5g. Establish the true size of the parallelogram ABCE and compute its area.

5h. Determine the magnitude of the angle MON. Show the bisector of this angle in all views.

5i. Line AB is one side of a regular hexagon lying in plane EGJK. Complete the views.

5j. Locate the views of the center of a 1m diameter pulley oriented to carry a belt from M to N around the turn near point O. If assigned, also plot the views of the pulley, neglecting thickness.

5k. Determine the true sizes of the airplane windshield ABCE and the side glass CEJK so that correct patterns may be cut. Compute the glass areas involved, neglecting waste.

5m. Locate the views of a branch pipeline from point O to connect with the main line AB with a standard 45° Y fitting.

5n. Find the views of the center of a circle passing through points X, Y, and Z.

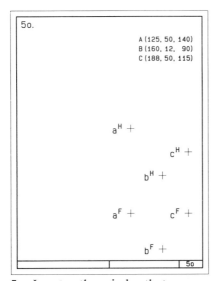

5o.

A (125, 50, 140)
B (160, 12, 90)
C (188, 50, 115)

a^H +

c^H +

b^H +

a^F + c^F +

b^F +

5o

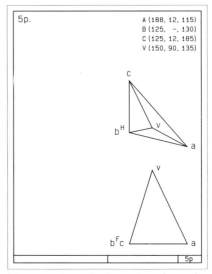

5p.

A (188, 12, 115)
B (125, -, 130)
C (125, 12, 185)
V (150, 90, 135)

5p

5o. Locate the circle that passes through points A, B, and C. Plot the front and top views of the circle and indicate the major and minor axes in these views.

5p. Determine which of the two planes ABV and BCV has the greater area. Show calculations.

5q. Indicate whether the following statements are true or false. If assigned, provide written explanations or sketches to justify the answers.

(1) No more than two successive auxiliary views are essential to obtain a normal view of a plane surface.

(2) Two successive projection planes are always at right angles.

(3) A point on the lateral surface of a cylinder always appears true distance from the center line of the cylinder.

(4) Any view of a sphere appears as a true circle.

(5) The true diameter of a circle is available in any view of the circle.

(6) The true length of any line in a plane is available in an edge view of the plane.

(7) In order to obtain a point view of a line, a line of sight may be selected parallel to any view of the line.

(8) The true angle between two lines is observed in a view that shows either one of the lines in true length.

SELF-TESTING PROBLEMS

5A. Add a complete auxiliary view of the tetrahedron that includes an edge view of surface M. Then add a complete auxiliary view that includes an edge view of surface N.

5B. Determine the angle formed by line AB and a line from point O to the midpoint of line AB.

5A.

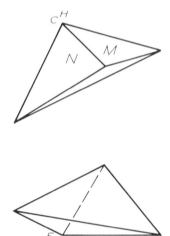

5B.

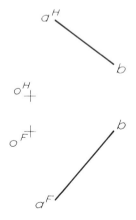

6 Piercing Points

Unless a line is in or parallel to a plane, it must intersect the plane. This intersection point, called a *piercing point*, may be within limits of the line segment or plane as given, or it may be necessary to extend one or both, in which case the piercing point can be considered imaginary. Such a point may prove useful in certain constructions. A line or a plane is considered indefinite in extent unless specific information to the contrary is available.

6.1 Piercing Point—Auxiliary-View Method

In Figure 6.1(a) let it be required to find by means of an auxiliary view the piercing point in plane ABC of line EG. The intersection of a line with a plane is a point common to both. An edge view of a plane contains all points in the plane. Therefore, in a view that shows the given plane in edge view, the point at which the given line intersects the edge view of the plane is the point common to both—the piercing point.

In Figure 6.1(a), line BC appears true length in the top view. Line of sight 1 is therefore assumed parallel to b^Hc, the result being the edge view acb^1. In this auxiliary view the piercing point p^1 appears as the intersection of e^1g and the edge view of the plane.

The top and front views of piercing point P, Figure 6.1(b), are found by projecting from p^1 to e^Hg to establish p^H and thence to the front view to establish p^F. The accuracy of the location may be checked by the distance D_1 as indicated. The visibility of line EG in the front and top views is then determined to complete the drawing (see §1.13).

Alternative Solution. The piercing point may also be determined by an edge view of the plane projected from the front view. This is not a different method but merely an alternative manner of applying the same method. In Figure 6.2 frontal line BF is added to procure a true-length line in the front view of the plane, and the views of the piercing point are found in a manner similar to that of Figure 6.1.

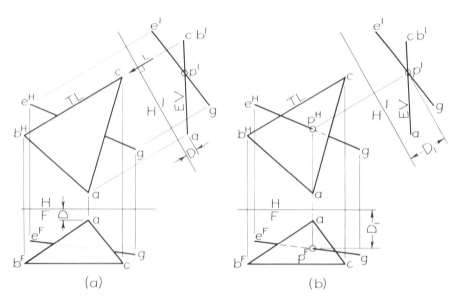

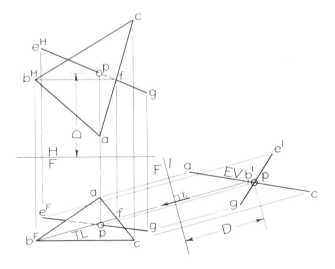

FIGURE 6.1   Piercing point—auxiliary-view method

FIGURE 6.2   Piercing point—auxiliary-view method (alternative solution)

6.2   Piercing Point—Two-View Method

The piercing point of line EG with plane ABC may be found using only the given views as follows, Figure 6.3:

- (a) Any convenient cutting plane containing line EG is introduced, Figure 6.3(b). A cutting plane perpendicular to one of the principal planes is convenient because it appears in edge view in a principal view. This simplifies the following step.
- (b) The line of intersection 1-2 between this cutting plane and plane ABC is determined.
- (c) Since lines EG and 1-2 both lie in the cutting plane, they intersect, locating point P.
- (d) Since line 1-2 also lies in plane ABC, point P is the required piercing point of line EG with the plane ABC.

In the multiview construction, Figure 6.3(c), a vertical cutting plane N containing line EG is introduced, with its top view an edge view containing e^Hg. The line of intersection 1-2 of the cutting plane and plane ABC coincides with the cutting plane in the top view. Since line 1-2 lies in plane ABC, its front view is located by projection of points 1 and 2 to the corresponding lines of the plane in the front view. The intersection of 1-2 and e^Fg locates the front view of the piercing point.

The top view of the piercing point is determined by a vertical projection line from the front view, Figure 6.3(d). The determination of the visibility of the line EG, if desired, completes the solution.

Alternative Solution. The foregoing problem may also be solved by introducing a cutting plane M that contains line EG and appears in edge view in the front view, Figure 6.4. Its line of intersection 3-4 with plane ABC when projected to the top view determines the top view of the piercing point. A vertical projection line then locates the front view of the piercing point, and the solution is completed as in the preceding example.

A Special Case. In Figure 6.5 the two-view method is shown applied to a vertical line and an oblique plane. Since the top view of the given vertical line XY is a point view, any cutting plane containing the line appears in edge view in the top view. A cutting plane M is selected that produces a line of intersection 1-2 meeting x^Fy at an angle large enough to establish the piercing point accurately. The top view of the piercing point coincides with x^Hy. In this case the piercing point is outside triangle ABC and may be considered imaginary. Imaginary piercing points, however, are often useful in certain constructions.

While the two-view method has the advantage of brief construction and minimum space requirements, the auxiliary-view method is often the easier of the two methods for the beginning student to comprehend. The auxiliary-view method is also frequently advantageous when several lines pierce the same plane (see, for instance, §16.4).

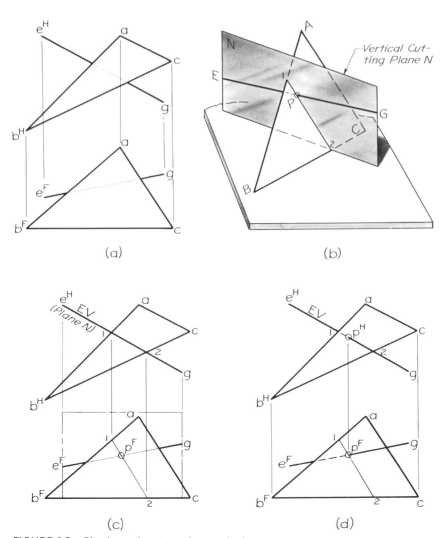

FIGURE 6.3   Piercing point—two-view method

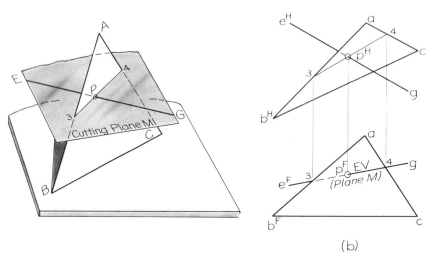

(b)

FIGURE 6.4   Piercing point—two-view method (alternative solution)

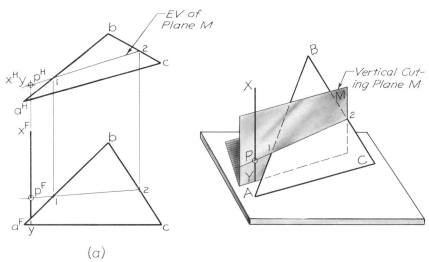

(a)

FIGURE 6.5   Piercing point of a vertical line and an oblique plane—two-view method

6.3 Pictorial Intersection of Line and Plane

An initial introduction to a series of pictorial intersection concepts was presented in §4.7. The solution to the problem of Figure 6.6(a) required a combination of those previous concepts and the use of a cutting plane for establishing the intersection of line AB and plane MNK.

Since lines MN and AT both exist in the same plane P of the isometric pictorial outline, Figure 6.6(b), these lines must intersect or be

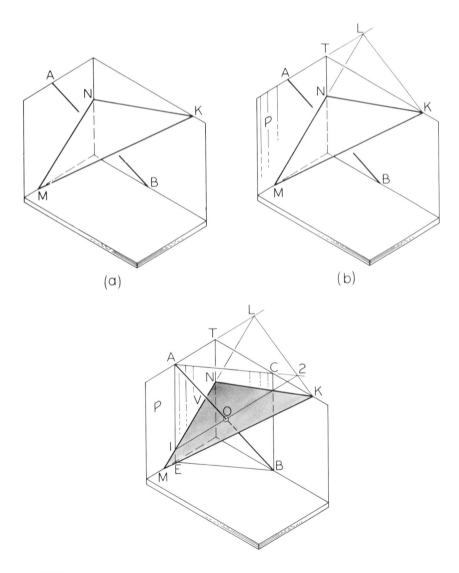

FIGURE 6.6   Pictorial intersection of line and plane

parallel. The two lines are extended as shown to locate their intersection point L. Now since points L and K both lie in the same imaginary upper horizontal surface of the pictorial outline, line LK represents a horizontal line of plane MNK.

Next a vertical cutting plane ACBE is introduced containing line AB, Figure 6.6(c). Since lines AE and MN both lie in plane P, their intersection point 1 represents one point common to plane MNK and cutting plane ACBE. Similarly, since lines LK and AC both lie in the same upper horizontal plane of the pictorial, their crossing point 2 locates a second point common to plane MNK and the vertical cutting plane ACBE. Now since lines 1-2 and AB each lie in the same cutting plane, their crossing point O is their actual intersection point. But since line 1-2 also lies in plane MNK, point O provides the piercing point of line AB with the oblique plane MNK.

Visibility of line AB and plane MNK is determined by observation of the relative orientation of lines AB and MN in the pictorial at the apparent crossing point V.

PROBLEMS

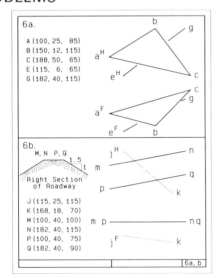

6a.
A (100, 25, 85)
B (150, 12, 115)
C (188, 50, 65)
E (115, 6, 65)
G (182, 40, 115)

6b.
M, N P, Q
Right Section of Roadway
J (115, 25, 115)
K (168, 18, 70)
M (100, 40, 100)
N (182, 40, 115)
P (100, 40, 75)
Q (182, 40, 90)

6a, b

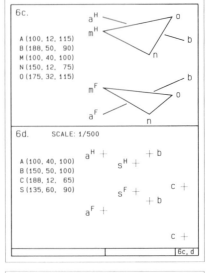

6c.
A (100, 12, 115)
B (188, 50, 90)
M (100, 40, 100)
N (150, 12, 75)
O (175, 32, 115)

6d. SCALE: 1/500
A (100, 40, 100)
B (150, 50, 100)
C (188, 12, 65)
S (135, 60, 90)

6c, d

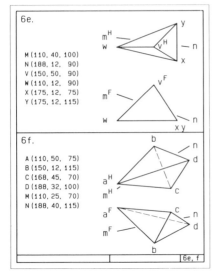

6e.
M (110, 40, 100)
N (188, 12, 90)
V (150, 50, 90)
W (110, 12, 90)
X (175, 12, 75)
Y (175, 12, 115)

6f.
A (110, 50, 75)
B (150, 12, 115)
C (168, 45, 70)
D (188, 32, 100)
M (110, 25, 70)
N (188, 40, 115)

6e, f

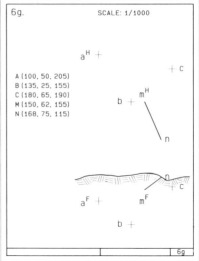

6g. SCALE: 1/1000
A (100, 50, 205)
B (135, 25, 155)
C (180, 65, 190)
M (150, 62, 155)
N (168, 75, 115)

6g

6a. Find the intersection of line EG with the plane ABC. Show complete visibility.

6b. Locate the views of the points at which the center line JK of a culvert pierces the embankment of the 1.5:1 fill of the roadway.

6c. Complete the views of the intersecting line AB and the plane MNO.

6d. Determine the depth of a vertical shaft from point S on the earth's surface to a thin ore vein ABC.

6e. Establish the intersection of line MN with the surfaces of the pyramid. Omit that portion of the line within the solid.

6f. Determine the intersection of line MN and the surfaces of the tetrahedron.

6g. Find the true length of the extension of inclined shaft MN needed to reach an ore vein determined by points A, B, and C.

95

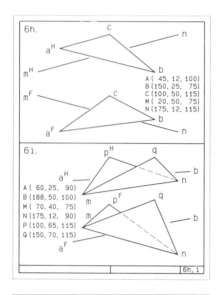

6h.

C
n
a^H
m^H
b
m^F
C
b
a^F
n

A (45, 12, 100)
B (150, 25, 75)
C (100, 50, 115)
M (20, 50, 75)
N (175, 12, 115)

6i.

p^H
q
a^H
b
n
m
p^F
q
m
b
a^F
n

A (60, 25, 90)
B (188, 50, 100)
M (70, 40, 75)
N (175, 12, 90)
P (100, 65, 115)
Q (150, 70, 115)

6h, i

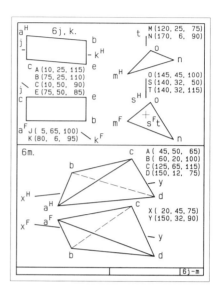

6j, k.

a^H
j
b
k^H
t
M (120, 25, 75)
N (170, 6, 90)
O
n
m^H

C A (10, 25, 115)
B (75, 25, 110)
C (10, 50, 90)
E (75, 50, 85)

e
O (145, 45, 100)
S (140, 32, 50)
T (140, 32, 115)
s^H
O

j
C
e
m^F
s^F
t
a^F J (5, 65, 100)
K (80, 6, 95)
k^F
n

6m.

C
A (45, 50, 65)
B (60, 20, 100)
C (125, 65, 115)
D (150, 12, 75)
b
y
x^H
d
a^H
C
X (20, 45, 75)
Y (150, 32, 90)
x^F
a^F
y
b
d

6j-m

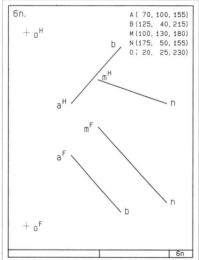

6n.

A (70, 100, 155)
B (125, 40, 215)
M (100, 130, 180)
N (175, 50, 155)
O (20, 25, 230)

+ o^H
b
m^H
a^H
n
m^F
a^F
b
n
+ o^F

6n

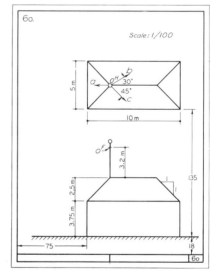

6o.

Scale: 1/100

5 m
Q
o^H
30°
b
45°
C
10 m

o^F
3.2 m
135
2.5 m
3.75 m
75
18

6o

6h. Determine the intersection of line MN and the plane ABC.

6i. Establish the piercing points of line AB with the given planes, which have edge MN in common.

6j. Locate the intersection of line JK with the plane ABCE.

6k. Locate the intersection of line ST with the plane MNO.

6m. Find the piercing points of line XY with the surfaces of the tetrahedron.

6n. Draw a line through point O and intersecting the skew lines AB and MN.

6o. Locate the intersections of guy wires OA, OB, and OC with the surfaces to which they are attached if each wire makes an angle of 30° with the vertical mast.

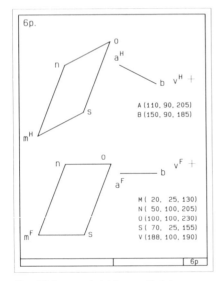

6p.

O
n   a^H
    b   v^H +

A (110, 90, 205)
B (150, 90, 185)

m^H S

n   O
         b   v^F +
    a^F

M (20, 25, 130)
N (50, 100, 205)
O (100, 100, 230)
S (70, 25, 155)
m^F S
V (188, 100, 190)

6p

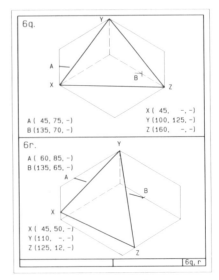

6q.

Y
A
    B +
X   Z

X (45,   -, -)
Y (100, 125, -)
Z (160,   -, -)

A (45, 75, -)
B (135, 70, -)

6r.

A (60, 85, -)
B (135, 65, -)

Y
A
    B
X
            Z

X (45, 50, -)
Y (110, -, -)
Z (125, 12, -)

6q, r

6p. Using point V as a light source, determine the shadow of line AB on the plane MNOS.

6q. In the isometric pictorial, find the piercing point of line AB and plane XYZ including proper visibility.

6r. In the isometric pictorial, find the piercing point of line AB and plane XYZ including proper visibility.

6s. Indicate whether the following statements are true or false. If assigned, provide written explanations or sketches to justify the answers.

(1) A straight line can intersect a plane surface at only one point.

(2) A straight line can intersect the surfaces of a regular prism at not more than two points including imaginary points.

(3) A line parallel to a plane surface does not intersect that surface.

(4) For the most convenient construction, a cutting plane used for finding the intersection of a line with a plane should be an edge-view cutting plane.

(5) The visibility of a line intersecting a limited plane surface changes at the piercing point of the line and plane.

(6) The intersection of a cutting plane with a plane surface is a single straight line.

SELF-TESTING PROBLEMS

6A. Determine the intersection of line CE and the surfaces of the pyramid. Omit the portion of line CE that is inside the pyramid.

6B. Determine the intersection of line MN and the surfaces of the tetrahedron. Omit the portion of line MN that is inside the tetrahedron.

6A.

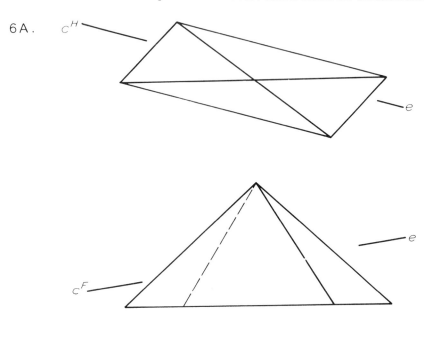

6 B.

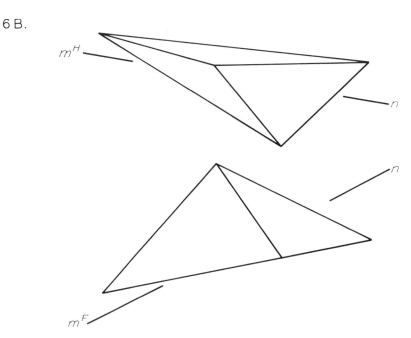

6C. In the isometric pictorial, obtain the piercing point of line AB in the oblique plane MNKS.

6D. Locate the piercing points of the line AB with the surfaces of the prism.

6C.

6D.

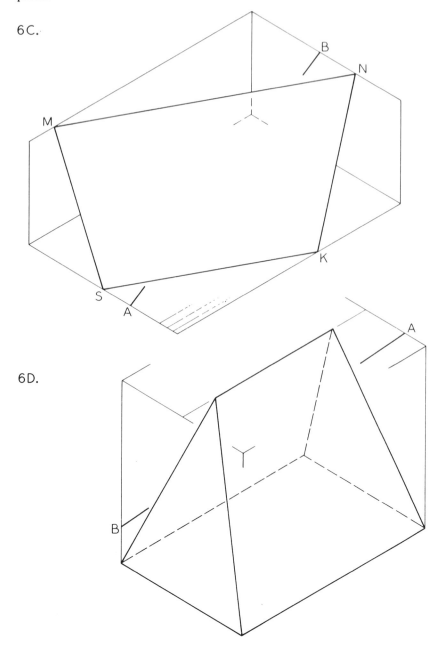

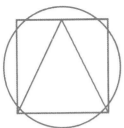

7 Intersection of Planes

Any two planes either must be parallel or they must intersect, even if the intersection falls beyond the limits of the planes as given. The intersection of two planes is a straight line common to the planes and its position is therefore determined by any two points common to both planes. For graphical accuracy the two points should be separated by a reasonable distance (see Appendix A).

Since a plane is unlimited in extent unless otherwise specified, the line of intersection of two planes is, in general, of indefinite length. It frequently occurs, however, that a given plane is bounded by a closed figure, as in the case of a plane surface of a solid object, in which instance the plane may be regarded as limited. The "real" portion of the line of intersection of such a plane and another plane would be limited correspondingly.

Points common to any two planes may be found by any one of three general methods: the *auxiliary-view* method; the *two-view*, *piercing-point* method; or the *cutting-plane* method.

7.1   Intersection of Two Planes—Auxiliary-View Method

The point in which a line in one plane pierces another plane is a point common to the two planes. As was discussed in §6.1, the point in which a line pierces a plane is shown in any view that shows the plane in edge view. Consequently, if it is required to find the line of intersection of two planes, any view showing one of the planes in edge view secures the piercing points of the lines of the other plane. Any two piercing points, if a sufficient distance apart for graphical accuracy, locate the required line of intersection. Of course, if a line is parallel to a plane, it does not pierce the plane, or if a line is nearly parallel to the plane, the piercing point may fall outside the working area of the drawing.

For the two planes in Figure 7.1(a), plane EGJK is arbitrarily selected to be shown in edge view. Horizontal line JH is added to plane EGJK as shown, and the desired edge view is projected from the top view. In the resulting view 1, the piercing points X and Y of lines AB and BC, respectively, appear as x and y. The required line of intersection (LI) of the two planes passes through the points X and Y and therefore coincides with the edge view of plane EGJK in view 1.

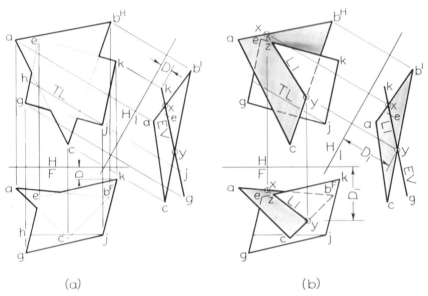

(a)   (b)

FIGURE 7.1   Intersection of two limited planes—auxiliary-view method

The top and front views of line XY are located by projecting from x and y to ab^H and b^{H}c in the top view, and then to ab^F and b^{F}c in the front view, Figure 7.1(b). In this problem the given planes are limited and the line of intersection may therefore be regarded as terminating at points Y and Z. The visibility of the remaining lines is then determined and the drawing completed as shown.

7.2   Intersection of Two Planes—Two-View, Piercing-Point Method

Since the piercing point of a line in a plane may be determined by using only the given two views (§6.2), the line of intersection of two planes may be located by applying this piercing-point method twice, or more times if necessary for accuracy.

In Figure 7.2(a) line DE of plane DEG is first selected as convenient for this method. Accordingly, an imaginary edge-view cutting plane is assumed coinciding with de^F. Its line of intersection 1-2 with plane ABC is then located in the side view by projecting points 1 and 2 to lines ac^P and bc^P, respectively. Since lines DE and 1-2 both lie in this plane, they are intersecting lines, and the point of intersection X, determined in the side view and projected to the front view, is the piercing point of line DE in plane ABC. Point X is therefore common to the given planes ABC and DEG and is one point on the required line of intersection.

The foregoing process is repeated for some other line of either plane—in this example, line EG, Figure 7.2(b). An edge-view cutting plane is

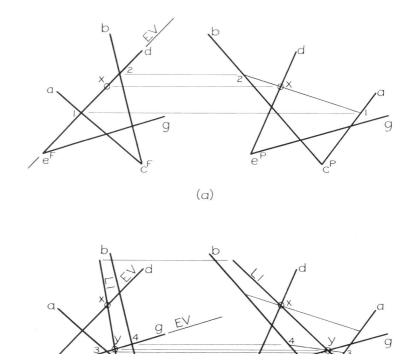

(a)

(b)

FIGURE 7.2   Intersection of two planes—two-view, piercing-point method

assumed coinciding with line e^{F}g to secure a second point Y common to both planes. The views of the line of intersection (LI) pass through the respective projections of X and Y as shown. It will be noted that the given planes in this problem are not limited. Visibility is· usually not considered in such cases, and the line of intersection may be drawn to any selected length.

If the two piercing points determined by the foregoing method are relatively close together, a third piercing point may be secured to assure the accurate location of the line of intersection. Since the line of intersection is common to the planes, it must intersect or be parallel to all lines in the planes. This fact may also be used to check the accuracy of the solution as suggested by intersection point 5 in Figure 7.2(b).

7.3 Intersection of Two Planes—Cutting-Plane Method

If two nonparallel planes are intersected by a third plane (not parallel to the intersection of the first two), the resulting lines of intersection meet at a point common to all three planes. Two or more such cutting planes may be employed to secure a corresponding number of points common to the two given planes, thus establishing the line of intersection of these two planes.

In Figure 7.3(a) the front and top views of two planes are given, and it is required to find their line of intersection through the use of cutting planes according to the preceding principle.

For convenience, edge-view cutting planes are used. These may be horizontal, frontal, profile, or inclined planes, or any combination of

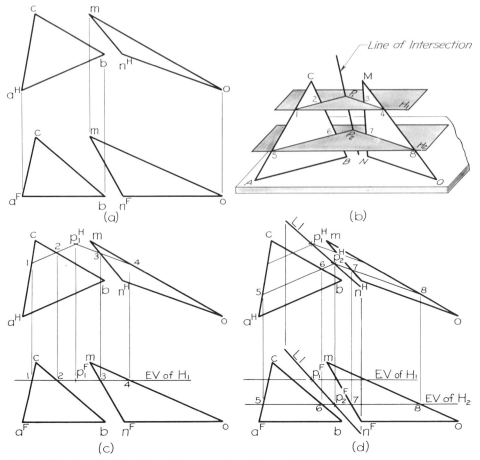

FIGURE 7.3 Intersection of two planes—cutting-plane method

these, whichever seems appropriate in a particular problem. In Figure 7.3(b), two horizontal cutting planes used in the solution are shown in pictorial form. It will be noted that cutting plane H_1 intersects planes ABC and MNO in two lines, which in turn intersect each other at point P_1 on the required line of intersection. This construction is shown in multiview form in Figure 7.3(c). The horizontal cutting plane H_1 appears as a horizontal line in the front view. Lines 1-2 and 3-4 coincide with the horizontal cutting plane H_1 in this view, but when projected to the top view they intersect, when extended, at point $p_1{}^H$. A vertical projection line from $p_1{}^H$ establishes $p_1{}^F$ and completes the location of a point common to the two given planes.

In similar fashion a second horizontal cutting plane H_2, Figure 7.3(b), intersects the given planes in lines that, when extended, locate point P_2 on the line of intersection of the given planes. This construction is shown in multiview form in Figure 7.3(d). Lines drawn through points $p_1{}^F$ and $p_2{}^F$, and $p_1{}^H$ and $p_2{}^H$, respectively, represent the front and top views of the line of intersection (LI) of the planes ABC and MNO. The length of the line of intersection is arbitrary but the ends of the lines should lie on common projectors.

7.4   Pictorial Intersection of Planes

Some constructions dealing with the topic of pictorial intersections were presented in §§4.7 and 6.3. A slight extension of these earlier constructions, together with the ideas of Chapter 7, are involved in the solution of the problem shown in Figure 7.4(a). Here the line of intersection and related visibility of the limited oblique planes ABC and MNK are desired.

To obtain a horizontal line in plane ABC, a vertical cutting plane ADES, which contains line AC, is introduced, Figure 7.4(b). Since AC and SE both lie in this cutting plane, these two lines extended intersect at point O. Owing to the fact that point O and point B both lie in the horizontal base plane of the pictorial, line BO represents a horizontal line of plane ABC.

Line BO of plane ABC and line MK of plane MNK each exist in the same horizontal base plane. They therefore intersect at point 2, which represents a point on the extended line of intersection of the two planes, Figure 7.4(c). A second point on the line of intersection is obtained by the introduction of horizontal line N3 in plane MNK parallel to MK, and then a horizontal line A3 in plane ABC is added parallel to BO. These two lines exist in the same upper horizontal plane of the pictorial outline, and therefore crossing point 3 locates the desired second point on the extended line of intersection of the given oblique planes.

Line of intersection 2-3 is drawn as shown, with the bold portion of this line terminated at the limiting outlines of the given planes. Appropriate visibility of the intersecting planes may be established by a pictorial analysis of the relative location of lines AC and NK at position V.

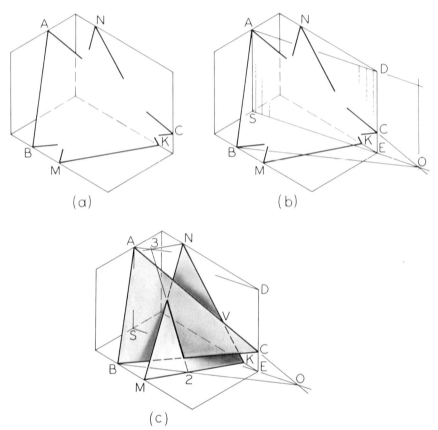

(a)

(b)

(c)

FIGURE 7.4   Pictorial intersection of planes

PROBLEMS

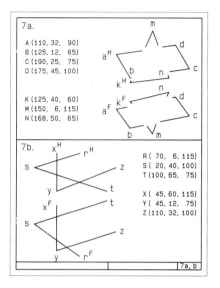

7a.

A (110, 32, 90)
B (125, 12, 65)
C (190, 25, 75)
D (175, 45, 100)

K (125, 40, 60)
M (150, 6, 115)
N (168, 50, 65)

7b.

R (70, 6, 115)
S (20, 40, 100)
T (100, 65, 75)

X (45, 60, 115)
Y (45, 12, 75)
Z (110, 32, 100)

| | 7a, b |

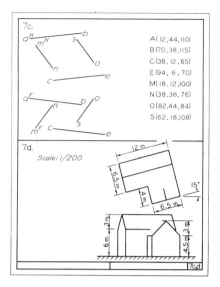

7c.

A (12, 44, 110)
B (70, 38, 115)
C (38, 12, 65)
E (94, 6, 70)
M (18, 12, 100)
N (38, 38, 76)
O (82, 44, 84)
S (62, 18, 108)

7d.

Scale: 1/200

| | 7c, d |

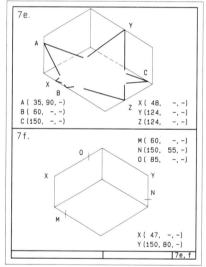

7e.

A (35, 90, -)
B (60, -, -)
C (150, -, -)

X (48, -, -)
Y (124, -, -)
Z (124, -, -)

7f.

M (60, -, -)
N (150, 55, -)
O (85, -, -)

X (47, -, -)
Y (150, 80, -)

| | 7e, f |

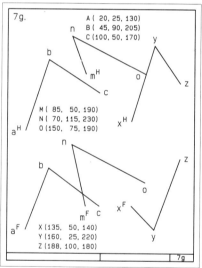

7g.

A (20, 25, 130)
B (45, 90, 205)
C (100, 50, 170)

M (85, 50, 190)
N (70, 115, 230)
O (150, 75, 190)

X (135, 50, 140)
Y (160, 25, 220)
Z (188, 100, 180)

| | 7g |

7a. By the auxiliary-view method determine the intersection of the planes. Show visibility.

7b. Using the auxiliary-view method, locate the intersection of the planes.

7c. By the auxiliary-view method find the line of intersection of the planes determined by the sets of parallel lines.

7d. Complete the plan and elevation views.

7e. In the isometric pictorial, find the line of intersection of planes ABC and XYZ. Show visibility.

7f. In the isometric pictorial, find the line of intersection of plane MNO with the prism. Show visibility.

7g. Locate the point common to the three given planes.

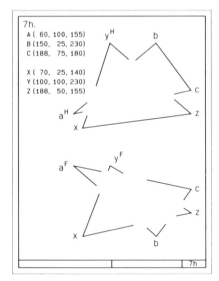

7h.
A (60, 100, 155)
B (150, 25, 230)
C (188, 75, 180)

X (70, 25, 140)
Y (100, 100, 230)
Z (188, 50, 155)

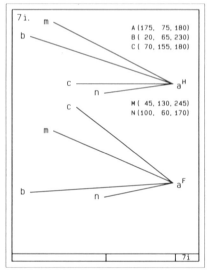

7i.
A (175, 75, 180)
B (20, 65, 230)
C (70, 155, 180)

M (45, 130, 245)
N (100, 60, 170)

7h. Determine the intersection of the given planes and show the correct visibility.

7i. Determine a point common to the planes ABC and AMN and 50 mm from point A.

7j. Indicate whether the following statements are true or false. If assigned, provide written explanations or sketches to justify the answers.

(1) Two nonparallel planes intersect in a single straight line.

(2) Three nonparallel planes have a single point in common.

(3) If two parallel planes are intersected by a third plane, the lines of intersection are parallel.

(4) Two perpendicular planes intersect in a single point.

(5) A line of intersection of any two planes appears true length in either a top, front, or side view.

(6) The line of intersection of two planes is visible if no other planes are present.

SELF-TESTING PROBLEMS

7A. By the auxiliary-view method, determine the intersection of the planes. Show correct visibility.

7B. Using only the given views, determine the intersection and show the correct visibility of the planes.

7A.

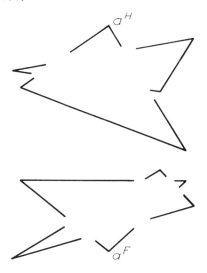

7B.

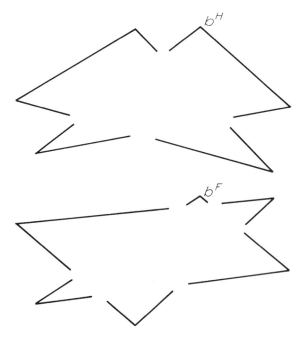

7C. Establish the line of intersection and the appropriate visibility of the intersecting planes ABC and MNK.

7D. Locate the intersection of unlimited plane MNK and the surfaces of the prism in the oblique pictorial. Show the appropriate visibility.

7C.

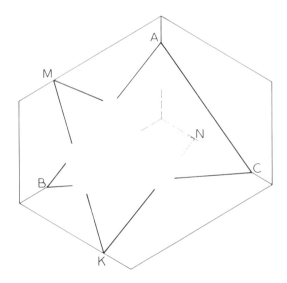

7D.

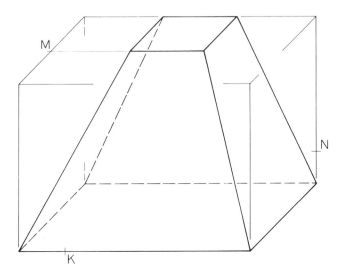

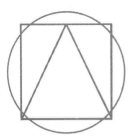

8 Angle Between Planes

The angle formed by two intersecting planes is called a dihedral angle, Figure 8.1(a). The true size of the dihedral angle is observed in a view in which each of the given planes appears in edge view, Figure 8.1(b). Since a plane is seen edgewise in any view in which a line in the plane appears as a point (see §5.4), a view showing a point view of a line common to two planes, the line of intersection, produces an edge view of each of the planes.

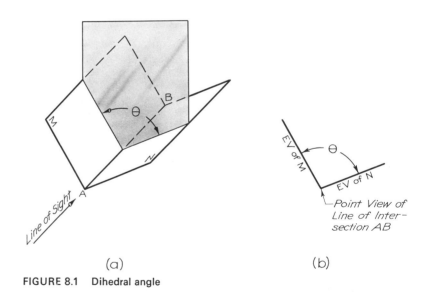

(a)   (b)

FIGURE 8.1   Dihedral angle

8.1   Dihedral Angle—Line of Intersection Given

In Figure 8.2, an application of the foregoing principles to an actual problem is illustrated. In order to manufacture or fabricate the *transition piece*, it is necessary to know the dihedral angles for the special bent-plate angles used at the corners. The construction shown illustrates the method of obtaining the angle for one of the bent-plate angles—the angle between surfaces A and B of the transition piece. A complete working drawing would, of course, include views showing all necessary angles.

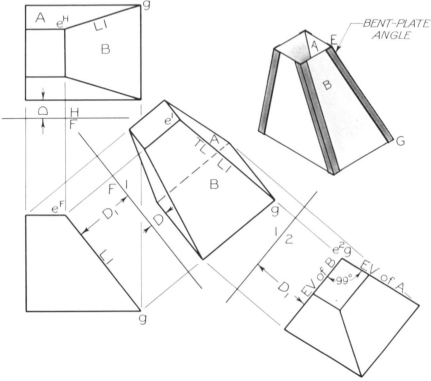

FIGURE 8.2   Dihedral angle—line of intersection given

In this problem, as is the case with many practical problems, the line of intersection EG of the two planes is known. To produce a point view of line EG, it is first necessary to construct view 1, which includes the true length of line EG. From view 1, view 2 is then projected, with a line of sight assumed parallel to e^1g. The resulting view 2 shows line EG as a point and consequently shows planes A and B in edge view. The desired dihedral angle may then be measured as indicated.

It will be noted in this example that the desired angle is greater than 90°. Theoretically, two intersecting planes form two dihedral angles which are supplementary (totaling 180°). In the absence of evidence to the contrary, it is customary to dimension the acute angle, Figure 8.3(c). In most practical examples, however, as in Figure 8.2, it is evident which of the two angles is desired.

8.2   Dihedral Angle—Line of Intersection Not Given

Occasionally it is necessary to find the angle between two planes for which the line of intersection is not a part of the original drawing, Figure 8.3(a). In such cases, one method of solution requires the determination of the line of intersection.

In Figure 8.3(b) the two-view, piercing-point method of §7.2 is employed to obtain the line of intersection XY of planes ABC and KMNO. Successive auxiliary views are then used to obtain the true length and point views of XY and the angle θ between the planes, Figure 8.3(c).

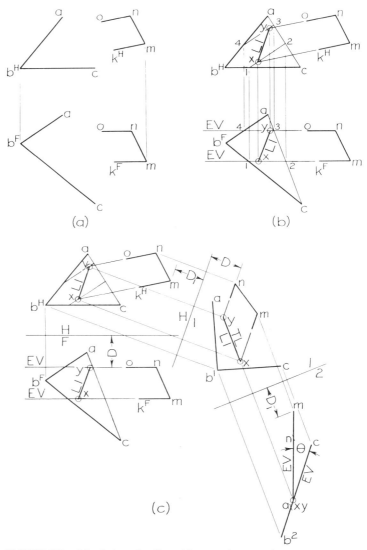

FIGURE 8.3   Dihedral angle—line of intersection not given

Alternative Solution. Figure 8.4 illustrates an alternative method for determining the dihedral angle in which it is not necessary to find the line of intersection.

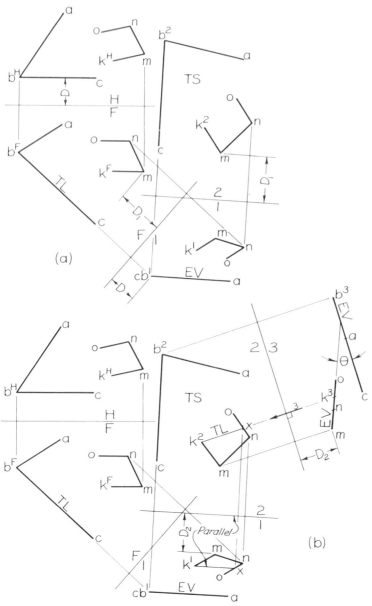

FIGURE 8.4   Dihedral angle—line of intersection not given (alternative method)

Since line BC appears true length in the front view, auxiliary view 1 may be added as shown to obtain an edge view of plane ABC, Figure 8.4(a). Auxiliary view 2 is then constructed, showing plane ABC in true size. Plane KMNO is also projected to these auxiliary views.

In Figure 8.4(b), a line KX is added to plane KMNO, with k^1x drawn parallel to folding line 1/2 so that k^2x will be true length. A line of sight for auxiliary view 3 is assumed parallel to k^2x. The resulting view 3 shows both planes in edge view, and the dihedral angle may be measured as indicated. This method is particularly appropriate when the line of intersection of the given planes falls outside the working area of the drawing.

8.3   Angle Between Oblique Plane and Principal Plane

A special application of the determination of the angle between two planes occurs when one of the planes is a horizontal, profile, or frontal plane. In Figure 8.5 it is desired to find the angle between plane ABC and a frontal plane. Any frontal plane appears as a line horizontal in the top view on the drawing paper. If such a plane is introduced, passing through a^H in Figure 8.5, its line of intersection with plane ABC is the frontal line AF.

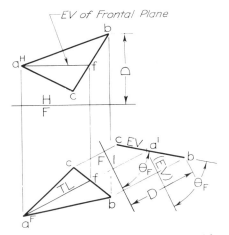

FIGURE 8.5   Angle between plane and frontal plane

Since line AF appears in true length in the front view, auxiliary view 1 shows the frontal plane and plane ABC in edge view, and the required angle is shown as indicated. Of course *any* frontal plane would appear in view 1 as a line parallel to folding line F/1. Hence the angle θ_F may be measured between the edge view of the plane, a^1bc and F/1, without the necessity of introducing an additional frontal plane.

In Figure 8.6 construction is shown for obtaining the angle θ_P between plane ABC and a profile plane. In this case it is necessary to add the side view to secure the true length of the profile line of intersection CP. Otherwise the construction corresponds to that of Figure 8.5.

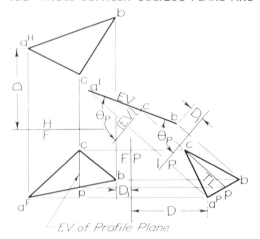

FIGURE 8.6   Angle between plane and profile plane

The angle between a plane and a horizontal plane can be found by adding a horizontal line to the oblique plane and proceeding in a similar fashion. In geology and mining engineering the angle between a sloping plane and a horizontal plane is called the *dip angle* of the sloping plane (see §12.4).

PROBLEMS

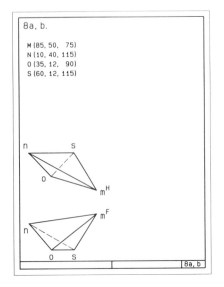

8a, b.

M (85, 50, 75)
N (10, 40, 115)
O (35, 12, 90)
S (60, 12, 115)

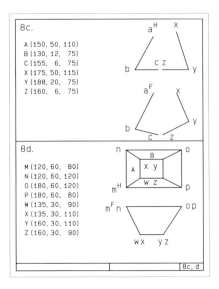

8c.

A (150, 50, 110)
B (130, 12, 75)
C (155, 6, 75)
X (175, 50, 115)
Y (188, 20, 75)
Z (160, 6, 75)

8d.

M (120, 60, 80)
N (120, 60, 120)
O (180, 60, 120)
P (180, 60, 80)
W (135, 30, 90)
X (135, 30, 110)
Y (160, 30, 110)
Z (160, 30, 90)

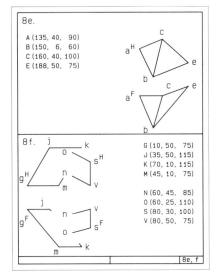

8e.

A (135, 40, 90)
B (150, 6, 60)
C (160, 40, 100)
E (188, 50, 75)

8f.

G (10, 50, 75)
J (35, 50, 115)
K (70, 10, 115)
M (45, 10, 75)

N (60, 45, 85)
O (60, 25, 110)
S (80, 30, 100)
V (80, 50, 75)

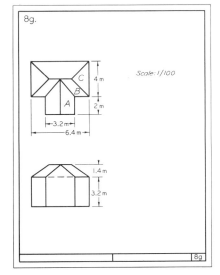

8g.

Scale: 1/100

4 m
2 m
3.2 m
6.4 m

1.4 m
3.2 m

8a. Find the true size of the dihedral angle formed by planes MNO and MNS.

8b. Find the true size of the dihedral angle formed by planes MNS and MOS.

8c. Find the line of intersection of, and the dihedral angle between, the two planes.

8d. Determine the true size of the angle formed by planes A and B so that a corner reinforcing plate can be correctly bent.

8e. Find the angle formed by the planes ABC and BEC.

8f. Find the line of intersection of, and the dihedral angle between, the given planes GJKM and NOVS.

8g. Determine the dihedral angles formed by roof planes A and B and the roof planes B and C.

116

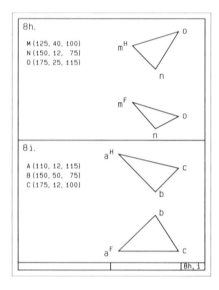

8h.

M (125, 40, 100)
N (150, 12, 75)
O (175, 25, 115)

8i.

A (110, 12, 115)
B (150, 50, 75)
C (175, 12, 100)

8h, i

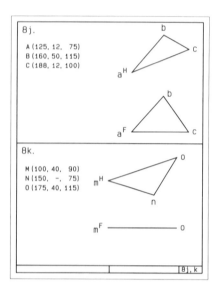

8j.

A (125, 12, 75)
B (160, 50, 115)
C (188, 12, 100)

8k.

M (100, 40, 90)
N (150, -, 75)
O (175, 40, 115)

8j, k

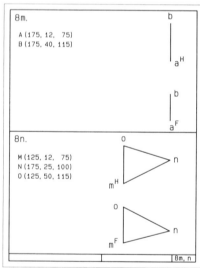

8m.

A (175, 12, 75)
B (175, 40, 115)

8n.

M (125, 12, 75)
N (175, 25, 100)
O (125, 50, 115)

8m, n

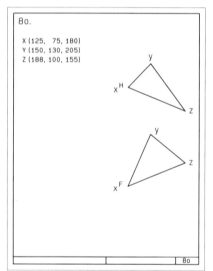

8o.

X (125, 75, 180)
Y (150, 130, 205)
Z (188, 100, 155)

8o

8h. Determine the angle that plane MNO makes with a horizontal plane.

8i. Find the angle that plane ABC makes with a frontal plane.

8j. Determine the angle formed by plane ABC and a profile plane.

8k. Complete the front view of plane MNO that makes an angle of $30°$ with a horizontal plane.

8m. Line AB lies in a plane that makes an angle of $30°$ with a profile plane. Locate a point S on the plane that lies 12 mm above B and 20 mm to the right of B.

8n. Determine the angle that plane MNO makes with a horizontal plane. Locate a point K on the plane that lies 24 mm to the left of N and 18 mm above N.

8o. Determine the angles that plane XYZ makes with the principal planes H, F, and P.

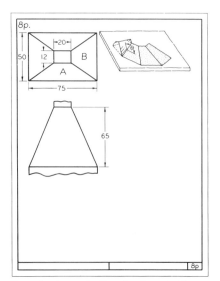

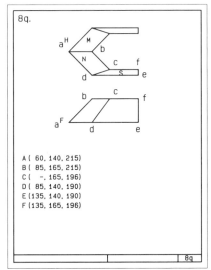

A (60, 140, 215)
B (85, 165, 215)
C (-, 165, 196)
D (85, 140, 190)
E (135, 140, 190)
F (135, 165, 196)

8p. The layout of a sheet-metal hopper is illustrated in its developed flat position and also with one face bent to its final angular position.

Find the true size of the bend angle θ formed by faces A and B.

8q. In order that support clips may be formed for the airplane windshield sections, determine the angles formed by planes M and N and by N and S.

8r. Indicate whether the following statements are true or false. If assigned, prepare written explanations or sketches to justify the answers.

(1) A view having a direction of sight parallel to a true-length view of the intersection of two planes shows the true size of the dihedral angle.

(2) The adjacent dihedral angles formed by two intersecting planes are complementary.

(3) Dihedral angles cannot be observed in principal views.

(4) Two intersecting oblique planes cannot both appear edgewise in the same view.

(5) The angles formed by an oblique plane with a horizontal and with a frontal plane are always complementary.

(6) If the line of intersection of two nonparallel planes is inaccessible in the given views, the dihedral angle cannot be found.

SELF-TESTING PROBLEMS

8A. Determine the dihedral angle formed by planes A and B of the tetrahedron.

8B. Complete the front view of plane MNO which makes an angle of 40° with a frontal plane.

8 A.

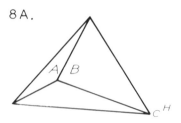

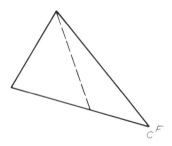

8B.

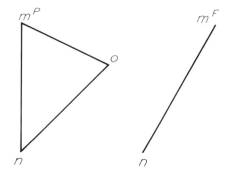

9 Parallelism

Parallelism of lines is a property that is preserved in orthographic projections. Thus lines parallel in space project as parallel lines in any view except in those views in which they coincide or appear as points—situations that do not alter the fact that the lines are parallel.

9.1 Parallel Lines

Oblique lines that appear parallel in two or more principal views are parallel in space. This principle is illustrated in Figure 9.1(a), in which three principal views of two lines AB and CD are drawn. Since the lines in the views are respectively parallel, the lines themselves are parallel in space, Figure 9.1(b).

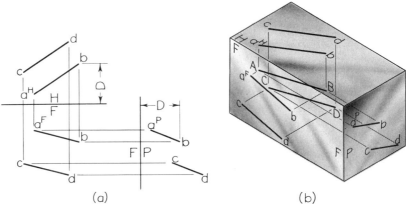

(a)   (b)

FIGURE 9.1   Parallel oblique lines

Two horizontal, two frontal, or two profile lines that appear to be parallel in two principal views may or may not be actually parallel in space. For example, in Figure 9.2(a) the two horizontal lines MN and

OS appear parallel in their front and side views. Without further study it might be concluded that the lines are parallel in space; but when the top view is added, Figure 9.2(b), it is apparent that the two lines are not parallel. The true spatial relationship of the two lines is shown pictorially in Figure 9.2(c). Nonintersecting, nonparallel lines are called *skew* lines.

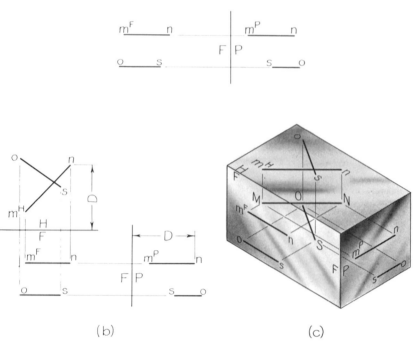

(b)   (c)

FIGURE 9.2   Check of parallelism of principal lines

As another example of the special requirements involved in parallel principal lines, let it be required to construct a line containing point C and parallel to profile line AB, Figure 9.3(a). If the side view is added, Figure 9.3(b), the true inclination of line AB becomes apparent, and the side view of the required line CE, of any appropriate length, may then be drawn parallel to aᴾb, Figure 9.3(c). The front and top views of point E are then established to complete the solution. It should be noted that a random location of point E in only the front and top views would not guarantee parallelism, since there would be no assurance that the inclination of such a line would be the same as that of AB.

The true distance between two parallel lines may be obtained either by constructing a view showing the lines as points, Figure 9.3(c), or by obtaining a normal view of the plane of the two lines (§5.4).

Any two lines in a plane must either intersect or be parallel. In Figure 9.4(a) it is evident that line AB cannot intersect line ON since their top views are parallel. If it is known that line AB is in plane MON, lines AB and ON must therefore be parallel in space. Consequently, the front view of AB may be established by projecting point A to the front view and by drawing aᶠb parallel to oᶠn, Figure 9.4(b).

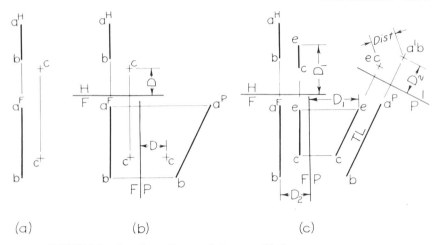

FIGURE 9.3   Drawing a line parallel to a profile line

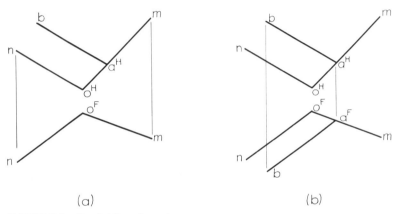

FIGURE 9.4   Parallel lines in a plane

9.2   Parallel Planes

If two intersecting lines in one plane are parallel respectively to two intersecting lines in a second plane, the planes are parallel in space, Figure 9.5. If the two planes are parallel, any line in one plane is parallel to the other plane, since it cannot intersect the other plane.

If two planes are parallel, any view showing one of the planes in edge view must also show the other plane as a parallel edge view. This principle may be used to check or to establish parallelism of planes represented by nonparallel lines, Figure 9.6. Parallelism of two such planes can also be checked by investigating the possibility of drawing a pair of intersecting lines in one plane parallel respectively to two lines in the other plane.

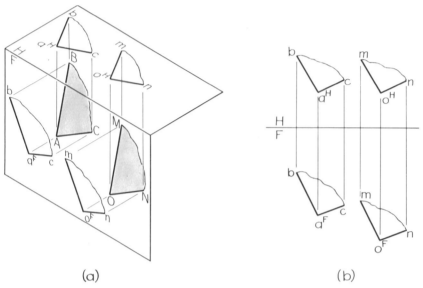

FIGURE 9.5   Parallel planes

(a)   (b)

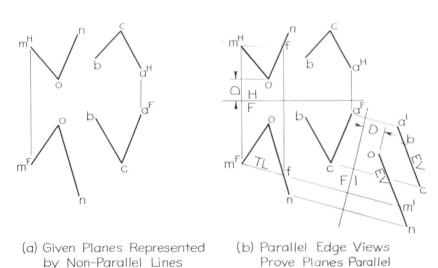

(a) Given Planes Represented   (b) Parallel Edge Views
    by Non-Parallel Lines   Prove Planes Parallel

FIGURE 9.6   Checking parallelism of planes by edge views

9.3   Lines Parallel to Planes; Planes Parallel to Lines

If two lines are parallel, any plane containing one of the lines is parallel to the other line (or, as a special case, contains the other line). Hence a line may be drawn parallel to a given plane by making it parallel to any appropriate line in the plane. Two such lines, if intersecting, establish a plane parallel to the given plane, Figure 9.5.

Conversely, a plane may be drawn parallel to a given line merely by having the plane contain a line parallel to the given line. In Figure 9.7,

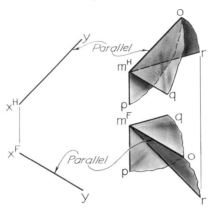

FIGURE 9.7   Planes parallel to a line

each of the planes OMP, OMQ, and OMR is parallel to line XY, since line OM is parallel to XY. In fact, there is an infinite number of planes containing OM which are parallel to XY. On the other hand, there is only one plane containing, say, line MR and which is parallel to XY, namely, plane OMR.

This same principle is applied to another case in Figure 9.8. Lines AB and CE as given are nonintersecting and thus do not lie in the same

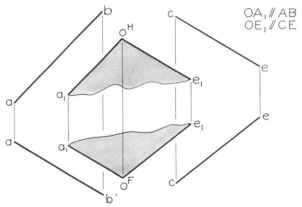

FIGURE 9.8   Plane parallel to two skew lines

plane; but it is possible to establish a plane parallel to both lines. If through the given point O, line OA_1 is established parallel to AB, and line OE_1 parallel to CE, the plane A_1OE_1 is parallel to both AB and CE.

9.4   Parallel Pictorial Lines

For lines to be parallel in the given isometric pictorial, Figure 9.9, such lines must not only have the appearance of being parallel but must also exist in the same or parallel planes of the pictorial. For example, lines AB and CE are parallel since they appear parallel and lie in the same top surface of the isometric. Lines AB and MN are parallel since they appear parallel and lie in parallel planes, namely the top and lower parallel surfaces of the pictorial. In contrast, lines AB and OK are not parallel in space; although they have the appearance of parallelism. Line OK does not exist in a plane that is parallel to the top surface that contains line AB.

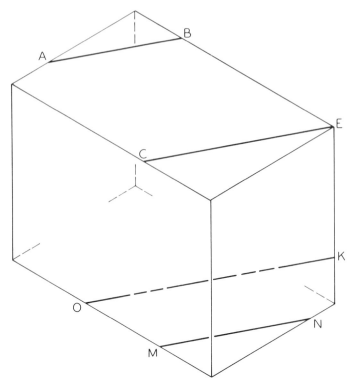

FIGURE 9.9   Pictorial parallel lines

9.5   Construction of Parallel Pictorial Lines

In Figure 9.10(a) the construction is needed for the representation of a line through point X parallel to line AB. A plane such as BCAE is first introduced containing line AB. Then in Figure 9.10(b), a plane MNSG is added that includes point X. This new plane is parallel to plane BCAE since MN appears parallel to CB and also lies in the same top surface of the pictorial; similarly SG is parallel to AE. Then line XY is drawn in plane MNSG parallel to line AB to fulfill the project specifications.

In contrast, observe that although alternate line OK appears parallel to AB, these two lines are not parallel in space because they do not lie in parallel planes.

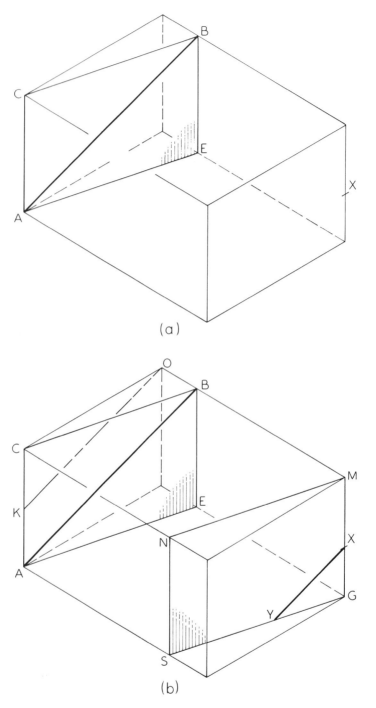

FIGURE 9.10   Construction of pictorial parallel lines

9.6 Parallel Pictorial Planes

Some of the previous concepts can be applied to establish pictorial parallel planes. For example, in Figure 9.11(a) let us provide a plane through point M parallel to plane ABC. In Figure 9.11(b) a line MN is drawn in the lower surface of the pictorial parallel to AB which exists in the parallel upper surface. A second line intersecting MN is now required to complete the parallel plane. To accomplish this

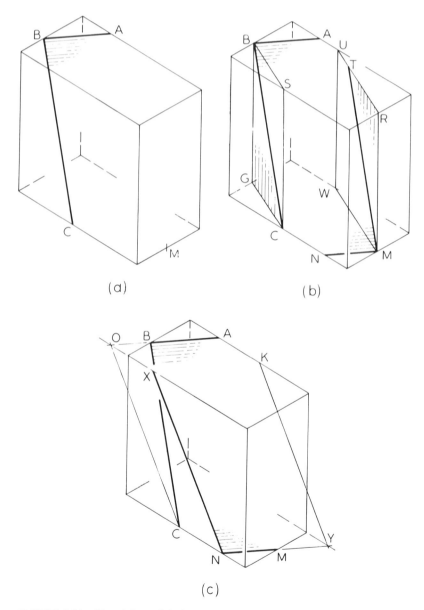

(a)　　　　　　(b)

(c)

FIGURE 9.11 Pictorial parallel planes

objective, a plane such as BSCG is provided first. Then a parallel plane MRUW is introduced, and MT is drawn parallel to BC in this new plane.

Another perhaps more convenient solution for this example is available by extending AB of the given plane to point O and then drawing NX or YK parallel to CO, Figure 9.11(c).

PROBLEMS

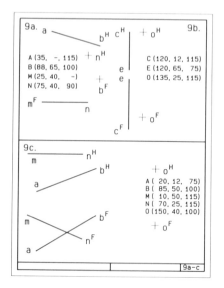

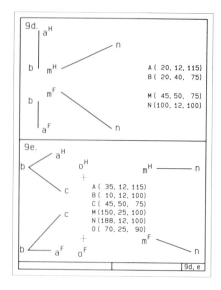

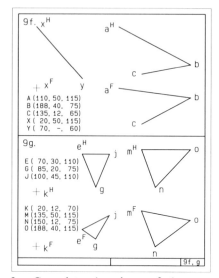

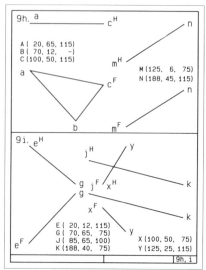

9a. Complete the views of the two parallel lines AB and MN.

9b. Complete the views of a 40 mm line OS that is parallel to line CE.

9c. Through point O, pass a plane parallel to the skew lines AB and MN.

9d. Represent a plane that contains line MN and is parallel to line AB.

9e. Draw a line that contains point O, is parallel to plane ABC, and intersects line MN.

9f. Complete the front view of line XY that is parallel to plane ABC.

9g. Pass a line through point K parallel to the planes EGJ and MON.

9h. Plane ABC is parallel to line MN. Complete its top view.

9i. Pass a line parallel to line EG and intersecting the skew lines JK and XY.

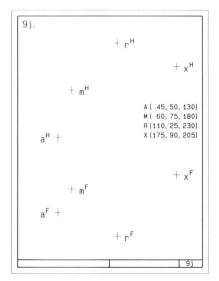

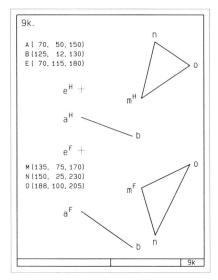

9j. Through the four given points, represent four equidistant and parallel planes, each plane to contain only one of the points.

9k. Draw a line through point E parallel to plane MON and intersecting line AB.

9m. Indicate whether the following statements are true or false. If assigned, prepare written explanations or sketches to justify the answers.

(1) Through a given point only one plane can be passed parallel to a specified plane.

(2) A plane can be passed through a point parallel to two nonintersecting lines.

(3) For a line to be parallel to a plane, the line must be parallel to a line in the plane.

(4) For a plane to be parallel to a horizontal line, it is essential that the plane be horizontal.

(5) Three parallel planes must also be equidistant.

(6) A line can be drawn through any point parallel to two nonparallel planes.

(7) A plane parallel to two non-parallel frontal lines is a frontal plane.

(8) A horizontal line can be drawn through any point parallel to an oblique plane.

(9) All horizontal lines are parallel.

SELF-TESTING PROBLEMS

9A. Pass a plane through point O parallel to the edges AC and BC of the tetrahedron. Then complete the front view of line MN parallel to this plane.

9B. Add the front and top views of line XY parallel to line AB and contacting the edges OM and VN of the tetrahedron.

9A.

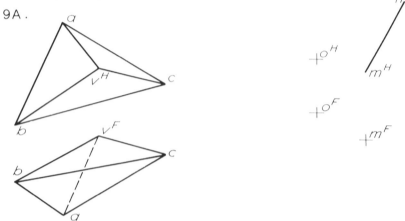

9B.

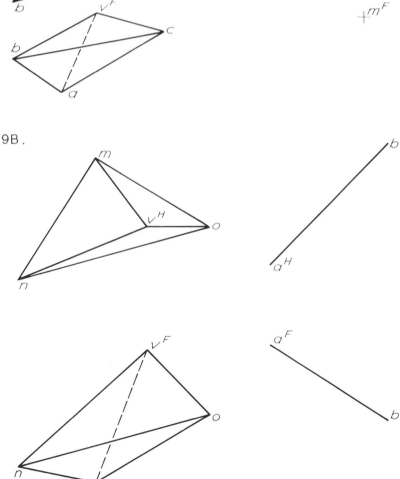

9C. Through point O establish a line that is parallel to plane MNK and that intersects line CE.

9D. Provide a line parallel to MN that intersects skew lines AB and OK.

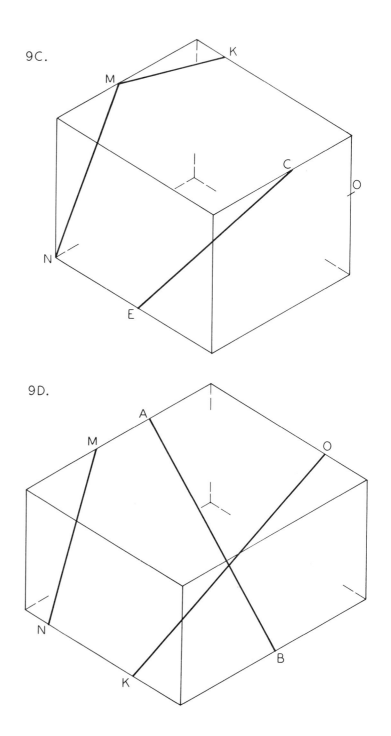

10 Perpendicularity

In solid geometry an important theorem is stated thus: If a line is perpendicular to a plane, it is perpendicular to every line in the plane through the foot of the perpendicular. In descriptive geometry, for reasons to be developed later, it is useful to broaden this as follows: *If a line is perpendicular to a plane, it is perpendicular to every line in the plane.* In Figure 10.1 lines GJ and XY are both considered to be per-

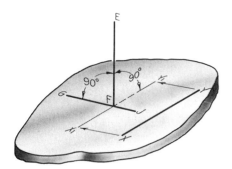

FIGURE 10.1   Perpendicular lines—intersecting and nonintersecting

pendicular to line EF, since they lie in a plane perpendicular to EF. Thus, in descriptive geometry, perpendicular lines are *not necessarily* intersecting lines; that is, they do not necessarily lie in the same plane.

10.1   Perpendicular Lines

A useful characteristic of perpendicular lines in orthographic projection is: *If two lines are perpendicular, they appear perpendicular in any view showing at least one of the lines in true length.* Conversely, if two lines appear perpendicular in a view, they are actually perpendicular in space *only* if at least one of the lines is true length in that same view.

133

Exceptions occur when one line is shown as a point or when the plane of intersecting perpendicular lines appears in edge view. In these cases an adjacent view, given or constructed, shows the true right angle.

In Figure 10.2 front and top views are shown of a 45° triangle in various positions. At (a) both legs of the 90° angle are true length and

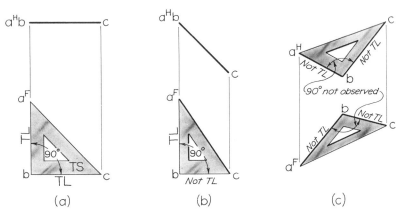

FIGURE 10.2   Views of perpendicular lines—a 45° triangle

the true 90° angle appears in the front view. At (b) the true 90° angle still appears in the front view because one of the legs $a^F b$, is true length. At (c), however, legs AB and BC of the 90° angle are both foreshortened, and the true 90° angle is not observed in either the top view or front view. It is suggested that the student view a triangle in various positions to verify these principles.

In Figure 10.3 each of the lines CD, CD_1, CD_2, and CD_3 is perpendicu-

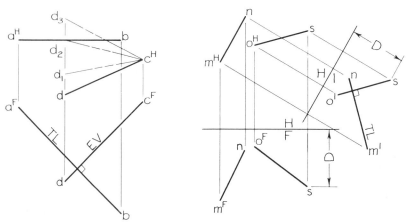

FIGURE 10.3 Perpendicular noninter-   **FIGURE 10.4 Auxiliary-view test for**
secting lines—one shown true length   **perpendicularity**

lar to line AB, since their coinciding front views are perpendicular to
true-length a^{F}b. Note that all these lines lie in an EV of a plane perpendicular to line AB. The directions of the top views of the lines are
immaterial.

The same principle may be used to test (or establish) perpendicularity
of oblique lines. In Figure 10.4 the true angular relationship of lines MN
and OS is not apparent in the front and top views; but when an auxiliary
view is added showing one of the lines in true length, line MN in this
case, it becomes evident that the two lines are perpendicular in space.

10.2 Plane Perpendicular to Line

Two-View Method. A plane is perpendicular to a line if the plane contains two intersecting lines each of which is perpendicular to the given
line. Thus a plane may be drawn containing a given point and perpendicular to a given line as follows, Figure 10.5: Let the given point be X

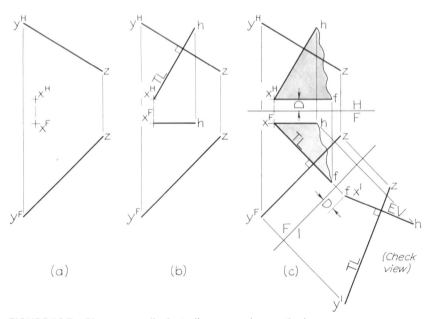

FIGURE 10.5   Plane perpendicular to line—two-view method

and the given line YZ, Figure 10.5(a). Horizontal line XH is drawn with
its true-length view x^{H}h perpendicular to y^{H}z, Figure 10.5(b). Frontal
line XF is drawn with its true-length view x^{F}f perpendicular to y^{F}z,
Figure 10.5(c). Since lines XH and XF are thereby made perpendicular
to line YZ in space, plane HXF is perpendicular to line YZ. This is
substantiated in the auxiliary "check" view, which shows plane HXF
in edge view and line YZ in true length.

Auxiliary-View Method. As suggested by the check view in Figure 10.5(c), a plane also may be established through a given point E perpendicular to a given line GJ by an auxiliary view showing line GJ in true length, Figure 10.6(a), since in this true-length view the required plane appears in edge view and at a right angle to g¹j. All lines in the required plane are perpendicular to line GJ. The front and top views may therefore be completed by projecting any random pair of points in the plane

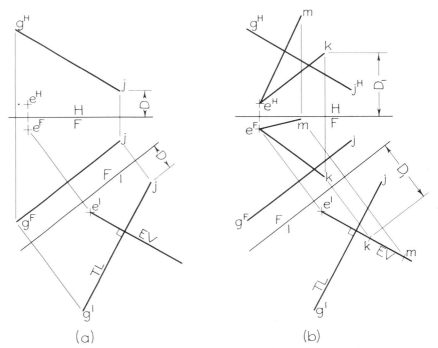

(a)   (b)

FIGURE 10.6   Plane perpendicular to line—auxiliary-view method

such as K and M, Figure 10.6(b), back to the given views. With k and m assumed on the edge view of the plane as shown, their front views may be placed anywhere along the projection lines from k and m to the front view. The points K and M are thus established at definite locations in space, and their top views are now located in the usual manner as indicated.

EXAMPLE INVOLVING PLANE PERPENDICULAR TO LINE

Problem

Draw the front and top views of a right square pyramid (see Appendix C.1) having its axis along line XY, its vertex at point A, and one corner of its base at point B, Figure 10.7(a).

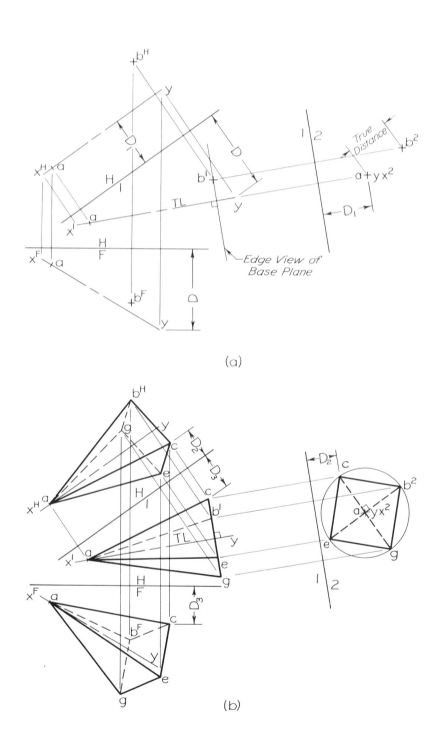

(a)

(b)

FIGURE 10.7 Construction of right square pyramid on given oblique axis

Analysis

A view showing line XY in true length will show the plane of the base in edge view and perpendicular to the axis because of the right pyramid specification. An additional view showing line XY as a point will thus show the base in true size and shape.

Graphic Solution

Auxiliary view 1 is added showing line XY in true length, Figure 10.7(a). The base of the pyramid must appear in edge view and contain point b^1 in this view. Auxiliary view 2 is then constructed showing line XY as a point and showing the true distance from B to XY.

Since the base of the pyramid must be centered at x^2y, line b^2e, Figure 10.7(b), is a diagonal of the base and the square may be constructed as shown (see Appendix B.2). Points c, e, and g are then established in the edge view of the base in view 1, from which the corresponding top and front views may be located in the usual manner. Note that the sides of the square must be parallel in the top and front views. Drawing the lateral edges with proper visibility from the base corners to vertex A completes the solution.

10.3   Line Perpendicular to Plane

Two-View Method. As has been stated, a line perpendicular to a plane is perpendicular to all lines in the plane. Therefore, a line perpendicular to a plane appears perpendicular to a line in the plane in any view in which the line in the plane is true length. For example, let it be required to draw a line from point A perpendicular to plane MNO, Figure 10.8(a). Since there are no true-length lines given in the plane, horizontal line

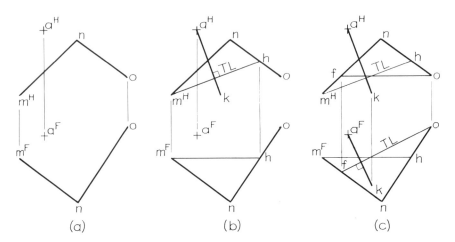

FIGURE 10.8   Line perpendicular to plane—two-view method

MH is added, Figure 10.8(b). The top view a^Hk of the required perpendicular line AK may now be drawn perpendicular to m^Hh as shown. The length of line AK is immaterial in this problem. It is important to realize that this construction establishes only the *direction* of the *top* view of the perpendicular. The front view must now be located by an additional construction.

A frontal line OF is added, Figure 10.8(c). Front view a^Fk is then drawn perpendicular to the true-length view of line OF with K located on the projection line from its top view. The two views of lines a^Hk and a^Fk thus drawn define a line AK of arbitrary length perpendicular in space to plane MNO. If it is required that a line be perpendicular to a given plane and also terminate in that plane, the point in which the line pierces the plane must also be located by one of the methods of Chapter 6 (see Figure 10.9).

Auxiliary-View Method. If an auxiliary view is drawn showing the given plane in edge view, Figure 10.9(a), the required line may be drawn perpendicular to the edge view. It may be noted that the true distance from point A to plane MNO is also apparent in this auxiliary view. If the measurement of this distance is a requirement of a particular problem, the auxiliary-view method is therefore more convenient than the two-view method.

Since view a^1k of the line is true length, top view a^Hk of the line must be parallel to folding line H/1 (or perpendicular to the true-length view m^Hh), Figure 10.9(b). The front view a^Fk is then established by projecting from the top view and transferring distance D_1.

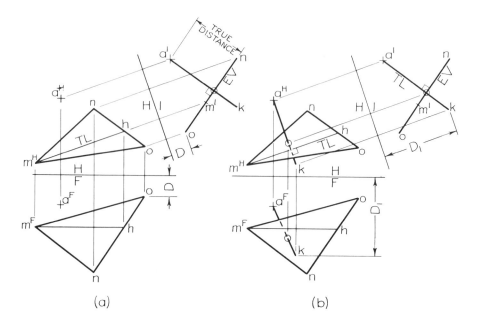

FIGURE 10.9   Line perpendicular to plane—auxiliary-view method

10.4   Common Perpendicular—Point-View Method

Connecting two skew (nonintersecting, nonparallel) lines is one and only one line which is perpendicular to both—the *common perpendicular*. Since the shortest distance from a point to a line is measured along the perpendicular from the point to the line, it follows that the shortest distance between two skew lines is measured along the line that is perpendicular to each of the skew lines.

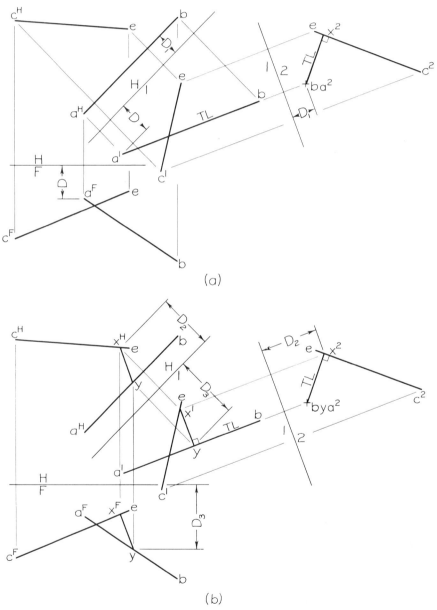

(a)

(b)

FIGURE 10.10   Common perpendicular—point-view method

In order to locate the common perpendicular between skew lines AB and CE, Figure 10.10, use is made again of the principle that perpendicular lines appear perpendicular in any view showing one of them true length. Since neither of the given lines appears true length in the front and top views, auxiliary view 1 is added, Figure 10.10(a), showing one of the lines true length, in this case AB. In view 1 any line perpendicular to line AB must appear perpendicular to a^1b, but at this stage the point at which to draw the common perpendicular has not been established. Accordingly, view 2 is added, showing line AB as a point. Although ec^2 is not true length, any line perpendicular to AB must appear true length. The common perpendicular may therefore be drawn in view 2 extending from a^2b and at a right angle to ec^2, establishing x^2. The length of this perpendicular is the shortest distance between lines AB and CE.

To locate the other views of the common perpendicular, x^1 is established by projection from x^2, Figure 10.10(b). The common perpendicular line XY does not appear true length in view 1; but since a^1b is true length, x^1y is drawn perpendicular to a^1b, establishing point y in view 1. By projection from x^1 and y to the top view and then to the front view, x^Hy and x^Fy are located. It is good practice to check the accuracy of location of the views by transfer distances such as D_2 and D_3.

10.5   Common Perpendicular—Plane Method

Another method of finding the common perpendicular between two skew lines is the *plane method*, which is particularly useful if only the shortest distance between the lines, and not the views of the perpendicular, is needed.

If a plane is passed through one of the two skew lines and parallel to the other (§9.3), the distance between the plane and the second line is the shortest distance between the lines. Through point E of line CE in Figure 10.11(a), line EK is drawn parallel to given line AB to establish a plane CEK parallel to AB. Auxiliary view 1 showing plane CEK in edge view is then drawn. Since a^1b must appear parallel to edge view c^1e, the shortest distance between lines AB and CE may be measured in view 1 as indicated.

If the views of the common perpendicular are desired, further construction is necessary since the location of the common perpendicular is not established in view 1. A true-size view of plane CEK will show as a point any line perpendicular to plane CEK. Consequently, true-size view 2 is added, Figure 10.11(b), and the common perpendicular line XY appears as the point of intersection of a^2b and c^2e. The remaining views of the common perpendicular line XY are then established by projection.

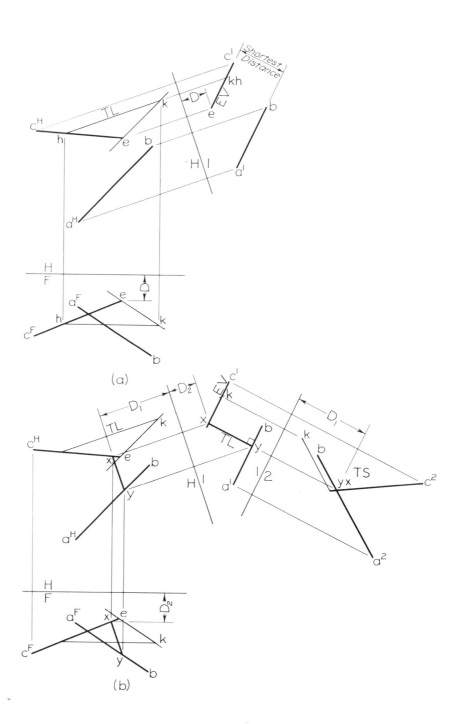

FIGURE 10.11   Common perpendicular—plane method

10.6 Shortest Horizontal Line Connecting Two Skew Lines

Although it does not involve perpendicularity, the problem of finding the shortest horizontal line connecting two skew lines is included here because of the similarity of its solution to that of the plane method of §10.5. The first step, Figure 10.12(a), is again the passing of a plane CEA_1 through one of the lines and parallel to the other. View 1 is then added showing plane CEA_1 in edge view.

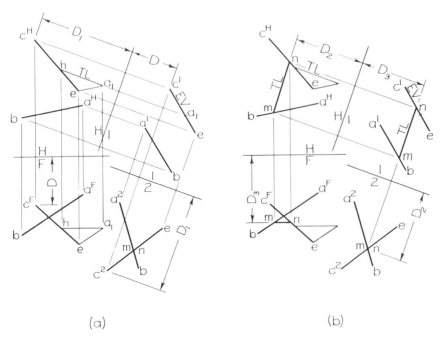

(a)   (b)

FIGURE 10.12   Shortest horizontal line connecting two skew lines

It should be carefully noted that the auxiliary view 1 is projected from the *top* view because in this auxiliary view horizontal lines will appear parallel to the folding line (H/1). The shortest horizontal *distance* between lines AB and CE may then be measured between a_1b and c_1e in a direction parallel to folding line H/1.

To locate the shortest horizontal *line* connecting the two given lines, view 2 is constructed, for which the line of sight is taken parallel to folding line H/1 in order to show a point view of the shortest horizontal line. This point view appears as intersection point mn of a^2b and c^2e. The other views of line MN are then established by projection, Figure 10.12(b).

A similar construction may be used to establish the shortest frontal line or shortest profile line connecting two skew lines.

10.7   Shortest Line at Specified Grade Connecting Two Skew Lines

The approach of the preceding section can be applied to the problem of Figure 10.13, where it is required to find the shortest connecting line having a downward grade of 15% from line MN to line OP. The first portion of the construction is the same as for view 1 of Figure 10.12. It is essential that auxiliary view 1, Figure 10.13(a), be projected from the top view, since percent grade is measured with respect to horizontal in *space* (§3.4). In auxiliary view 1 a line is drawn from any convenient point such as m¹, parallel to folding line H/1, in order that the line will be horizontal

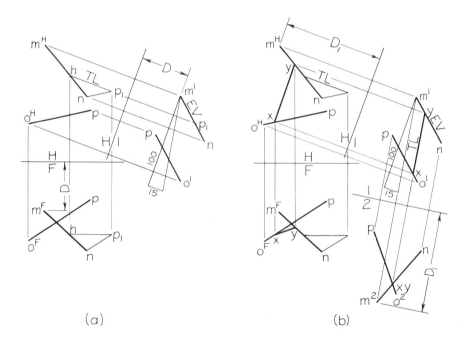

(a)   (b)

FIGURE 10.13   Shortest line at specified grade connecting two skew lines

in space. Along this line 100 units at any appropriate scale are set off and 15 units are then set off downward, or away from and perpendicular to folding line H/1. This establishes the direction and length of the line at the specified downward grade of 15%, but it does not locate the line. With this established direction used for the line of sight, view 2 is then added, Figure 10.13(b). The crossing point of m^2n and o^2p is a point view of the required connecting line XY. Projection back to the other views, as shown, completes the solution.

10.8   Projection of a Line on a Plane

The orthographic projection of a line on a plane is the line connecting the projection on the plane of the endpoints of the line. The projection of a point on a plane is the point in which a perpendicular from the point to the plane pierces the plane (see Figure 10.14). With these

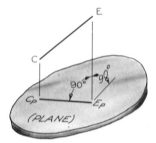

FIGURE 10.14   Projection of a line on a plane

definitions in mind, either of the following methods may be used to find the projection of a given line on a given oblique plane.

Two-View Method. Lines are drawn from points A and B perpendicular to plane MNO, Figure 10.15(a), by drawing their views respectively

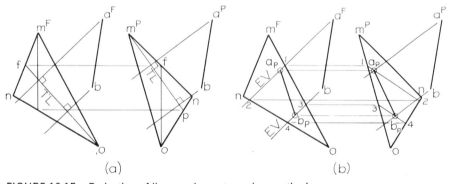

(a)   (b)

FIGURE 10.15   Projection of line on plane—two-view method

perpendicular to the true-length views of frontal line OF and profile line MP (see §10.3). The piercing points A_P and B_P of these lines in plane MNO are then found by the two-view method, Figure 10.15(b). Line A_PB_P is the required projection of line AB on plane MNO.

Auxiliary-View Method. View 1 is constructed showing plane MNO in edge view, Figure 10.16. Lines are drawn perpendicular to this edge view from a^1 and b, establishing projections a_P and b_P. Since the projection lines are true length in view 1, their side views must appear parallel

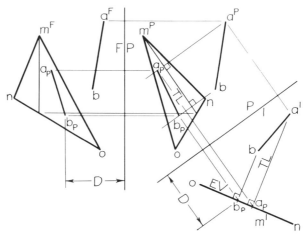

FIGURE 10.16   Projection of line on plane—auxiliary-view method

to folding line P/1, as shown in view P. Projection from a_P and b_P to view P then locates a_P and b_P in the side view, from which their front views are established by projection and transfer of distances as indicated.

PROBLEMS

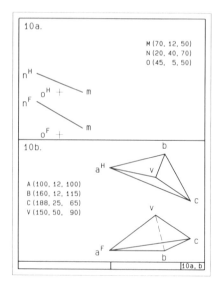

10a.

M (70, 12, 50)
N (20, 40, 70)
O (45, 5, 50)

10b.

A (100, 12, 100)
B (160, 12, 115)
C (188, 25, 65)
V (150, 50, 90)

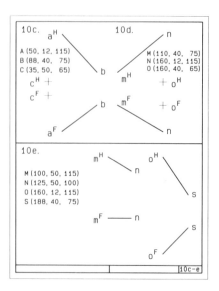

10c.

A (50, 12, 115)
B (88, 40, 75)
C (35, 50, 65)

10d.

M (110, 40, 75)
N (160, 12, 115)
O (160, 40, 65)

10e.

M (100, 50, 115)
N (125, 50, 100)
O (160, 12, 115)
S (188, 40, 75)

10a, b

10c-e

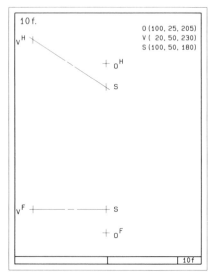

10f.

O (100, 25, 205)
V (20, 50, 230)
S (100, 50, 180)

10f

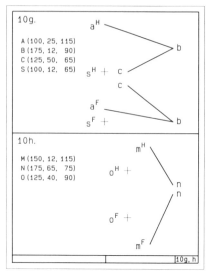

10g.

A (100, 25, 115)
B (175, 12, 90)
C (125, 50, 65)
S (100, 12, 65)

10h.

M (150, 12, 115)
N (175, 65, 75)
O (125, 40, 90)

10g, h

10a. Show the true length and views of a line from point O perpendicular to and intersecting line MN. Use the auxiliary-view method.

10b. Find the views and true length of the altitude of the pyramid having the vertex V. Use the auxiliary-view method.

10c. Through point C, draw a 25 mm horizontal line perpendicular to line AB. Use only the given views.

10d. Represent a plane that contains point O and is perpendicular to line MN. Use the two-view method.

10e. Locate the center of a circle that passes through points O and S and has its center on line MN. Use the two-view method.

10f. Complete the given views of a right square pyramid that has the vertex V on its axis SV and that has a corner of the square base at point O.

10g. Locate the orthographic projection of point S on the plane ABC.

10h. Use the two-view method to locate a line from O perpendicular to and intersecting MN. Check the solution by means of an auxiliary view.

147

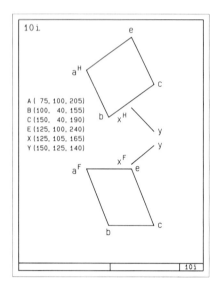

10 i.

A (75, 100, 205)
B (100, 40, 155)
C (150, 40, 190)
E (125, 100, 240)
X (125, 105, 165)
Y (150, 125, 140)

10i

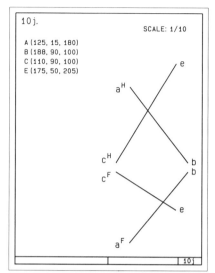

10 j.   SCALE: 1/10

A (125, 15, 180)
B (188, 90, 100)
C (110, 90, 100)
E (175, 50, 205)

10j

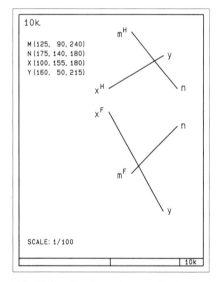

10k.

M (125, 90, 240)
N (175, 140, 180)
X (100, 155, 180)
Y (160, 50, 215)

SCALE: 1/100

10k

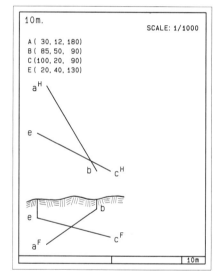

10m.   SCALE: 1/1000

A (30, 12, 180)
B (85, 50, 90)
C (100, 20, 90)
E (20, 40, 130)

10m

10i. Using the two views only, locate the reflected ray having given the mirror surface ABCE and the light ray YX.

10j. Determine the clearance between the two control cables AB and CE.

10k. Locate the views of the shortest branch connecting pipes MN and XY.

10m. Find the following: (1) the true length and projections of the shortest shaft connecting the shafts AB and CE; (2) the true length of a vertical connecting shaft; (3) the true length and bearing of a horizontal connecting tunnel originating at point E.

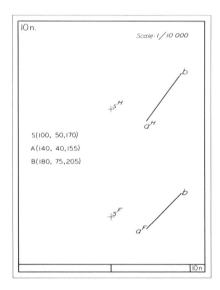

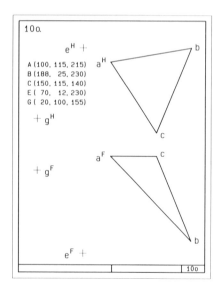

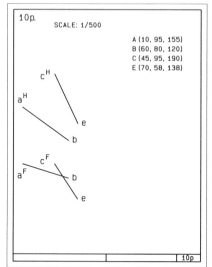

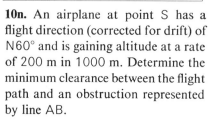

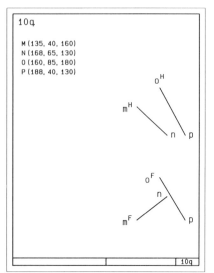

10n. An airplane at point S has a flight direction (corrected for drift) of N60° and is gaining altitude at a rate of 200 m in 1000 m. Determine the minimum clearance between the flight path and an obstruction represented by line AB.

10o. Determine the locus of points equidistant from points E and G and lying in the limited plane ABC.

10p. Find the bearing and length and show the views of the shortest horizontal tunnel connecting shafts AB and CE.

10q. Show the views of the shortest profile line connecting lines NM and OP.

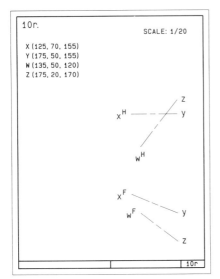

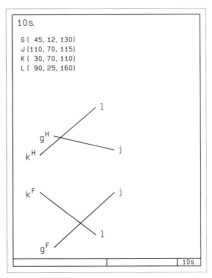

10r. Pipes XY and WZ are to be connected with a branch pipe having a downward grade of 20% from pipe XY to pipe WZ. Find the length and show the views of the center line of the branch pipe.

10s. Find the views of a tube connecting tubes GJ and KL and having an upward slope of 30° from tube GJ to tube KL.

10t. Indicate whether the following statements are true or false. If assigned, prepare written explanations or sketches to justify the answers.

(1) Perpendicular lines do not necessarily intersect.

(2) The locus of points 50 mm from the ends of a 25 mm line is a circle.

(3) A sphere can be passed through any four points not in a straight line.

(4) The locus of points equidistant from three points not in a straight line is a line perpendicular to the plane of the three points and passing through the center of a circle circumscribing the points.

(5) The locus of all points equidistant from two given points is a perpendicular plane containing the two points.

(6) A line can be drawn perpendicular to each of two nonintersecting lines.

(7) Only one line can be drawn perpendicular to a given line through a specified point on the given line.

(8) Two lines perpendicular in space will appear perpendicular in any orthographic view.

SELF-TESTING PROBLEMS

10A. Determine the distance between the parallel planes and show the front and top views of a line representing this distance.

10B. Determine the true length and views of the shortest connector joining the given skew lines.

10A.

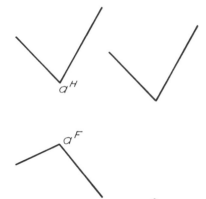

10B.

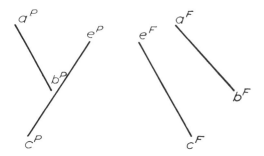

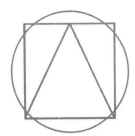

11 Angle Between Line and Oblique Plane

The angle between a line and a plane lies in a plane that is perpendicular to the given plane and contains the given line. This angle is also defined as the angle between the given line and its projection upon the given plane, Figure 11.1. The three methods for finding this angle presented in this chapter are based upon the use of successive auxiliary views; for another variation, see §13.4. Problems involving the angle between a line and a principal plane are presented in Chapters 3 and 13.

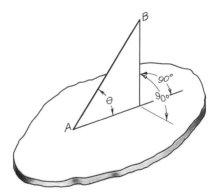

FIGURE 11.1   Angle between line and plane

11.1   Angle Between Line and Plane—Plane Method

In order to show in its true magnitude the angle between a line and a plane, it is necessary to show in the same view the line in *true length* and the plane in *edge view*.

For the general case of an oblique plane and line as in Figure 11.2(a), three successive auxiliary views of the plane are constructed to achieve the desired view. In Figure 11.2(b), the first auxiliary view shows the

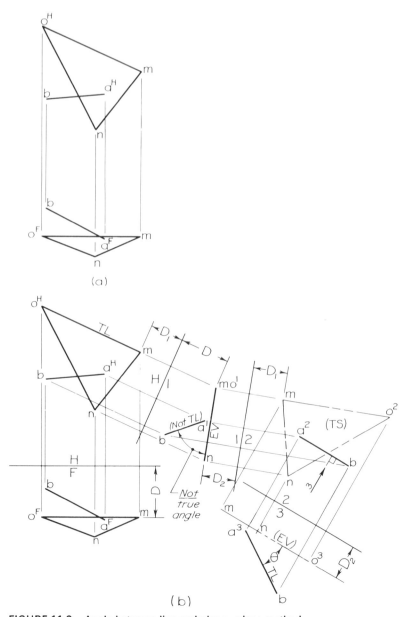

FIGURE 11.2 Angle between line and plane—plane method

plane OMN in edge view as well as line AB. Since the line AB is not shown true length, the required angle does not appear true size. An additional view showing the line AB and the true size of plane OMN is then drawn. Any auxiliary view projected from a true-size view of a plane produces an edge view of the plane. Therefore, if a line of sight 3 is introduced perpendicular to a^2b, the resulting view 3 shows line AB in true length and plane OMN in edge view. The required angle is then measured as indicated.

Unless there are reasons for obtaining data relative to the line and plane other than the angle between the two, the construction necessary for the angle can be simplified in the second and third auxiliary views, Figure 11.2(b). After an edge view of the plane is obtained, it is not necessary to project the plane into the successive views beyond the auxiliary view containing the edge view of the plane. Since in view 3 the edge view of the plane OMN must appear parallel to the folding line 2/3, the angle between the line and the plane may thus be measured between a^3b and folding line 2/3. Hence for the purpose at hand it is not necessary to show plane OMN in views 2 and 3.

11.2   Angle Between Line and Plane—Line Method

Another method of determining the true size of the angle between a line and a plane consists of a series of views first showing the given line in true length, then as a point, and finally in true length with the given plane in edge view.

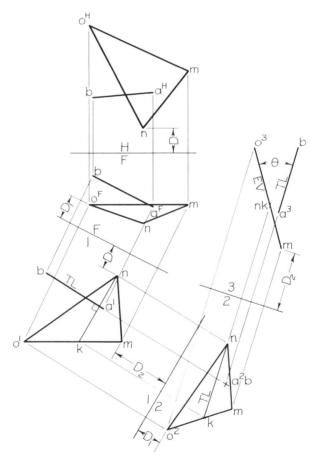

FIGURE 11.3   Angle between line and plane—line method

In Figure 11.3, the front and top views of the plane MNO and line AB are given. The first auxiliary view shows line AB in true length and the second auxiliary view shows it as a point, while plane MNO is projected to each view. Any view projected from view 2 will then show line AB in true length. A line NK is introduced in plane MNO in views 1 and 2 in such position as to appear true length in view 2. The third auxiliary view may then be drawn showing this line as a point and plane MNO as an edge view with line AB in true length. The true angle between this line and the plane is then measured as shown.

This method is particularly advantageous in those problems in which the given line appears true length or as a point in one of the given views, situations that occur frequently in practice.

11.3 Angle Between Line and Plane—Complementary- Angle Method

The angle between a line and a plane is also defined as the angle between the line and its projection on the plane, Figure 11.4. Therefore, the true size of the plane determined by the given line and its projection would produce the required angle in true size. This procedure, however, involves considerable construction when applied to orthographic views. Further study of Figure 11.4 reveals a simpler method. The triangle

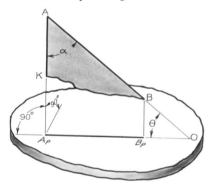

FIGURE 11.4   Complement of angle between line and plane

AOA_P is a right triangle and hence angle α is the complement of the required angle θ. The angle α may be formed by constructing a line from any point on the given line perpendicular to the given plane.

In Figure 11.5(a) line BK is constructed perpendicular to plane MNO by the two-view method of §10.3. Auxiliary views 1 and 2 are then added, Figure 11.5(b), to show plane ABK and angle α in true size (see §5.4). It must be remembered at this point that the angle at B, angle α, is *not* the required angle between line AB and plane MNO. The required

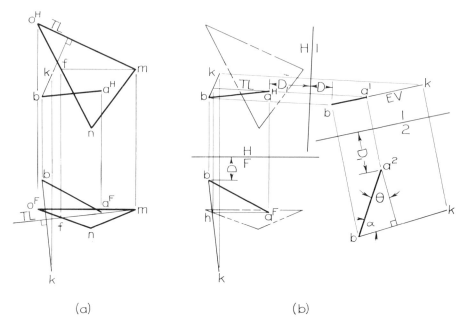

(a)   (b)

FIGURE 11.5   Angle between line and plane—complementary-angle method

angle is the complement of angle α. To obtain this complement, a right
triangle is constructed as shown, with angle α as one of its acute angles.
The other acute angle is then the required angle θ.

PROBLEMS

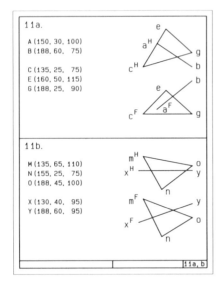

11a.

A (150, 30, 100)
B (188, 60, 75)

C (135, 25, 75)
E (160, 50, 115)
G (188, 25, 90)

11b.

M (135, 65, 110)
N (155, 25, 75)
O (188, 45, 100)

X (130, 40, 95)
Y (188, 60, 95)

11a, b

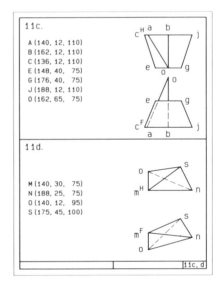

11c.

A (140, 12, 110)
B (162, 12, 110)
C (136, 12, 110)
E (148, 40, 75)
G (176, 40, 75)
J (188, 12, 110)
O (162, 65, 75)

11d.

M (140, 30, 75)
N (188, 25, 75)
O (140, 12, 95)
S (175, 45, 100)

11c, d

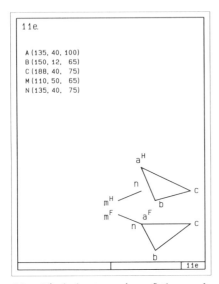

11e.

A (135, 40, 100)
B (150, 12, 65)
C (188, 40, 75)
M (110, 50, 65)
N (135, 40, 75)

11e

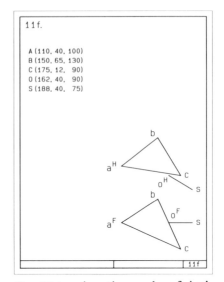

11f.

A (110, 40, 100)
B (150, 65, 130)
C (175, 12, 90)
O (162, 40, 90)
S (188, 40, 75)

11f

11a. Find the true size of the angle between line AB and plane CEG.

11b. Find the true size of the angle between line XY and plane MON.

11c. Determine the angles formed by the guy wires OA and OB with the roof plane CEGJ.

11d. For the given tetrahedron find the angle between edge SO and base MNO.

11e. Determine the angle of incidence of light ray MN and the polished surface ABC.

11f. Determine the angle formed by line OS and plane ABC. Select the method that requires only two auxiliary views.

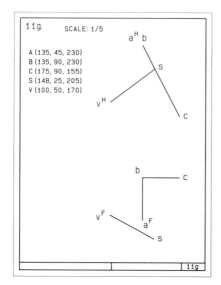

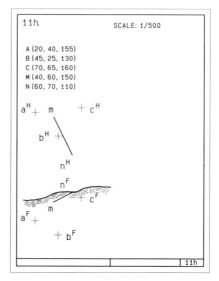

11g. Line VS is an element of a right cone having its circular base in plane ABC. Find the true angle formed by line VS and the base plane, and determine the diameter of the base circle.

11h. What distance must tunnel MN be extended to meet an ore vein determined by points A, B, and C? What angle is formed by tunnel MN and the vein ABC?

11i. Indicate whether the following statements are true or false. If assigned, prepare written explanations or sketches to justify the answers.

(1) The corresponding angles formed by a line intersecting two parallel planes are equal.

(2) Supplementary adjacent angles are formed by a line intersecting a plane.

(3) The angle formed by a line and a given plane is contained in a plane through the line and perpendicular to the given plane.

(4) The true angle formed by a line and a plane is observed in any view in which the plane appears edgewise.

(5) The true angle formed by a frontal line and an oblique plane may be measured in a front view.

(6) Only a single plane can be passed through a line and perpendicular to a given plane.

SELF-TESTING PROBLEM

11A. Determine the true angle formed by edge EA and plane ABC of the tetrahedron. Use complete auxiliary views.

11A.

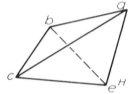

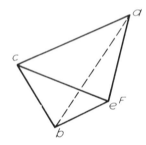

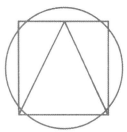

12 Mining and Civil Engineering Problems

Although graphical methods may be used to advantage in all fields of engineering, the natures of many problems in mining and civil engineering particularly fit them for graphical representation and solution. The mining or highway engineer as well as the military strategist has frequent occasion to prepare or use *topographic maps*, which are graphical means of representing the irregularities of the earth's surface in single views. A topographic map is based on the principles of *horizontal projection*.

FIGURE 12.1   Chicago Circle Expressway Interchange (courtesy Illinois Division of Highways)

12.1   Horizontal Projection

Horizontal projection is a method of indicating the position of a point
in space by means of its projection on a horizontal plane (top view)
together with an accompanying symbol or number specifying the elevation of the point in relation to a horizontal datum plane. For topographic
maps the most common datum plane is the mean level of the sea, which
is used as zero elevation.

Figure 12.2 shows in pictorial form the representation by horizontal
projection of a point A in space. The height of point A above the datum
plane is indicated by the subscript.

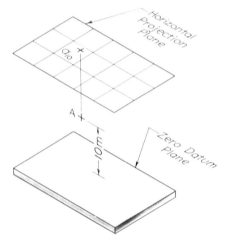

FIGURE 12.2   Horizontal projection

12.2   Topographic Map

On a topographic map a series of points at a selected elevation may be
connected with a line called a *contour line*. Thus a contour line approximately represents a continuous series of points of a designated elevation
on a terrain. If a portion of the terrain were in the form of a cone, the
contour lines would be a series of circles, Figure 12.3. The cone is shown
intersected by a series of horizontal cutting planes at 10 m intervals in
height, resulting in equally spaced, concentric contour circles in the top
view or map. The high point or vertex of the cone is indicated in the top
view and is called a *topographic crest*.

The surface of the earth, unlike a cone, is irregular, and therefore contour lines on the earth's surface are irregular. The plan view of Figure
12.4 shows typical contour lines. If a large enough area is included,
contour lines will be continuous and closed and will not cross each other
unless an *overhang* is involved.

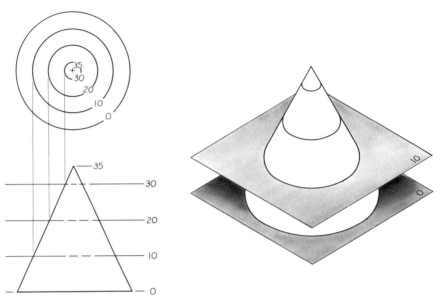

FIGURE 12.3   Contour lines on a cone

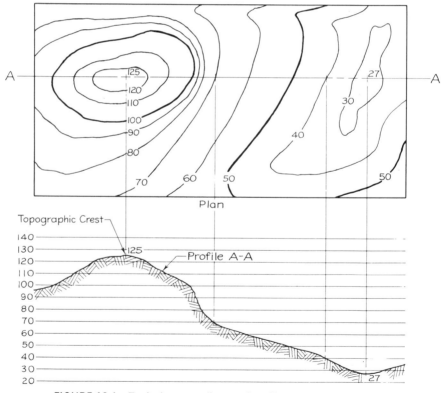

FIGURE 12.4   Typical contour lines and profile

The elevation view of Figure 12.4 is called a *profile* (section) of the terrain, resulting from the vertical cutting plane A-A shown in the plan view. It should be noted that the term "profile" in this usage does not refer to the profile (side) view. A profile in this context is the line of intersection of the earth's surface and any vertical cutting plane. A study of profile A-A together with the contour lines in the plan view reveals the fact that contour lines closely spaced indicate a relatively steep slope while the opposite condition suggests a gentle slope.

12.3   Preparation of Contour Map Using Grid Survey

Before a topographic map can be prepared, a survey of the area must be made to determine the elevation of an adequate number of strategically selected points. The points chosen are dependent on the relative

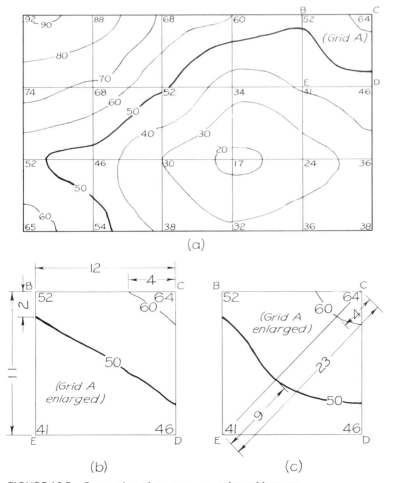

FIGURE 12.5   Preparation of contour map using grid survey

irregularity of the terrain involved and the proposed utilization of the map.

One frequently used method for locating points is the grid system shown in Figure 12.5(a). After the field survey has established the elevation of each intersection on the grid, the map maker has the task of plotting the contour lines. Assuming that the contour lines are to be plotted at 10 m intervals, the following procedure would be used to locate the contour lines in grid A, which is shown enlarged in Figure 12.5(b). The difference in elevation of points B and C is 12 m. Based on the supposition that the slope between the points B and C is constant, contour line 60 passes through a point four-twelfths or one-third of the distance from C to B. Contour line 50 passes through a point two-elevenths of the distance from B to E. These points may be approximated by eye or by a graphical method of proportion similar to Appendix B.1, depending on the accuracy necessary.

Points are obtained similarly for the sides of each grid. Contour lines are then drawn freehand through all points of the same elevation as shown. Although helpful only under special circumstances, additional points on the contour lines may be secured by interpolation along one diagonal of each grid, preferably that diagonal which is more nearly perpendicular to the contour lines. This procedure is indicated in Figure 12.5(c), which shows the altered contour lines 50 and 60 resulting from this additional interpolation.

12.4 Mining and Geology

The following definitions cover the pertinent technical terms among the many employed in mining and geology. They are illustrated in Figure 12.6.

1. *Strike.* The bearing of a horizontal line in a plane, usually measured from north, for example, N50°W or N310°.

2. *Dip.* The slope of a plane or angle between the plane and a horizontal plane (§8.3) plus the *general* direction of downward slope of the plane, for example, 30°SE. The direction of the dip is always at a right angle to the strike line. The dip is indicated on the map (top view) as illustrated.

3. *Stratum.* A layer of sedimentary rock. *Strata* usually lie below the earth's loose surface but may be exposed by weathering. Since they were formed by sedimentation in ancient seas or rivers, their bounding surfaces usually may be considered as parallel planes within limited areas. The term *seam* is sometimes used in place of stratum, as in "a seam of coal."

4. *Bedding Plane.* A bounding surface of a stratum.

5. *Vein.* A deposit of mineral or ore formed in a fissure in rock frequently bounded by two bedding planes. A vein is sometimes called a *lode.*

6. *Fault.* A discontinuity or break in a stratum or vein involving a shifting of one portion with respect to the other. The term *fault plane* may be applied if the break and displacement take place along a plane.

7. *Outcrop.* An area of a stratum or vein exposed (or only lightly covered) at the earth's surface. The bounding edges of the area, which are the intersections of the bounding surfaces within the earth's surface, are called the *outcrop lines.*

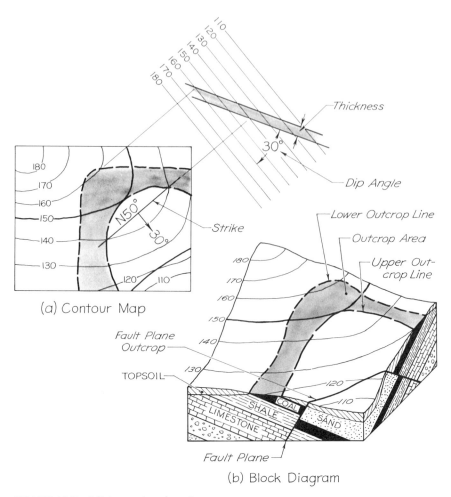

(a) Contour Map

(b) Block Diagram

FIGURE 12.6   Mining and geology terms

12.5   Strike, Dip, Thickness, Outcrop

A bedding plane is theoretically located by establishing three points in the plane. In practice, more than three points are usually located to allow for irregularities and measurement inaccuracies. These points may be on the outcrop line if it exists or may be determined by drilling.

EXAMPLE

Problem

Points X, Y, and Z are given in the upper bedding plane of an ore vein and point W in the lower bedding plane. It is required to find the strike, dip, thickness, and outcrop of the vein.

Graphic Solution

The elevations of the points are given in the top view, Figure 12.7(a), and from this information the front view is constructed according to the given scale. A horizontal line ZH is drawn in the plane XYZ, and its bearing is measured as N295°. This is the strike of the vein.

An auxiliary view projected from the top view with the line of sight parallel to the true-length strike line is used to obtain the dip angle of the vein. The size of the dip angle, 45°, is measured in the auxiliary view. The complete description of the dip includes in addition to the angle the general direction of downward slope of the vein. This direction is always at right angles to the strike line, and thus the possible directions are reduced to two if the strike is known. In Figure 12.7(a) the strike is found to be N295°, so that immediately the general direction of the dip (which is all that need be given) must be either northeast or southwest. Study of the front view and/or the auxiliary view reveals that point Y is lower than the strike line ZH. Thus in the top view the arrow representing the direction of downward slope is drawn pointing from zh in the general direction of y, which is northeast. Consequently, the dip is recorded as "45°NE."

On the customary assumption that the upper and lower bedding planes of the vein are parallel, the edge view of the lower plane is drawn parallel to the upper plane through point w, and the thickness of the vein is measured on a perpendicular between the edge views.

In Figure 12.7(b) a contour map has been superimposed on the plan view of plane XYZ. Points on the outcrop lines are located as follows: Since the contour lines lie in horizontal planes, they appear as a series of straight lines perpendicular to the line of sight in any elevation view. Accordingly, these lines are spaced in auxiliary view 1 as shown using the given scale. [The elevations of the given points on the upper and lower bedding planes of the vein are given in the top view of Figure 12.7(a).] A typical point nm, at which the 15 m contour intersects the edge view of the upper plane of the vein, represents the point view of a horizontal line in this upper plane at a 15 m elevation. This line

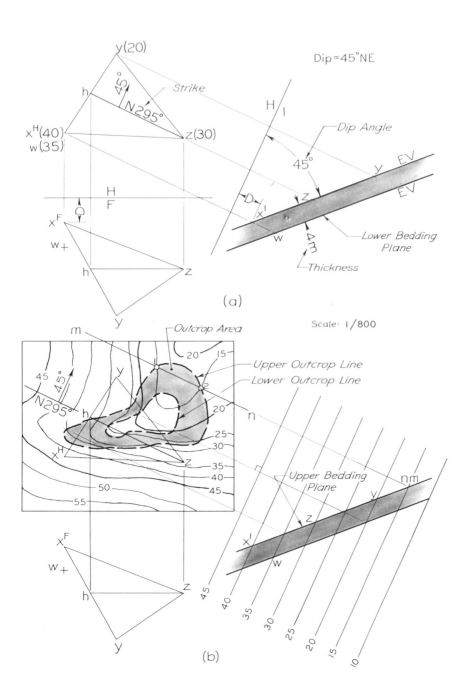

FIGURE 12.7   Strike, dip, thickness, and outcrop of a vein

projected to the plan view intersects the 15 m contour line at points 1 and 2. These are points on the upper plane of the vein and on the surface of the earth and thus are points on the upper outcrop line. Additional points are similarly located on other contour lines until all available points are secured on both the upper and lower outcrop lines. The curves are then drawn through these points as shown, establishing the outcrop area.

If a problem is encountered in which the outcrop lines are given and the strike, dip, and thickness are required, the above procedure is reversed as follows: Any two points in which an outcrop line intersects the same contour line, such as points 1 and 2, Figure 12.7(b), determine a horizontal line and thus the strike of the vein. This in turn establishes the direction of sight to produce the edge views of the planes of the vein.

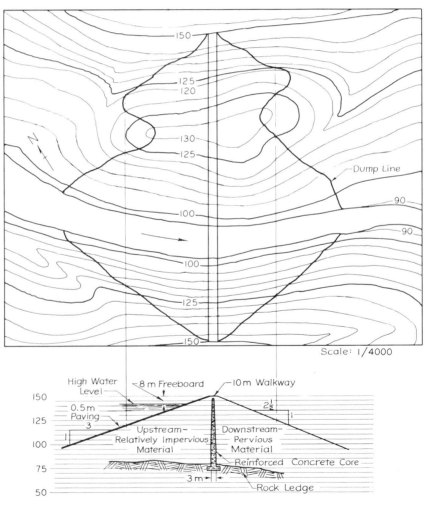

FIGURE 12.8   Dump lines for earth-fill dam

12.6 Cut and Fill or Dump Lines

The problem of locating the outlines of earth *fills* or *cuts* is similar to the preceding construction for outcrop lines. As an example, Figure 12.8, the construction for the so-called *dump* lines (fill lines) is shown for an earth dam. As indicated, the dam is designed with a 1 on 3 slope on the upstream side and a 1 on $2\frac{1}{2}$ slope on the downstream side. Since the front view is a vertical section of the dam, the given slopes establish edge views of the plane surfaces of the dam in the front view as shown. The intersection points of these edge views with the several horizontal planes are projected to the corresponding contour lines in the plan view to establish points on the dump lines just as in plotting outcrop lines. If it were desired to continue the dump lines across the stream area, it would of course be necessary to establish contour lines on the stream bed.

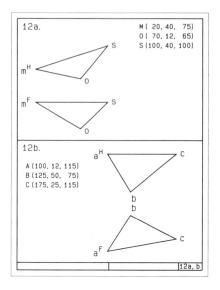

12a.

M (20, 40, 75)
O (70, 12, 65)
S (100, 40, 100)

12b.

A (100, 12, 115)
B (125, 50, 75)
C (175, 25, 115)

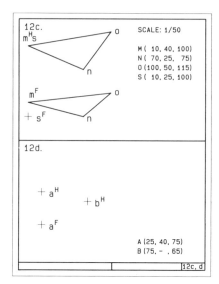

12c.

SCALE: 1/50

M (10, 40, 100)
N (70, 25, 75)
O (100, 50, 115)
S (10, 25, 100)

12d.

A (25, 40, 75)
B (75, – , 65)

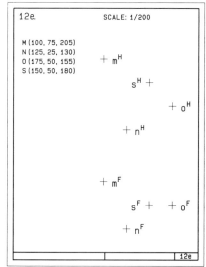

12e.

SCALE: 1/200

M (100, 75, 205)
N (125, 25, 130)
O (175, 50, 155)
S (150, 50, 180)

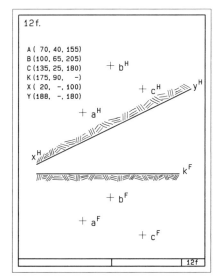

12f.

A (70, 40, 155)
B (100, 65, 205)
C (135, 25, 180)
K (175, 90, –)
X (20, –, 100)
Y (188, –, 180)

12a. Determine the strike and dip of the oblique plane MOS.

12b. Determine the strike and dip of the oblique plane ABC.

12c. Points M, N, and O are located on the upper bedding plane of an ore vein. Point S is located on the parallel lower bedding plane. Determine the strike, dip, and thickness of this vein.

12d. A stratum that contains points A and B has a strike of N75° and a dip of 40°SE. Locate the front view of point B.

12e. Points M, N, and O are located on the upper bedding plane of an ore vein. Point S is on the parallel lower bedding plane. Determine the strike, dip, and thickness of this vein.

12f. Find the strike and dip of a thin vein determined by points·A, B, and C. Show the outcrop of this vein on the level ground surface at the elevation through point K and on the vertical cliff passing through points X and Y.

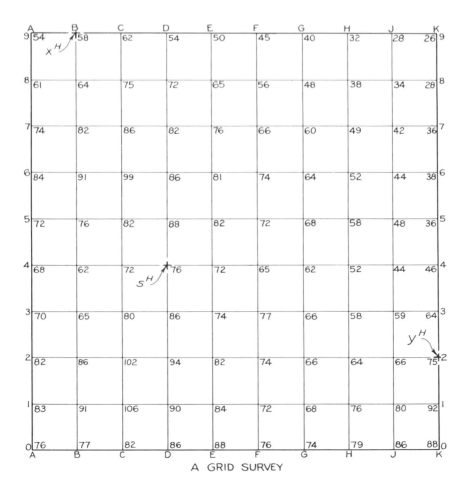

A GRID SURVEY

In each of problems **12g–12j** contour lines for either 5 m or 10 m intervals, as assigned, are to be drawn for the given grid survey. The grids are 12 mm apart. Scale 1/500. For each problem the grid is to be placed in the upper left-hand corner 12 mm from the borders of the drawing paper, which is to have the long side horizontal.

12g. Plot the profile along grid line 6-6. On the profile, label the topographic crest and shade the *defiladed area* (space not visible from topographic crest). Label the *military crest* (that crest from which the previous defiladed area is visible).

12h. Show the outcrop of a thin vein that passes through point S at an ele-

vation of 65 m and has a strike of N285° and a dip of 15°NE.

12i. Show dump lines for a 25 m wide horizontal roadway extending from point X to point Y at a 70 m elevation with the side slopes 1:1 for both the cut and fill.

12j. Plot the profiles along the grids assigned by the instructor.

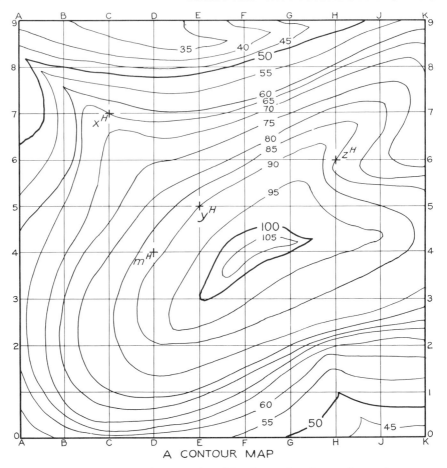

A CONTOUR MAP

The contour map may be reproduced by the student by tracing or by using a cross-section grid paper. The grids shown are 12 mm apart. If the instructor so desires, the map may be simplified by using only the contour lines at 10 m intervals. For each of problems **12k–12o** the map should be placed in the upper left-hand corner 12 mm from the borders of the drawing paper, which is to have the long side horizontal. Scale: 1/2000.

12k. Show the outcrop line of a thin vein that passes through point M at an elevation of 70 m and that has a strike of N300° and a dip of 20°NE.

12m. Points X, Y, and Z are on the surface of the terrain. Vertical test drills to the upper bedding plane of a vein are respectively 5 m, 10 m, and 40 m deep. The drill at point X is continued through the vein and breaks through the lower bedding

plane at a total depth of 25 m from the surface of the terrain. Find the strike, dip, and thickness of the vein and show the outcrop area.

12n. In the plan view plot the dump lines for a 20 m horizontal roadway at an 80 m elevation along grid line 5-5. The cut and fill side slopes are 1:1.

12o. Plot the grid profiles as assigned by the instructor.

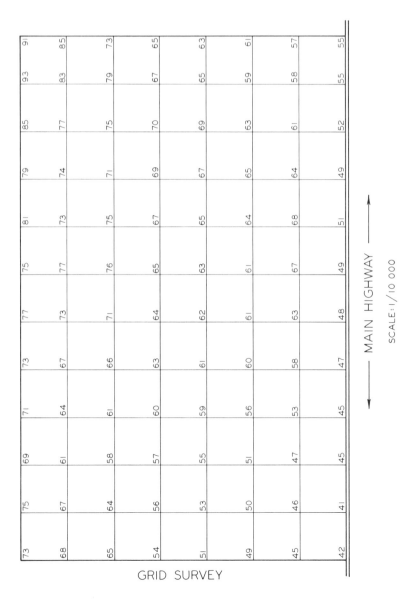

GRID SURVEY

12p. Duplicate the given grid survey using 20 mm spacing for the grids and plot contour lines for each even 2 m interval.

(1) Prepare an original layout for streets, parks, and residential lots superimposed on this contour map or on a transparent overlay.

(2) Primarily for sales purposes, prepare in addition a three-dimensional model of the contour lines and your plans, utilizing sheet stock or such material as Styrofoam, plaster, or clay.

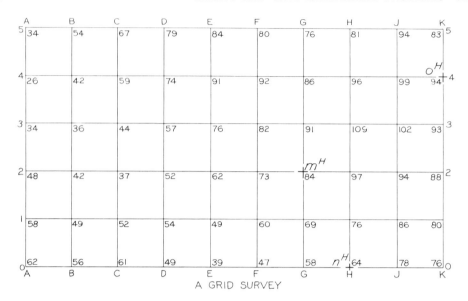

A GRID SURVEY

For the given grid survey, plot contour lines at 5 m or 10 m intervals as assigned. The grids are 12 mm apart. Scale: 1/500. The grid is to be placed in the upper, right-hand corner 12 mm from the borders of the drawing paper, which is to have the long side horizontal.

12q. Plot the profile along grid line 3-3. On the profile, label the topographic crest and shade the *defiladed area* (the space in air or above earth's surface not visible from the topographic crest). Label the *military crest* (that crest from which the previous defiladed area is visible).

12r. Points M, N, and O are located on the outcrop of a thin vein. Determine the strike and dip of this vein and show its complete outcrop line.

12s. Plot additional profiles along the grids assigned by the instructor. Along each grid show the center line of a roadway with a uniform grade which does not exceed 10% and which is such that the areas of cut and fill along the center line are approximately minimized and balanced.

12t. Indicate whether the following statements are true or false. If assigned, prepare written explanations

or sketches to justify the answers.

(1) If the strike lines of two planes are parallel, the planes are parallel.

(2) The strike of a plane is the direction of a horizontal line in the plane and is always measured in a top view.

(3) A contour line represents points of approximately equal elevation on the surface of the earth.

(4) Contour lines cannot cross.

(5) The dip of a plane is observed in any view that shows the plane edgewise.

(6) The thickness of a vein is always measured in a front view.

(7) A dump line is a horizontal line.

(8) An outcrop of a vein is its intersection with a bedding plane.

(9) The relative space between contour lines on a map is an indication of the relative slope of the terrain.

SELF-TESTING PROBLEMS

12A. Determine the strike and dip of the given plane.
12B. Unlimited plane NOM has a strike of N35° and unlimited plane ABC has a dip of 30°SW. Complete the given views and determine the line of intersection of these planes.

12 A.

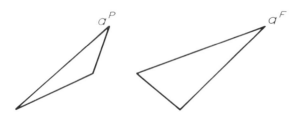

12 B.

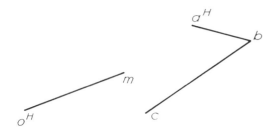

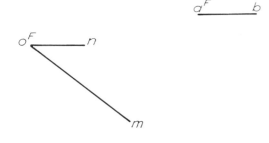

13 Revolution

Drawing-board problems are usually solved by the addition of principal or auxiliary views. This procedure is sometimes called the *change-of-position method* because the successive views are obtained by successive changes in the position of the observer (or of the line of sight), with the object remaining stationary. It is also possible to consider the observer stationary and the object revolved to whatever position results in a new or altered view showing the desired information. As will be seen, this *revolution method* frequently requires less construction and less working space on the drawing. On the other hand, it often results in crowded or overlapping views with increased confusion. Consequently, in practice the revolution method is used only when it possesses a distinct advantage over the change-of-position method for a particular problem. Because of this latter possibility, the engineer and technician should be familiar with both methods.

13.1   Revolution of a Point

For the purposes at hand, a point is considered as revolving only about a straight-line axis and only in a circular path lying in a plane perpendicular to the axis. These are the conditions present in rotation of familiar objects such as wheels, pulleys, and hand-cranks, as illustrated in Figure 13.1(a). The axis of revolution is the center line AB of the shaft, and points on the pulley such as point C rotate in circular paths lying in planes perpendicular to AB. Other types and positions of paths of revolution are, of course, theoretically possible but are of no utility in the solutions of problems illustrated in this chapter. Reduced to bare essentials and drawn in multiview arrangement, the rotation is demonstrated in Figure 13.1(b), where point C is revolved through an

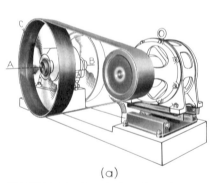

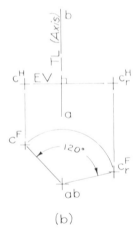

(a)   (b)

FIGURE 13.1   Revolution of a point

angle of 120° in a clockwise direction. Two important characteristics of any revolution of this type may be observed in Figure 13.1(b):

1. *In a view showing the axis of revolution as a point, the path of revolution of any point not on the axis appears as a circle* or as an arc of a circle in case the revolution is less than 360°. The center of the circle is the point view of the axis, and the radius is the distance from this point view to the corresponding projection of the point to be revolved. This is illustrated in the front view of Figure 13.1(b).
2. *In a view showing the axis of revolution in true length, the plane of the path of revolution of any point appears in edge view and perpendicular to the axis.* See the top view in Figure 13.1(b). Experience shows that one of the most common errors in revolution constructions is the violation of this perpendicularity principle, particularly in rotation about an inclined or an oblique axis.

13.2   Revolution of a Line

As a line revolves about an axis, all points of the line revolve *through the same angle*. Otherwise the length of the line is altered and the end product becomes a different line instead of merely a new position for a given line.

As an example, Figure 13.2(a), let it be required to revolve line CE about axis XY until line CE lies in a horizontal plane above axis XY. The first step is the construction of view 1 showing axis XY as a point. In that view arcs of revolution about axis XY show in their true circular shape. A line is then drawn from yx^1 perpendicular to c^1e, locating k. As c^1e revolves, it remains tangent to the circular arc drawn through k.

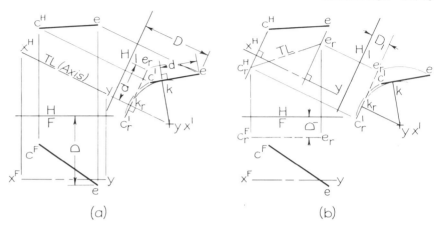

FIGURE 13.2   Revolution of a line

In view 1 any horizontal plane appears as a line parallel to folding line H/1. Hence, in this case, point k is revolved to k_r in which position line CE in space becomes horizontal and above axis XY. Points $c_r{}^1$ and e_r may be located as the intersections of their respective arcs of revolution with the horizontal plane through k_r or, more accurately, by transferring line segments kc^1 and ke to positions $k_rc_r{}^1$ and k_re_r, respectively. The revolved line is drawn as a phantom line to make it readily distinguishable from the line in its original position.

Since axis XY appears true length in the top view, any arcs of revolution about XY must appear in the top view as straight lines perpendicular to the axis. Points $c_r{}^H$ and e_r are therefore located by drawing construction lines from c^H and e perpendicular to x^Hy and projecting from $c_r{}^1$ and e_r, Figure 13.2(b). The front view $c_r{}^Fe_r$ is then established by projection from $c_r{}^He_r$ and transfer of distance D_1.

13.3   True Length by Revolution

The true length of a line may be obtained by revolving the line about an axis that is parallel to a projection plane until the line itself is parallel to that projection plane. In Figure 13.2(b) line CE was revolved into a horizontal plane. The new top view $c_r{}^He_r$ is therefore true length.

If, however, it is desired merely to obtain the true length of a line OA, Figure 13.3, by any appropriate revolution, a more convenient axis may be chosen, resulting in considerably less construction. Since it is necessary to have both true-length and point views of the axis, the simplest arrangement is an axis that is parallel to one principal projection plane and perpendicular to an adjacent principal projection plane. Further simplification results from assuming an axis containing one endpoint of the given line, Figure 13.3(a) and (b).

In this case the endpoint O on the arbitrarily chosen vertical axis remains stationary while all other points of the line, including point A, describe circular arcs, the whole effect being the generation of a right circular cone. For the position $o^H a_r$, line OA is frontal and this view of $o^F a_r$ is true length, Figure 13.3(b). Since the entire cone is not used, the construction may be simplified to that of Figure 13.3(c).

In its various positions the revolving line remains at a constant angle with the plane of the base of the cone. In Figure 13.3 this plane is horizontal, so that in the true-length position $o^F a_r$, the true slope, θ_H, may be measured as indicated. It should be emphasized that the angle θ_H is obtained because the chosen axis is perpendicular to a horizontal plane. For purposes of comparison an auxiliary view is included in Figure 13.3(c) showing the true length and slope of line AB as obtained in §3.3.

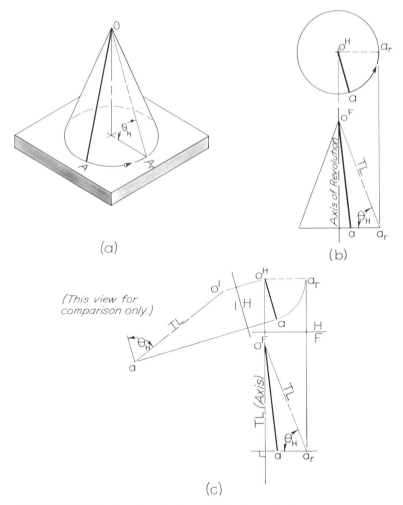

(a)

(b)

(c)

FIGURE 13.3   True length and slope of a line by revolution

If an axis perpendicular to a frontal plane is selected, Figure 13.4(a), a given line may be revolved to a horizontal position as shown. The true length and the angle with the frontal plane, θ_F, are then measured in the top view.

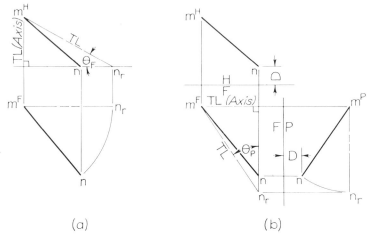

FIGURE 13.4   True length and angles with frontal and profile planes by revolution

If a given line is revolved about an axis perpendicular to a profile plane, Figure 13.4(b), the true length and angle with the profile plane, θ_P, may be found as shown.

13.4   Examples of Converse Problems (Establishing Views with Angles Known)

EXAMPLE 1

Problem

Complete the front view of line EG, Figure 13.5(a), if the line slopes upward from point E at an angle of 30° with a horizontal plane.

Graphic Solution

The top view is revolved about a vertical axis to position $e^H g_r$, Figure 13.5(b). The front view of the revolved line must then be true length and may be drawn at the slope of 30° as shown. A vertical projection line from the top view of point g_r establishes its front view. If line EG is then considered *counterrevolved* to its original position, point G moves horizontally in the front view from g_r to g, established by a vertical projection lines from the top view of G. Line $e^F g$ is then the required front view of line EG.

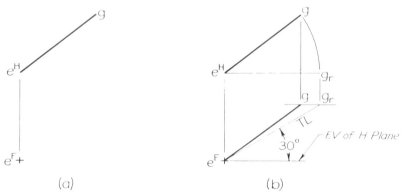

FIGURE 13.5   Establishing the front view of a line of given slope

EXAMPLE 2

Problem

Complete the front view of line AB, Figure 13.6(a), if the line slopes upward from point A and makes an angle of 25° with a frontal plane.

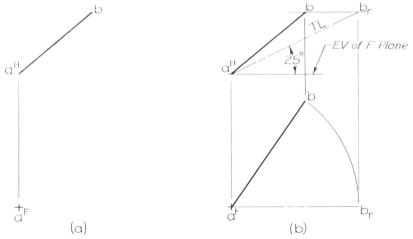

FIGURE 13.6   Establishing the front view of a line with angle with frontal plane given

Graphic Solution

If the unknown front view is considered revolved to a horizontal position about an axis through point A, the revolved top view must then be true length and may be drawn at 25° with a frontal plane as shown, Figure 13.6(b). The arc of revolution appears in the top view as a line parallel to the edge view of a frontal plane, establishing b_r, from which the front view is located by projection. Counterrevolution to the projection line from b in the top view establishes the front view of point B.

13.5   Angle Between Line and Oblique Plane by Revolution

In §13.3 the angles formed by a given line and the principal planes were found by revolution. As was emphasized therein, if it is desired to find by revolution the angle between a line and any particular given plane, *the axis of revolution must be perpendicular to that plane.* Revolution about an axis not perpendicular to the given plane alters the angle being sought. Violation of this principle is a common error in applying the revolution method, and the student should take special pains to avoid this mistake.

To find the angle between line GE and plane ABCD, Figure 13.7, auxiliary view 1 is first constructed showing the given plane in edge view. An axis of revolution perpendicular to the plane must appear true length together with the given line GE. An axis of revolution perpendicular to the plane must appear true length in this view as indicated. View 2

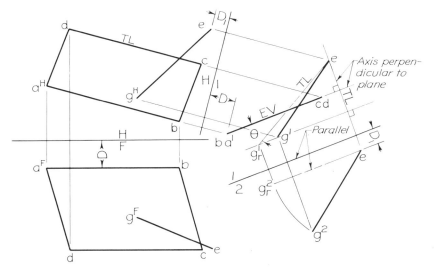

FIGURE 13.7   Angle between line and oblique plane

is then added, showing the axis of revolution as a point. View eg^2 is revolved to position eg_r^2 and the true-length line eg_r^1 is established. The angle θ between the line and plane may then be measured in view 1 as shown.

For other methods see Chapter 11.

13.6   Normal View of a Plane by Revolution

The true size and shape of a plane surface may be found by revolving the plane as a unit until it is parallel to a principal plane. In Figure 13.8(a) auxiliary view 1 is added, showing given plane ABCE in edge

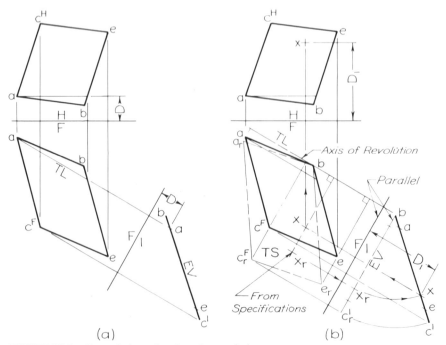

FIGURE 13.8 Normal view of a plane by revolution

view. The edge view of the plane is then revolved to the position bc_r^1, Figure 13.8(b), where it becomes frontal. The resulting revolved front view $a_rbe_rc^F$ is therefore a normal or true-size view.

In this true-size view, points or lines may be located from specifications and subsequent locations in the given views are obtained by counterrevolution and projection. Point X of Figure 13.8(b) is an example of this procedure.

13.7 Revolution of a Solid

A jig or fixture for holding a piece of work for machining must sometimes incorporate in its design provision for revolving the object to successive positions for each of several machining operations. Construction of the views of a revolved solid involves the principles discussed in the preceding material of this chapter. When a line, a plane, or any geometric form revolves, all points must revolve about the same axis and through the same angle. The following example illustrates a single revolution of a solid object.

EXAMPLE

Problem

After plane ABJG of the jig block has been surface-milled, Figure 13.9, through what angle must the piece be revolved to bring surface

ABCE into the same horizontal plane for milling? Draw the views of the
revolved object.

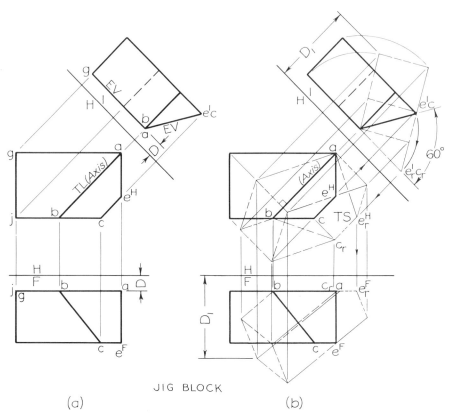

JIG BLOCK

(a)   (b)

FIGURE 13.9   Revolution of a solid

Graphic Solution

In Figure 13.9(a) auxiliary view 1 is added to secure edge views of
surfaces ABCE and ABJG including a point view of edge AB, which is
to serve as the axis of revolution. Figure 13.9(b) shows the revolution of
the object through an angle of 60° to bring surface ABCE into the
plane of the original position of surface ABJG. All other points must be
revolved through the same angle. The resulting new top and front views
are shown.

13.8   Establishing a Line at Given Angles with Two Principal Planes

Occasionally in practice it is necessary to establish a line forming a
compound angle with two planes. In more familiar terms, the line must

be located in such a position as to form given angles with each of two principal planes.

As an example of the solution of such a problem, let it be required, Figure 13.10, to establish a 19 mm line terminating at given point O and forming angles of 45° with a horizontal plane and 30° with a frontal plane. Since all of the elements of a right circular cone are at equal angles with the base plane, the intersection of two cones with proper base angles and in appropriate positions locates the required line, Figure 13.10(a). One cone is established as shown in Figure 13.10(b)

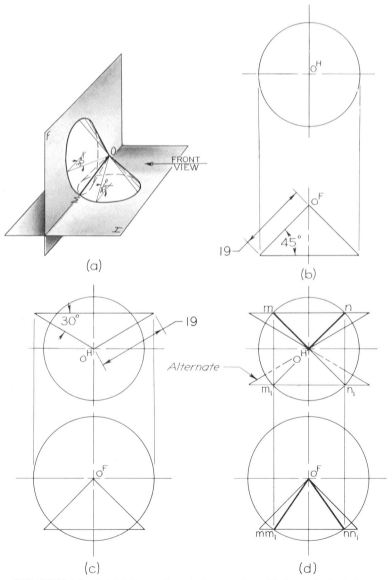

FIGURE 13.10 Establishing a line at given angles with two principal planes

with its vertex at given point O and 19 mm elements at 45° with its horizontal base. This conical surface is the locus of all lines extending downward from point O that are of 19 mm length and at 45° with horizontal.

The second cone is drawn, Figure 13.10(c), with its vertex also at point O, but with its 19 mm elements at 30° with a frontal plane. The surface is the locus of all the 19 mm lines extending backward from point O and forming 30° angles with the frontal plane. Since the elements of the two cones are of the same length, the base circles intersect as shown in Figure 13.10(d), and the two elements OM and ON are common to the two cones. Either of the two elements fulfills the requirements of the problem as stated.

It should be emphasized that the elements of the two cones *must* be drawn to the same length; otherwise the base *circles* will not intersect (although the base *planes* will). By reversing both or either of the cones in turn, it is possible to find eight different line segments (six in addition to OM and ON) which form the given angles. Four of these are aligned with the other four, respectively, so that there are actually only four different lines of indefinite length that answer the requirements. In Figure 13.10(d) the results of reversing the second cone (30°) are shown. In a practical application it is normally apparent which of the possible solutions is the one desired.

These same principles can be used to establish a line making given angles with *any* two intersecting planes (with certain exceptions as noted below). In practice a desired angle between the required line and a given plane may be larger than 90°, for example, 135°. The cone incorporating the angle would then be drawn with a base angle equal to the *supplement* of the given angle. In the example mentioned, the base angle of the cone would be 180° minus 135° or 45°.

It should be noted that a line can be established at specified angles with two planes only within certain limits. For example, if the two planes are perpendicular, the sum of the two angles must fall in the range of 0–90° or 270–360°; otherwise the problem is incapable of solution. For a sum more than 90° but less than 270°, the two cones do not intersect. For the particular case of the sum being 90° or 270°, the two cones are tangent and the *element of tangency* is the required line. If the latter condition is applied to a situation similar to that in Figure 13.10, the two possible elements of tangency are profile lines.

PROBLEMS

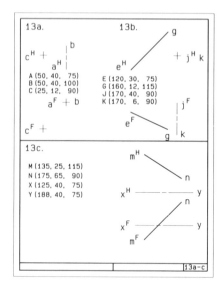

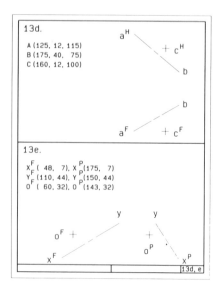

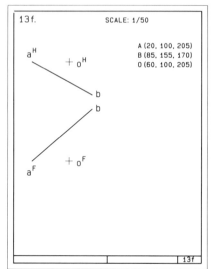

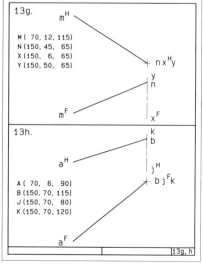

13a. Revolve point C 200° counter-clockwise about axis AB.

13b. Revolve line EG 90° clockwise about axis JK.

13c. Show the views of line MN revolved about axis XY until MN appears true length in the front view.

13d. Revolve point C 90° counter-clockwise about axis AB.

13e. Show the views of the path of point O revolved 360° about axis XY.

13f. Point O is the extreme tip of a lever revolving on shaft AB. Locate a frontal wall behind the lever that clears the lever by 100 cm.

13g. Revolve line MN about axis XY until MN appears true length in the front view.

13h. Revolve line AB about axis JK until AB appears true length in the top view.

187

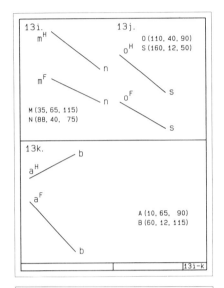

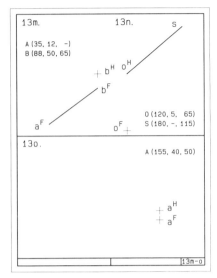

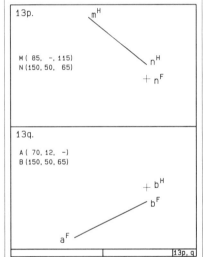

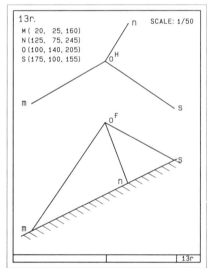

*Use revolution methods in problems **13i–13r**.*

13i. Find the true length and θ_F of line MN.

13j. Find the true length and θ_H of line OS.

13k. Find the true length and θ_P of line AB.

13m. Complete the top view of line AB that has a length of 75 mm.

13n. Complete the front view of line OS if θ_H is 30°.

13o. Complete the view of a 115 mm line AB that has a bearing of N60° and a downgrade of 30%.

13p. Complete the front view of the line MN for which θ_F is 30°.

13q. Complete the top view of a line AB that has a grade of 40%.

13r. Determine the true lengths of the three guy wires OM, ON, and OS.

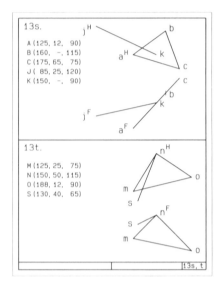

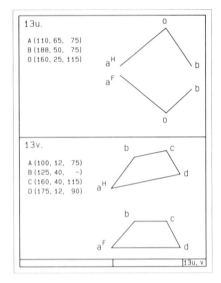

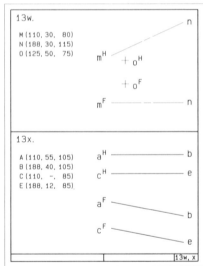

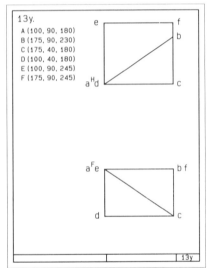

Use revolution methods in problems **13s–13y.**

13s. Find the angle that line JK makes with plane ABC.

13t. Find the angle that line NS makes with plane MNO.

13u. Show the views of the bisector of angle AOB.

13v. Show the true size and compute the area of the trapezoid ABCE.

13w. Locate the center line of a branch pipe from line O to the main

line MN if the branch is connected to main line MN with a 45° Y-fitting.

13x. Determine the distance between the parallel lines AB and CE.

13y. Through what angle must the block be revolved about AB in order to bring surface ABC into a horizontal plane for surface milling? Using phantom lines show the revolved position of the block in all the views. Omit hidden lines.

SELF-TESTING PROBLEMS

13A. By revolution, determine the true length and grade of line AB. If line AC has the same grade as AB, complete the top view of this line.
13B. Determine the angles formed by edges VB and VC with the base plane ABC.

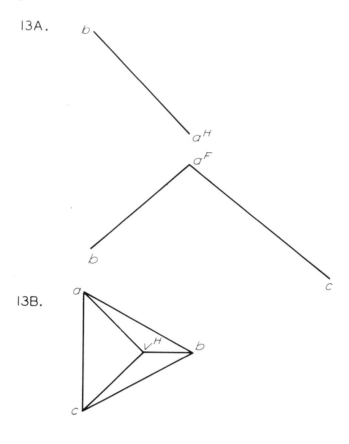

13A.

13B.

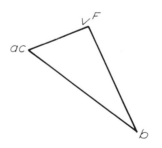

14 Concurrent Vectors

Many engineering and scientific data are graphical in nature—read from meters, gages, voltmeters, scales, and other measuring devices. Therefore, graphical methods utilizing such data can provide comparable accuracy for analysis of forces, velocities, accelerations, and directional quantities in structural and machine design, mechanics, and other physical sciences. In addition, graphical solutions are usually faster and are more easily visualized in comparison to analytical methods.

14.1 Definition of Terms

1. *Vectors.* Forces, velocities, and other directed magnitudes may be represented graphically by straight-line segments called vectors. These lines have definite lengths, relative positions, and directions in space. The true length of a vector, to an appropriate scale, represents its magnitude (a *scalar* quantity). An arrowhead indicates the direction (*sense*) of the vector.

    Vectors are usually identified in print with boldfaced type: **A**, **B**, etc. In writing they may be given a distinguishing symbol such as $\vec{A}$.
2. *Concurrent Vectors.* A system of vectors whose lines of action pass through a common point are known as concurrent vectors.
3. *Resultant.* The resultant of a system of vectors is a single vector that may be used to replace the system.
4. *Components.* When a vector is replaced by several vectors, these are known as components of the original vector. The vector is said to be *resolved* into components, a process reversing the procedure of obtaining a resultant.
5. *Equilibrant.* A vector that will balance a system of vectors is the equilibrant. It has the same magnitude as the resultant but has the opposite sense.

6. *Coplanar Vectors.* Vectors whose lines of action lie in one plane are known as coplanar vectors.

7. *Noncoplanar Vectors.* A system of vectors existing in more than one plane is noncoplanar.

8. *Vector Polygon.* If a system of vectors is in equilibrium, the vectors when laid end to end in any sequence but in continuous direction, form a *closed* figure called a vector polygon.

9. *Newton* (*N*). The basic unit of force, kg·m/s². A *mass* of one kilogram (1 kg) exerts a gravitational force of 9.8 N (theoretically 9.80665 N) at mean sea level.

14.2   Resultant of Concurrent Coplanar Vectors

In Figure 14.1(a) two forces, **B** and **C**, are shown acting through a given point. To find the direction and magnitude of a force that will replace these two concurrent forces, the *parallelogram method* is used, Figure 14.1(b).

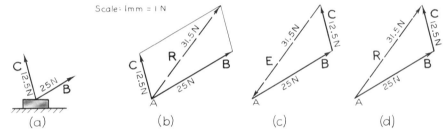

FIGURE 14.1   Resultant of two concurrent forces acting in plane of paper

If the two concurrent forces are represented by their vectors and both are acting either toward or away from their point of intersection, the diagonal of the completed parallelogram drawn through their intersection point represents the resultant of the two forces. A pushing or pulling force may be replaced by its opposite in order to satisfy the requirements for the parallelogram method. For example, if one of two given forces acts toward a point and the other force acts away from the point, the pushing force might be replaced by a pulling force of the same sense and magnitude. Then there would be two pulling forces and the parallelogram method could be applied.

Before the parallelogram is constructed, it is first necessary to establish the lengths of the vectors representing the given forces. Since the forces are acting in the plane of the paper, the vectors are shown in true length. The lengths of the vectors are determined directly from the given scale. The parallelogram is drawn, and the diagonal through the point of intersection of the forces represents the resultant **R**, which replaces the given forces. The magnitude of the resultant is established using the given scale, and the direction is indicated by the arrowhead.

By the vector-polygon method (§14.1, definition 8), the vectors representing the forces are laid out end to end in continuous direction and parallel to the given positions of the forces, Figure 14.1(c). The side or vector necessary to close the polygon represents the equilibrant force **E**, which maintains the given forces in equilibrium. The resultant **R**, which replaces the given forces, has the same magnitude as **E** but acts in the opposite direction, Figure 14.1(d). The magnitude of this force is determined according to the given scale.

If several vectors are acting through a point, Figure 14.2, the resultant vector can be more quickly determined by the vector-polygon method than by application of the parallelogram method to successive pairs of vectors. The vectors are laid end to end (tip to tail) in any sequence,

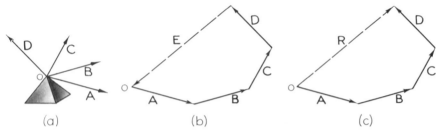

(a)   (b)   (c)

FIGURE 14.2   Resultant of more than two concurrent forces acting in plane of paper

Figure 14.2(b). The vector necessary to close the polygon represents the equilibrant capable of balancing the given vectors **A**, **B**, **C**, and **D**. The resultant **R**, Figure 14.2(c), acts in the opposite direction. In this illustration all the vectors are in the plane of the paper so that their magnitudes may be scaled directly.

In many instances vectors acting through a point lie in an oblique or an inclined plane, and hence the true magnitude of the vectors are not directly shown in principal views. For example, in Figure 14.3(a) two concurrent vectors **M** and **N** acting in an oblique plane are represented

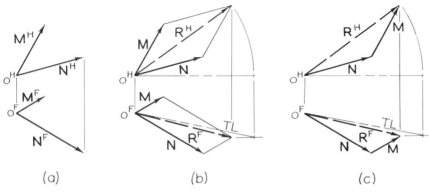

(a)   (b)   (c)

FIGURE 14.3   Resultant of two concurrent forces acting in an oblique plane

by their front and top views. To find the resultant vector, a parallelogram is constructed in the two given views, Figure 14.3(b). The diagonal
of the parallelogram through the intersection point of the vectors is the
resultant. The true magnitude of the resultant may be determined by
finding its true length by revolution as shown, or by the auxiliary-view
method.

The vector-polygon method may also be used in this instance to find
the resultant, as shown in Figure 14.3(c). The true magnitude of the
resultant is determined by scaling its true length as found by revolution.

Note that the views are treated separately, as if they were independent
vector systems. Actually these views represent horizontal and frontal
components of the space vectors. These components are combined in
their respective views to obtain horizontal and frontal components
(views) of the resultant. The vertical alignment of the components checks
the accuracy of construction.

14.3   Resolution of a Vector into Concurrent Coplanar Components

The resolution of a known vector into two concurrent coplanar vectors involves the reversal of the procedures used in securing the resultant.

In Figure 14.4(a) the directions and locations of a force and components 1 and 2 are known. If the magnitude of the force is given as
800 N, let it be required to resolve this force into components along 1
and 2. It is first necessary to revolve the force into a true-length position
and properly scale the 800 N length. To find a revolved true-length
position, any convenient point such as X is selected and revolved as

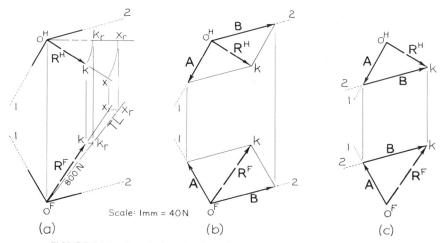

FIGURE 14.4   Resolution of a force into concurrent coplanar components

shown. The magnitude 800 N is then set off to scale on the true-length line o^{F}x$_r$, and the point k$_r$ thus determined is counterrevolved to the original position to establish the force **R**. From points o^H and k, Figure 14.4(b), the sides of the parallelogram are drawn parallel to the given views of components 1 and 2, respectively. The parallelogram thus formed determines the vector components **A** and **B** along 1 and 2. The magnitudes of the forces **A** and **B** may now be found if desired by additional revolutions. For simplicity this true-length construction is not shown.

The vector-polygon method, Figure 14.4(c), also furnishes a ready resolution of a force into two components. After laying out the force **R** as before, the polygon is closed by intersecting lines 1 and 2 drawn through points o^H and k parallel to the given views of Figure 14.4(a). Again, the magnitudes of the components **A** and **B** may be found if needed.

14.4 Resultant of Concurrent Noncoplanar Vectors

The resultant of three vectors whose lines of action lie in more than one plane can be found by combining two of the vectors into a resultant and then combining this resultant with the third vector into a resultant for all three vectors. By continuing this procedure, any number of concurrent vectors may be finally combined into one resultant.

Plane Method. In Figure 14.5(a) three vectors **A**, **B**, and **C** acting through point O determine three planes with point O in common. There-

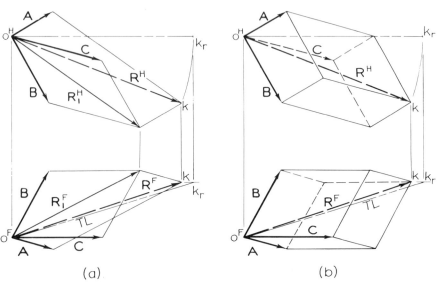

(a)   (b)

FIGURE 14.5   Resultant of three noncoplanar concurrent vectors—plane method and parallelepiped method

fore, to find the resultant of any two of the vectors, the appropriate plane is selected and the parallelogram is constructed. In this case the vectors **B** and **C** were selected, and their resultant is shown by the diagonal R_1.

The next step is to combine the resultant R_1 with the remaining vector **A**. These two vectors determine still another plane. The parallelogram in this plane produces the resultant **R** of the three noncoplanar forces **A**, **B**, and **C**. The true magnitude of the resultant **R** may be determined by revolution as shown.

Parallelepiped Method. A variation of the plane method utilizes the parallelepiped (see Appendix C.1), Figure 14.5(b). Since the three vectors establish three planes that intersect at point O, the construction of a parallelogram in each plane produces three sides of a parallelepiped. The opposite sides (parallelograms) are drawn parallel, respectively, to the three parallelograms established by the views of the vectors. The diagonal of the parallelepiped through the point of intersection of the vectors is the resultant of three vectors. The true magnitude of the resultant **R** is obtained by revolution.

Vector-Polygon Method. The resultant of concurrent, noncoplanar vectors also may be determined by application of the vector-polygon

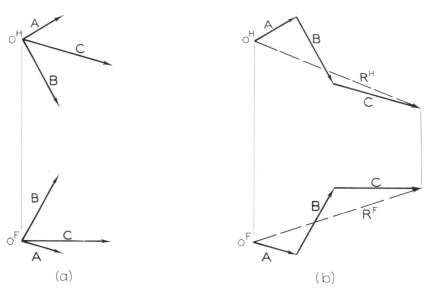

FIGURE 14.6   Resultant of three noncoplanar concurrent vectors—vector-polygon method

method in the separate views, Figure 14.6. The true magnitude of the resultant **R** could be obtained by revolution as in Figure 14.5.

The vector-polygon method has obvious advantages in clarity for a system involving multiple vectors.

14.5   Resolution of a Vector into Concurrent Noncoplanar Components

It is often necessary in engineering practice and design work to resolve a known vector into concurrent, noncoplanar components.

For example, in Figure 14.7(a) it is required to resolve the vertical force exerted by the 387.8 kg mass into components in the three legs of the supporting structure or tripod. The vector **B** representing the 3800 N force, Figure 14.7(b), is the diagonal of a parallelepiped of which three of the edges coincide with the legs of the tripod. To begin the

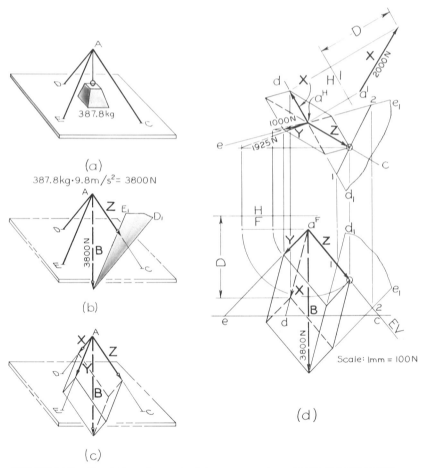

FIGURE 14.7   Resolution of a vector into concurrent noncoplanar components

construction of the parallelepiped, a plane is passed through the endpoint of the vector **B** and parallel to a plane determined by any two of the tripod legs. In this case the plane is passed parallel to the plane of legs AD and AE, Figure 14.7(b), thereby establishing a plane of an opposite face of the parallelepiped. In order to determine the length of the edges between the two parallel faces, the piercing point of leg AC is then found in the plane. This procedure establishes the length of one component vector **Z**. The parallelepiped can now be drawn and the remaining components thus established, Figure 14.7(c).

The foregoing procedure is shown in multiview form in Figure 14.7(d). The piercing point of leg AC in the plane established parallel to legs AD and AE is found by the two-view method of §6.2. The parallelepiped is completed with parallel lines in the two given views, and the magnitudes of the components are determined by revolution as shown for component vectors **Z** and **Y**, or by auxiliary views as shown for component vector **X**.

As illustrated in Figure 14.8, the foregoing problem may be solved through the means of an auxiliary view showing in edge view the plane of two of the components.

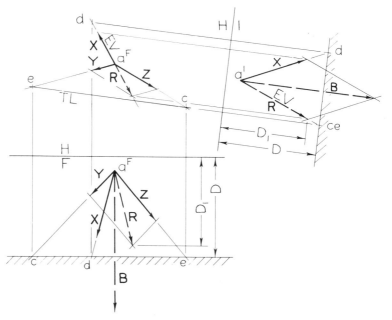

FIGURE 14.8   Resolution of a force into concurrent noncoplanar components—edge-view method

Plane ACE was selected as being convenient. In the top view leg ce is true length and establishes the line of sight for auxiliary view 1 showing plane ACE in edge view. In view 1 vector **B** is resolved into two components, **X** and **R**. These three vectors form a coplanar system. Hence in the top view, where vector **B** shows as a point, **R** must be aligned with **X**. (The coplanar system shows here in edge view.) The front view of **R** is established by projection and transfer of distance D_1. Vector **R** may now be resolved into components along legs AE and CE, establishing the views of vectors **Y** and **Z**. The magnitudes of the components could now be determined as in Figure 14.7(d).

14.6   Other Vector Problems

Thus far the vector examples have been largely forces. A great variety of graphical solutions has been devised for other force problems involving nonconcurrent forces, both parallel and nonparallel, *moments*, *couples*, and so on. Other vectors arising in connection with velocities and accelerations are also amenable to graphical treatment. These problems are encountered by the engineering student in later courses and so will not be treated here except for the following discussion of velocity vectors.

14.7   Velocity Vectors

Concurrent velocity vectors may be combined in exactly the same manner as forces. As an example, let it be required to find the true ground speed of an aircraft in level flight at point A if the indicated air speed is 510 knots on a compass course of N120° and there is a 130 knot wind blowing due north, Figure 14.9(a). (One knot is a velocity of 1 nautical mile per hour or 1852 m per hour.)

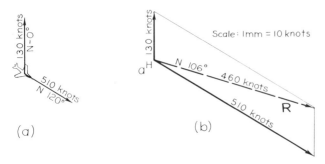

(a)

(b)

FIGURE 14.9   Sum of velocity vectors

The pilot's compass indicates the direction in which the aircraft is headed. The air-speed indicator tells him his velocity with respect to the air in which the aircraft is flying. If, however, the air is also moving with respect to the earth's surface (which is considered stationary for reference purposes), the absolute movement of the aircraft with respect to ground is represented by the vector sum of the two velocities. Thus in Figure 14.9(b) the aircraft and wind vectors are laid out in the prescribed directions and to the given scale. Only the top view is needed here because the aircraft is in level or horizontal flight. The vector sum **R** indicates that the aircraft is moving with respect to the earth's surface at a velocity of 460 knots on a course of N106°.

14.8   Relative Motion

Another example involving velocity vectors is illustrated in Figure 14.10. Here two ships represented by points X and Y are on their indicated courses of N45° at 15 knots and N330° at 19 knots, respectively. The problem is to find how close the ships will pass.

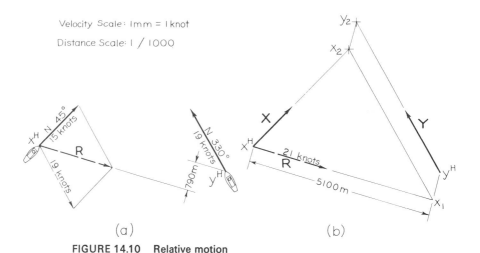

FIGURE 14.10   Relative motion

The given velocity vectors are specified with respect to the earth. The real concern here, however, is the motion of one of the ships relative to the other, which is the vector difference of the two motions. This involves reversing one vector before adding it to the other, which is equivalent to considering one ship stationary while the other moves in the composite direction represented by the vector difference. Hence in Figure 14.10(a) ship Y is temporarily considered stationary while the motion of ship X becomes that indicated by vector **R**. It is then evident that ship X will pass ship Y with a clearance of 790 m.

It must be remembered that the distance and velocity scales are independent. Thus, if the elapsed time to closest approach is to be calculated, the relative distance of travel of ship X is measured at the given distance scale, which gives 5100 m, Figure 14.10(b). The relative velocity of ship X is the length of vector **R**, 21 knots, at the vector scale. Thus the elapsed time becomes:

Time to closest approach = distance ÷ velocity

$$= \frac{(5100 \text{ m})(\text{hr})(\text{naut. mile})(60 \text{ min})}{(21 \text{ naut. mile})(1852 \text{ m})(\text{hr})}$$

= 7.868 min or 7 min 52 sec

If it is desired to show the location on the water at which this closest approach occurs, the clearance distance $y^H x_1$ is simply moved along the direction of vector $\mathbf{Y}$ until x_1 falls on the path of vector $\mathbf{X}$ at x_2, Figure 14.10(b). The actual positions of the ships at this moment would then be at x_2 and y_2.

The foregoing procedure may be employed for a three-dimensional relative motion problem. An example is the determination of how close a ship and a diving aircraft will pass. If the ship is temporarily considered stationary, then the aircraft moves in space in a direction and with a velocity represented by the vector difference of the original velocity vectors. The problem then becomes that of finding the shortest distance from a point to a line (§5.5).

PROBLEMS

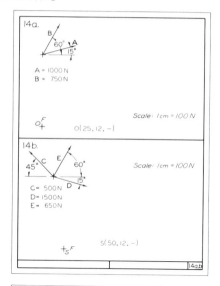

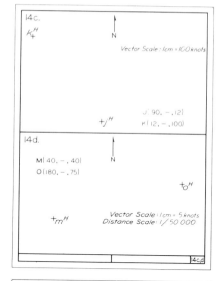

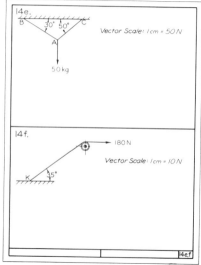

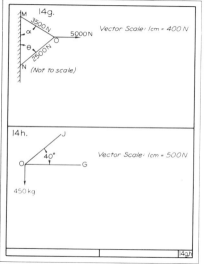

14a. Starting the vector polygon at point O, find the resultant of vectors **A** and **B** acting in a frontal plane.

14b. Starting the vector polygon at point S, find the resultant of the vectors **C**, **D**, and **E** acting in a frontal plane.

14c. An aircraft at point J is flying at an air speed of 250 knots on a compass course of N45°. Find the true flight direction and resulting ground speed if there is a 75 knot wind blowing due south. With the same air speed and wind force, in what direction should the aircraft

head to pass over point K? What is the resulting ground speed?

14d. A ship at point M is traveling at 25 knots and N30°. A ship at point O is traveling at 15 knots and N330°. How close will the ships pass? Find the time needed to reach this position.

14e. Find the forces in the members AB and AC acted upon by the given mass.

14f. Determine the horizontal and vertical components of the force on the cable anchorage at point K.

14g and **14h.** See next page.

14 i.

A = 350 N
B = 500 N
C = 300 N
D = 550 N

40° 15° 35°-60°

Vector Scale: 1cm = 50N

14i

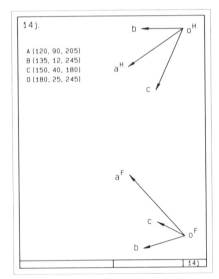

14 j.

A (120, 90, 205)
B (135, 12, 245)
C (150, 40, 180)
0 (180, 25, 245)

14j

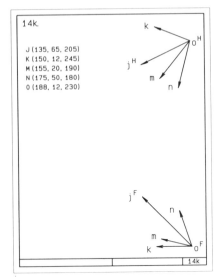

14k.

J (135, 65, 205)
K (150, 12, 245)
M (155, 20, 190)
N (175, 50, 180)
0 (188, 12, 230)

14k

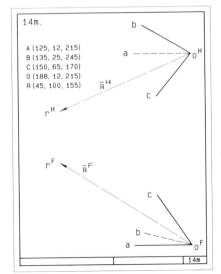

14m.

A (125, 12, 215)
B (135, 25, 245)
C (150, 65, 170)
0 (188, 12, 215)
R (45, 100, 155)

14m

14g. Find the angles α and θ when the 5000 N force at point O produces the given forces in cables OM and ON.

14h. Find the forces in the members OJ and OG.

14i. Determine the resultant of the forces **A**, **B**, **C**, and **D** and the horizontal and vertical components of the resultant.

14j. Find the views and the true

length of the resultant of the three forces OA, OB, and OC. Complete the parallelepiped.

14k. Find the views and the true length of the resultant of the forces OJ, OK, OM, and ON.

14m. Resolve the vector **R** into components along the members OA, OB, and OC.

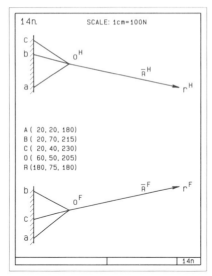

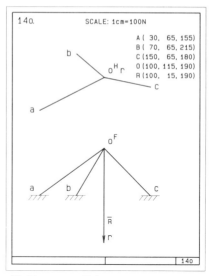

14n. For the applied force **R** find the force in each of the frame members OA, OB, and OC.

14o. For the vertical force **R** determine the force in each member of the tripod frame.

14p. Indicate whether the following statements are true or false. If assigned, prepare written explanations or sketches to justify the answers.

(1) The magnitude of the resultant of any two given vectors is always greater than either of the given vectors.

(2) Two concurrent vectors are always coplanar.

(3) Two parallel vectors are noncoplanar.

(4) The equilibrant has the same magnitude as the resultant but acts in the opposite direction.

(5) An inclined force is greater than either its vertical or horizontal component.

(6) A vector polygon always has an even number of sides.

(7) In problems involving relative motion of two objects, a vector may be introduced to put one of the objects at rest.

SELF-TESTING PROBLEMS

14A. Find the forces in boom AB and cable BC.
14B. The paths and velocities of two aircraft in level flight at the same elevation are indicated by vectors **M** and **N**. How close will the aircraft pass?

14 A.

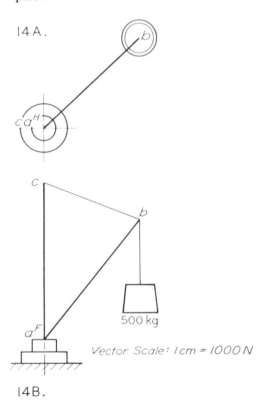

Vector Scale: 1cm = 1000 N

14B.

Distance Scale: 1/5 000

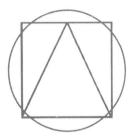

15 Plane Tangencies

A plane tangent to a ruled surface (see Appendix C.2) such as a cylinder or a cone contains one and only one straight-line element of that surface. A plane tangent to a double-curved surface such as a sphere contains one and only one point in that surface. Since all lines tangent to a curved surface at a particular point or at points along the same straight-line element lie in the plane tangent at that point or element, a tangent plane may be conveniently represented in one of the following manners, depending on the situation:

1. By two straight lines, one an element (the element of tangency) and the other a line tangent to the surface at a point on the element.
2. By two straight lines, both tangent to the curved surface at the same point.

15.1   Lines Tangent to Curved Surfaces

A line may be drawn tangent to a curved surface by drawing it tangent to a curve lying in that surface. This is a general statement that is most often applied in practice to surfaces in which the curve in the surface is a circle. Examples in this chapter will be confined to objects of that nature, such as a circular cylinder, a circular cone, a sphere, or a torus.

Although the engineering or technology student is probably familiar with the mechanics of drawing a line tangent to a circle, it should be realized that such constructions encountered previously have been *plane geometry* constructions. If the construction is understood to be confined to a plane, it is sufficient to define a line tangent to a circle as being a line that contains one and only one point of the circle. But if a circle or other plane curve is regarded as existing *in space* rather than on paper, it should be borne in mind that: A line tangent to a circle *must lie in the plane of the circle*. This is necessary because, mathematically speaking, a line tangent to a curve has the same *slope* as the curve at the point of tangency. In other words the tangent line must coincide with the curve for a very short (actually infinitely short) distance.

Thus, in Figure 15.1, while both lines AB and XY are *apparently* tangent to the circle in the auxiliary view 1, only line AB is *actually* tangent to the circle, since line XY does not lie in the plane of the circle.

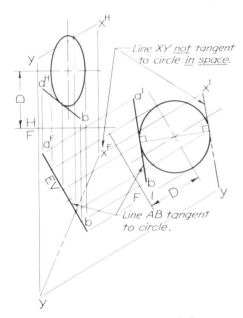

FIGURE 15.1   Line tangent to a circle

The truth of this statement becomes evident when the resulting top view is examined, since x^Hy crosses the ellipse, while a^Hb is tangent to the ellipse. The student should keep this principle constantly in mind as he studies the following constructions.

15.2   Plane Tangent to a Cone and Containing a Given Point on the Surface of the Cone

Let the top view of the point and the front and top views of the cone be given, Figure 15.2(a). The element VT through point A is drawn, Figure 15.2(b), by drawing v^Ht through a^H and projecting to t in the front view. The front view of point A is located by projecting from a^H to v^Ft. Element VT is then the element lying in the required tangent plane. Line HT is drawn tangent to the circular base of the cone by constructing ht tangent in the top view and drawing ht in coincidence with the front view of the base. Intersecting lines HT and VT both lie in the desired tangent plane and therefore represent that plane.

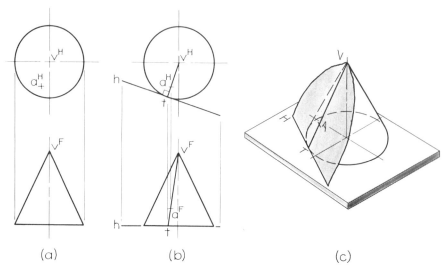

FIGURE 15.2   Plane tangent to cone and containing given point on the surface of the cone

15.3   Plane Tangent to a Cone and Containing a Given Point Outside the Surface of the Cone

Let the given cone and point be as in Figure 15.3(a). Since all elements of a cone pass through the vertex, the vertex V lies in any tangent plane. Line BV joining the given point and vertex must therefore lie in the required tangent plane, Figure 15.3(b). In this case a line tangent to the base must coincide with the top view of the base. In order to lie in a plane with line BV, the tangent line must intersect line BV, which could occur only at point P, the point at which line BV extended pierces the

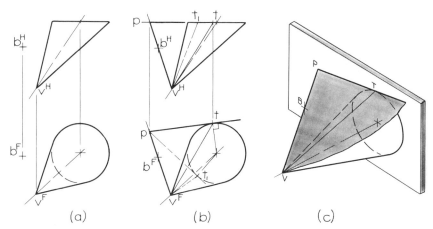

FIGURE 15.3   Plane tangent to cone and containing given point outside the cone

plane of the base. Line PT is then drawn tangent to the base, completing the representation of the tangent plane. If desired, the element of tangency VT may be added as indicated. There are actually two solutions to the problem, the alternative tangent plane being plane VPT_1. In a practical application it is normally apparent which tangent plane is consistent with other design features.

15.4   Plane Tangent to a Cone and Parallel to a Given Line

Let the given cone and line be as shown in Figure 15.4(a).

A plane is parallel to a given line if it contains a line parallel to the given line (§9.3). Thus, if a line VP is drawn through the vertex V of the cone and parallel to the line CE, Figure 15.4(b), a tangent plane containing line VP fulfills the problem requirements. Line VP pierces the base plane of the cone at point P, from which tangent line PT may be drawn as shown. An alternative solution may be secured by drawing line PT tangent on the opposite side of the base. This is not shown.

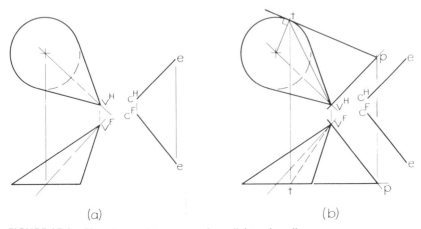

(a)   (b)

FIGURE 15.4   Plane tangent to cone and parallel to given line

15.5   Plane Tangent to a Cylinder and Containing a Given Point on the Surface of the Cylinder

Let the cylinder and the front view of the point be given, Figure 15.5(a).

Element AT is drawn in the front view and then in the top view, Figure 15.5(b). A line tangent to either base at the corresponding end of the element completes the representation of the tangent plane.

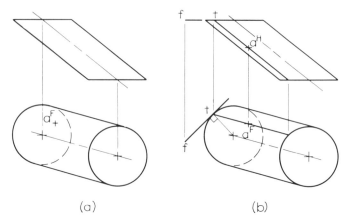

(a)   (b)

FIGURE 15.5   Plane tangent to cylinder and containing given point on surface of cylinder

15.6   Plane Tangent to a Cylinder and Containing a Point Outside the Cylinder

Let the cylinder and point B be given, Figure 15.6(a).

Since all elements of a cylinder are parallel, a line BP drawn parallel to the axis of the cylinder is parallel to the element of tangency and is thus in the same plane as the element of tangency. Point P is the piercing point of line BP in the plane of the lower base, and thus line PT may be drawn tangent to this base as shown. The element of tangency through

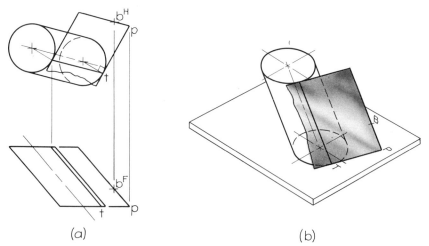

(a)   (b)

FIGURE 15.6   Plane tangent to cylinder and containing given point outside the cylinder

T may be added if desired. There is an alternative solution, since line PT may be drawn on the opposite side of the base.

15.7 Plane Tangent to a Cylinder and Parallel to a Line Outside the Cylinder

Let the cylinder and line EC be given as in Figure 15.7(a). (If the given line is parallel to the given cylinder, the solution is similar to that of Figure 15.6.)

Since at the beginning a point on the cylinder at which to start the construction is not known, this problem is solved indirectly. A plane that contains the given line and is parallel to the axis of the cylinder is parallel to the required tangent plane. This is true because a plane (in this case the tangent plane) that is parallel to another plane is parallel to all lines in the second plane. Consequently, the first step, Figure 15.7(b), is to construct plane ECH parallel to the axis of the cylinder by drawing line CH parallel to the axis. Since the bases of the cylinder in this case are horizontal, any line tangent to one of the bases is of necessity a horizontal line. Therefore, a horizontal line EH is drawn in plane ECH. This establishes in the top view the direction of all horizontal lines in

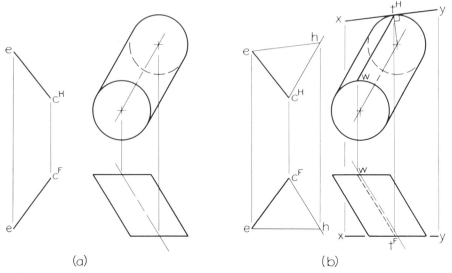

(a)   (b)

FIGURE 15.7 Plane tangent to cylinder and parallel to given line

plane ECH and in all planes parallel to ECH. Line XY is then drawn tangent to one base and parallel to line EH. This line XY plus the element of tangency TW completes the representation of a tangent plane that is parallel to plane ECH and thus to given line CE. Again there is an alternative solution that is omitted for simplicity.

15.8   Plane Tangent to a Sphere Through a Point on the Surface of the Sphere

Let a sphere and the front view of a point A on its surface be given, Figure 15.8(a). The top view of point A may be found by passing a horizontal plane through point A. This plane cuts from the sphere a horizontal circle that contains point A. This circle is then drawn in the top view and a^H is located on it as shown in Figure 15.8(a). A line AH tangent to this circle must be a horizontal line and is also tangent to the sphere.

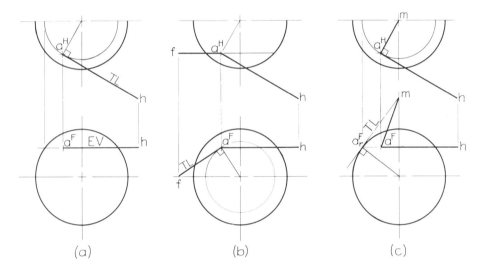

FIGURE 15.8   Plane tangent to sphere at a given point on the surface of the sphere

A frontal circle through point A and lying on the surface of the sphere is then added as in Figure 15.8(b). Frontal line AF drawn tangent to this circle at point A is also tangent to the sphere. The two lines AH and AF thus represent a plane tangent to the sphere at point A.

The same result may be obtained by utilizing the principle that a plane tangent to a sphere is perpendicular to the radius drawn to the point of tangency. This construction may be performed as demonstrated in §10.2, obtaining exactly the same lines, AH and AF.

A third line tangent at point A may be used in place of line AF or line AH, Figure 15.8(c). If the *meridian* (see Figure 22.1) through point A is resolved to a frontal position, it will coincide with the circle representing the front view of the sphere. A line drawn tangent to the revolved meridian will locate point M on the extended vertical axis. If the meridian is counterrevolved, point M will remain stationary and line AM will be the tangent to the meridian. The required tangent plane thus may be represented as MAH.

15.9 Plane Tangent to a Sphere and Containing a Given Line—Line Not Intersecting the Sphere

In this construction, Figure 15.9, an auxiliary view is added to show given line AB as a point. In this view any plane containing line AB must appear in edge view, and thus the required tangent plane may be drawn in edge view as a line tangent to the auxiliary view of the sphere either at t or w. Projection of any suitable points of the planes to the top and front views completes the representation. In this illustration it was chosen to project the actual points of tangency T and W, so that the alternative solutions would be plane ABT and plane ABW.

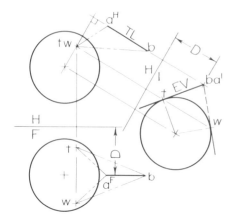

FIGURE 15.9   Plane tangent to sphere and containing given line

15.10 Plane Tangent to a Torus at a Given Point on its Surface

With the front view a^F of the point given, Figure 15.10(a), a horizontal plane may be passed through point A, cutting two circles that appear in their true circular form in the top view. Thus the top view of point A may be at a^H, $a_1{}^H$, $a_2{}^H$, or $a_3{}^H$. The tangent plane construction is shown in Figure 15.10(b) for a^H only and is similar to Figure 15.8(c). A line AH is drawn tangent to the horizontal circle through point A. The vertical right section of the torus through point A is a circle lying in a vertical plane. When this circle is revolved to a frontal position along with point A, line A_rO may be drawn tangent to the circle as shown, locating point O on the vertical axis of the torus. In counterrevolving to the original position of point A, point O remains stationary. Hence the tangent plane is represented by OAH.

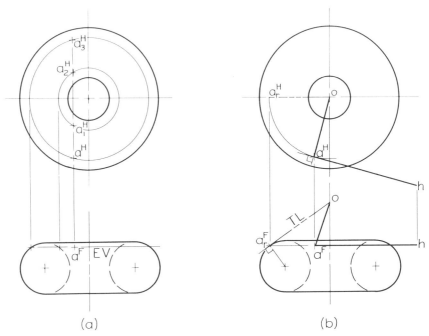

FIGURE 15.10   Plane tangent to a torus at a given point on the surface of the torus

15.11   Application of the Tangent Plane—Representation of a Plane Containing a Given Line and Making a Specified Angle with a Given Plane

All elements of a right circular cone form the same angle with the base plane of the cone. Therefore, all planes tangent to a given right circular cone form this same angle with the base plane. This principle may be used as the basis for the representation of a plane making a specified dihedral angle with a given plane when the line of intersection of the given and required planes is not known.

EXAMPLE 1

Problem

Represent a plane containing line EG and making an angle of 60° with a horizontal plane, Figure 15.11(a).

Analysis

If a right circular cone has its base in a horizontal plane and if its elements make a 60° "base angle" with the horizontal plane, any plane tangent to the cone forms a 60° dihedral angle with horizontal. In order to make it possible for the required plane both to contain line EG and be

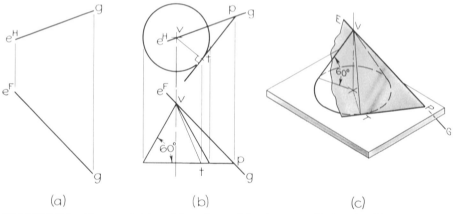

(a) (b) (c)

FIGURE 15.11 Plane containing oblique line and forming a specified angle with a horizontal plane

tangent to the cone, the vertex of the cone must be placed at some point such as point V on line EG, Figure 15.11(c).

Graphic Solution

The cone may be of any height consistent with accuracy and available space, Figure 15.11(b). The tangency construction becomes similar to Figure 15.3(b). The piercing point P of line EG and the base plane is located in the front view and is projected to the top view. From this point, line PT is drawn tangent to the base of the cone. There is an alternative solution, but in a practical application additional data permit selection of one of the two possible planes. Such an application may also require the addition of other lines to complete the views of an actual object as demonstrated in Example 3.

EXAMPLE 2

Problem

Given the front and side views of line XY, Figure 15.12(a), complete the representation of a plane XYZ if the plane is known to be at 60° with a profile plane.

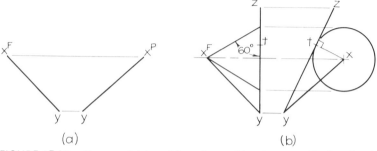

(a) (b)

FIGURE 15.12 Plane containing oblique line and forming a specified angle with a profile plane

Analysis

The introduced right circular cone must have its vertex on the oblique line XY and its base in a profile plane.

Graphic Solution

For simplicity of construction the 60° cone is placed with its vertex at point X and its base in the profile plane through point Y, Figure 15.12(b). In the right-side view a line may then be drawn from y tangent to the circular base of the cone. Selection of any suitable point Z on the tangent line completes the representation. There would, of course, be an alternative solution in which the tangent line is drawn on the opposite side of the base.

EXAMPLE 3

Problem

Given the side view and the incomplete front view of a *guide block*, Figure 15.13(a), and given that the dihedral angle between surfaces ABX and BCX is 65°, complete the front view.

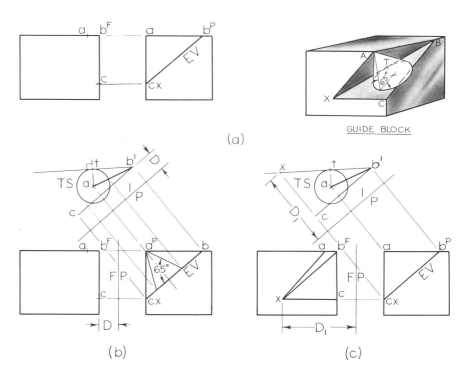

FIGURE 15.13   Application of tangent plane

Analysis

Since line AB is completely given and since the position of plane BCX is established by virtue of the fact that it appears in edge view in the side view, the solution involves the following requirement: Represent a plane containing line AB and making an angle of 65° with plane BCX. This rewording makes it apparent that the problem is merely another variation of the previously discussed Example 1.

Graphic Solution

A right circular cone having a base angle of 65° is introduced, Figure 15.13(b), with its vertex on line AB at point A and with its base in plane BCX. An auxiliary view showing plane BCX in true size and shape is then added. In this view the base of the cone appears as a circle, and since point B is the point at which line AB pierces the plane of the base, b¹t may be drawn tangent to the circle as shown. (Why is line BT not drawn tangent on the opposite side of the base?) A plane making the required angle of 65° with plane BCX is then represented by plane ABT. Since line BT is common to planes ABT and BCX, it is the line of inter-section of these planes and is therefore a segment of edge BX. A pro-jection line from the side view thus establishes x on b¹t extended, Figure 15.13(c). Transfer distance D₁ then locates x in the front view, and the remaining edges are drawn to complete the front view as shown.

15.12   Pictorial Solution of a Plane Tangent to a Cone and Containing a Point Outside the Surface of the Cone.

Some concepts presented in this chapter on tangent planes and some earlier material on pictorial parallelism and pictorial intersections can be combined to include pictorial tangencies involving cones and cylinders.

In this first example, a plane that is tangent to the cone and contains a point C outside the surface of the cone is to be established, Figure 15.14(a). Since the desired tangent plane must contain the vertex of the cone as well as point C, line VC is drawn to contain these two points, Figure 15.14(b). A second line intersecting line VC and also tangent to the circular base of the cone is needed next. It is critical that this second line exist in the circular base plane of the cone. With this fact in mind, line VC is extended to point P which lies in the same rear pictorial surface as does the circular base. Then line PT is added tangent to the circular base, providing the two intersecting lines VCP and PT that represent the specified tangent plane. Line VT may be added to show the element of tangency.

15.13   Pictorial Solution of a Plane Tangent to a Cone and Parallel to a Given Line

In this example of tangency, a plane is needed that is tangent to the conical surface and parallel to line AB, Figure 15.15(a). This desired

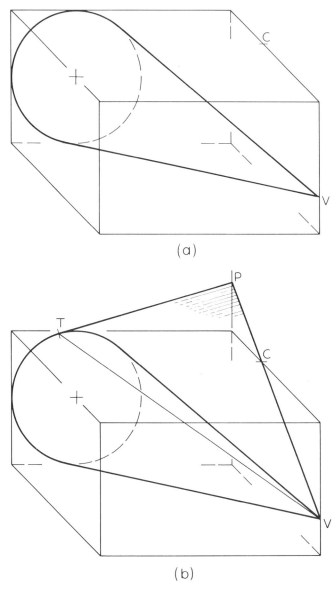

(a)

(b)

FIGURE 15.14   Pictorial solution of plane tangent to cone and containing a
point outside the cone

plane will require a line that must contain the vertex of the cone and
be parallel to line AB. For this construction a plane such as ACBE is
provided that contains line AB, Figure 15.15(b). Then a plane VXOY
is introduced through vertex V and parallel to plane ACBE. This new
plane entails a line VX parallel to line AC with both these lines existing
in the same top pictorial surface. Similarly, line OY is provided parallel
to line BE with both these lines lying in the same lower pictorial surface.

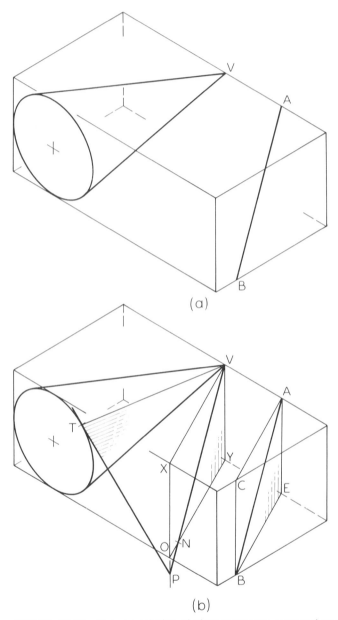

(a)

(b)

FIGURE 15.15   Pictorial solution of plane tangent to cone and parallel to a line

In plane VXOY, line VN is added parallel to line AB, and then line VN is extended to point P to provide a point on line VN that lies in the front pictorial surface that contains the base of the cone. Line PT is next drawn tangent to either side of this base circle, which of course appears elliptical in this isometric pictorial. The specified tangent plane consists of the two intersecting lines VNP and PT. The element of tangency TV may be added as a supplementary provision.

15.14   Pictorial Solution of a Plane Tangent to a Cylinder
and Containing an Outside Point

For this example a plane is required that is tangent to the cylindrical
surface and also contains a given point M, Figure 15.16(a). This specified
tangent plane must contain a line through point M that is parallel to
center line AC of the cylinder. This initial construction requires the
addition of a plane such as ABCE that includes center line AC,
Figure 15.16(b). Line AE of this plane lies in the top surface of the
pictorial and line BC lies in the lower surface. A plane parallel to
plane ABCE is added to include point M. This plane consists of
line SN parallel to line AE with both these lines existing in the top
pictorial surface and line OK parallel to BC with these two lines
existing in the lower surface. Line XY, which is parallel to the cylinder
center line AC as well as all elements of the cylinder, is then added to
plane SOKN.

Next the intersection of line XY with the base planes of the cylinder
is needed. Since these bases lie in the top and bottom horizontal surfaces
of the isometric frame, the intersection of line XY with horizontal
lines SN and OK provides the pertinent piercing points P and P_1.
Lines PT and P_1T_1 are drawn tangent to the cylindrical bases to
complete the presentation of the specified tangent planes. The tangent
points T and T_1 may be connected to represent the supplementary
element of tangency.

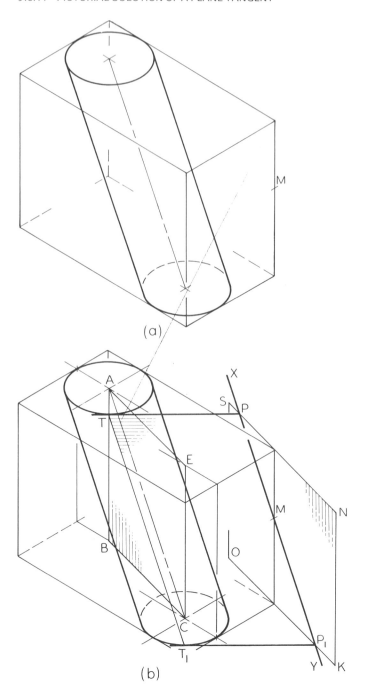

FIGURE 15.16 Pictorial solution of plane tangent to cylinder and containing an outside point

PROBLEMS

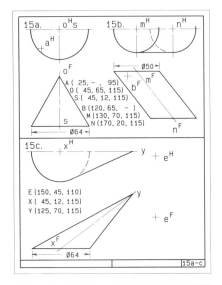

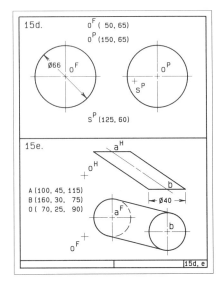

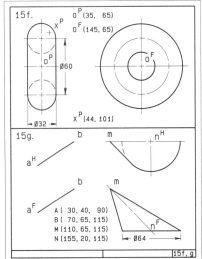

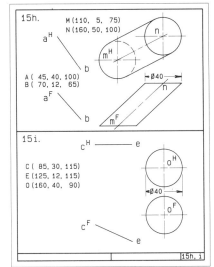

15a. Pass a plane tangent to the cone and containing point A on the surface of the cone.

15b. Pass a plane tangent to the cylinder and containing point B on the surface of the cylinder.

15c. Pass a plane tangent to the cone and containing point E.

15d. Pass a plane tangent to the sphere and containing point S on its surface.

15e. Pass a plane tangent to the cylinder and containing point O.

15f. Pass a plane tangent to the torus and containing point X on its surface.

15g. Pass a plane tangent to the cone and parallel to line AB.

15h. Pass a plane tangent to the cylinder and parallel to line AB.

15i. Pass a plane tangent to the sphere and containing line CE.

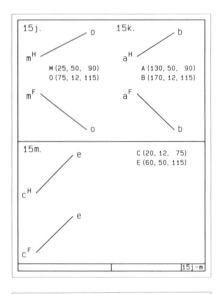

15j.

M (25, 50, 90)
O (75, 12, 115)

15k.

A (130, 50, 90)
B (170, 12, 115)

15m.

C (20, 12, 75)
E (60, 50, 115)

|15j-m|

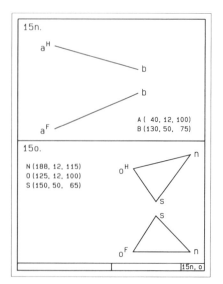

15n.

A (40, 12, 100)
B (130, 50, 75)

15o.

N (188, 12, 115)
O (125, 12, 100)
S (150, 50, 65)

|15n, o|

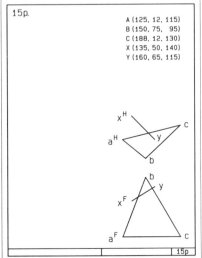

15p.

A (125, 12, 115)
B (150, 75, 95)
C (188, 12, 130)
X (135, 50, 140)
Y (160, 65, 115)

|15p|

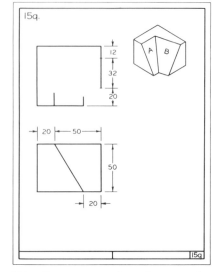

15q.

12
32
20

20 50

50

20

|15q|

15j. Represent a plane that contains line MO and makes an angle of 60° with horizontal.

15k. Represent a plane that contains line AB and makes an angle of 45° with a frontal plane.

15m. Represent a plane that contains line CE and makes an angle of 60° with a profile plane.

15n. Represent a plane that contains line AB and has a dip of 45°NW.

15o. Represent a plane that contains line OS and makes an angle of 60° with plane NOS.

15p. Pass a plane containing line XY and making an angle of 60° with plane ABC.

15q. The dihedral angle formed by planes A and B is 120°. Complete the top view.

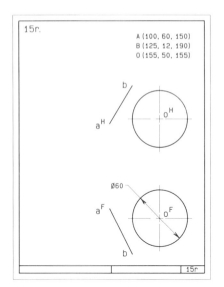

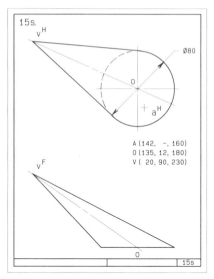

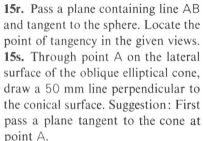

15r. Pass a plane containing line AB and tangent to the sphere. Locate the point of tangency in the given views.

15s. Through point A on the lateral surface of the oblique elliptical cone, draw a 50 mm line perpendicular to the conical surface. Suggestion: First pass a plane tangent to the cone at point A.

15t. Indicate whether the following statements are true or false. If assigned, prepare written explanations or sketches to justify the answers.

(1) A plane tangent to the lateral surface of a cone contains the vertex of the cone.

(2) A line tangent to a circle lies in the plane of the circle.

(3) A plane tangent to a cylinder contains a single element of the cylinder.

(4) Only one plane can be passed through an outside point and tangent to a cone.

(5) A plane can be passed through an outside line and tangent to a sphere.

(6) A plane containing the vertex of a cone is tangent to the lateral surface of the cone.

(7) A plane that contains an outside line and is tangent to a cylinder contains an element of the cylinder that is parallel to the outside line.

(8) A single plane may be passed tangent to any two oblique cones.

(9) A plane tangent to a sphere can be passed through a line that intersects the sphere.

SELF-TESTING PROBLEMS

15A. Pass a plane through point A and tangent to the cone.
15B. Pass a plane tangent to the sphere and containing point A on the surface of the sphere.

15A.

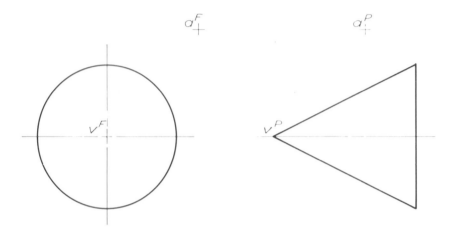

15B.

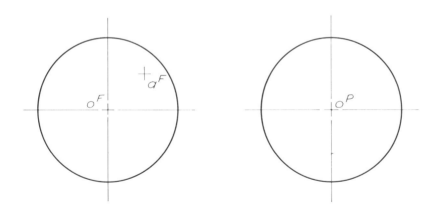

15C. Represent a plane parallel to line MN and tangent to the conical surface.

15D. Provide a plane through point K tangent to the cylindrical surface.

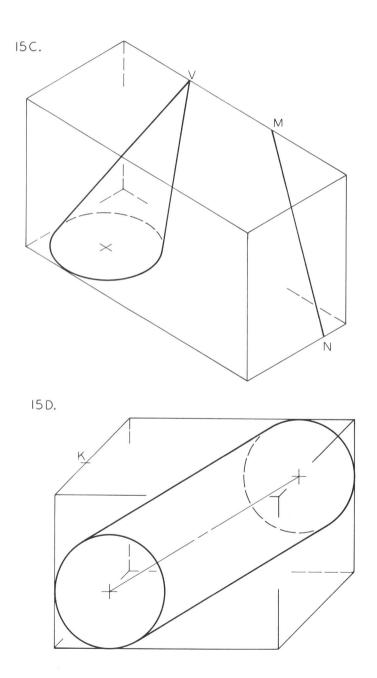

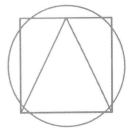

16 Intersections of Planes with Solids

Problems involving the intersections of two surfaces in general may be solved by one of two methods:

1. Lines in one surface are selected and their piercing points with the other surface are found. For practical reasons, the selected lines should be of a type convenient to manipulate, such as straight lines or circles.
2. Additional cutting surfaces are introduced, cutting pairs of lines from the given surfaces. The point of intersection of the two lines of one pair is a point common to the given surfaces and is therefore on their line of intersection. The additional cutting surfaces are usually planes, but may be spheres for certain problems (see §18.7).

These methods have already been applied in finding the line of intersection of two planes, the first method in §§7.1 and 7.2, and the second in §7.3. The methods will now be employed in finding the intersections of planes with the surfaces of solids and will be used again in Chapter 18.

16.1 Intersection of Plane and Pyramid

Figure 16.1 illustrates the use of the two-view method to secure the intersection of plane ABCD and the pyramid OEGJ. An edgewise cutting plane is passed through a^Hb. This cutting plane intersects the plane OGJ in the line 1-2 and intersects the plane OEG in the line 1-3. The intersections of the front views of these two lines with a^Fb produce the front views of the piercing points P_1 and P_2 of line AB with surfaces OGJ and OEG. The top views of these two points are then obtained by pro-

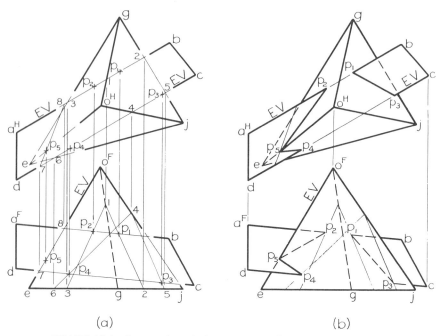

FIGURE 16.1   Intersection of plane and pyramid

jection to a^Hb. The piercing points P_3 and P_4 of line CD with the pyramid are obtained similarly by the use of a cutting plane passed through line CD that intersects the pyramid in the lines 4-5 and 4-6.

Plane ABCD intersects plane OGJ in the line P_1P_3, Figure 16.1(b), as both these piercing points lie on the same surface of the pyramid. Since points P_2 and P_4 lie on different surfaces of the pyramid, it is evident that a single line connecting these two piercing points lies *within* the pyramid and does not represent a line of intersection of the plane and the surface. In such a situation the lines of the intersection cannot be completed until an intermediate piercing point of a line of the solid with the given plane is obtained. To secure this missing point, an edgewise cutting plane is passed through o^Fe, Figure 16.1(a). This cutting plane intersects plane ABCD in the line 7-8. The intersection of line 7-8 with o^He in the top view locates the required point P_5. Since P_2 and P_5 are points common to the planes ABCD and OEG, the line connecting these two points represents the line of intersection of these two planes, Figure 16.1(b). Similarly, the line connecting points P_4 and P_5 represents the intersection of planes ABCD and EOJ. The solution is completed by indicating the correct visibility of the plane and pyramid as shown. Since, in practice, combinations of intersecting forms are usually portions of a single (one-piece) object, the hidden line segments of the plane that fall inside the pyramid are omitted.

16.2 Intersection of Plane and Oblique Cone

In the problem of Figure 16.2 the given plane is shown edgewise in the front view. Thus the front view of the intersection of the plane and oblique cone coincides with this edge view of the plane. Since this intersection must necessarily lie on the surface of the cone, its top view may be secured by projecting points on elements of the cone in the front view to the corresponding views of these elements in the top view. Piercing

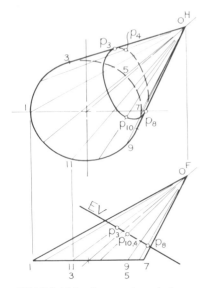

FIGURE 16.2   Intersection of plane and oblique cone

points P_4 and P_{10} on elements 4 and 10 illustrate this procedure. A sufficient number of points to assure an accurate line of intersection should be obtained. This does not always mean that equally spaced elements must be used or that equally spaced points on the intersection are necessarily desirable. Rather one should plan to secure points more closely spaced on the line of intersection, where the rate of change of curvature is greatest. Tangent points P_3 and P_8 are other points that are needed to produce an accurate intersection, which in this case is an ellipse.

If the base of the cone in Figure 16.2 had been circular rather than elliptical, this intersection could also have been found by the method of §16.3.

If the intersecting plane as given does not appear in edge view, an auxiliary view is added, showing this plane in edge view. Then the solution is obtained as before. The construction would be similar to that of Figure 16.5. In this situation both given views would show the intersection as ellipses, although it is possible under particular circumstances for the intersection to appear circular.

16.3　Intersection of Plane and Right Circular Cone

In Figure 16.3 the intersection of the plane and right circular cone could be obtained as in the preceding example by using elements of the cone. However, in the case of a right cone some of the elements would appear too nearly vertical to assure accurate direct projection of points on these elements from one view to the other.

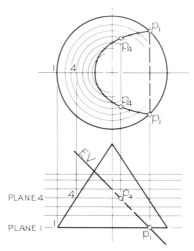

FIGURE 16.3　Intersection of plane and right circular cone

To avoid the inaccuracy that might occur by the use of elements of the cone, horizontal cutting planes are used in this illustration to produce concentric circles in the top view. Points such as P_1 and P_4 are then obtained on the line of intersection by projecting from the front view to the corresponding circle in the top view. The resulting curve in the top view is a portion of an ellipse, one of the *conic sections* that will be discussed in more detail in Chapter 21.

16.4　Intersection of Plane and Prism

Figure 16.4(a) shows the addition of an auxiliary view that includes an edge view of the given plane MON. In this view the piercing points 1, 2, and 3 of the edges of the prism with the plane MON are obtained. Figure 16.4(b) shows the projection of these points to the side and front views. The correct indication of the visibility of the lines connecting these points completes the solution. This problem also might be solved using only the given views, by the two-view, piercing-point method of Chapter 6. The auxiliary-view method shown, however, affords a direct check of accuracy by means of transfer distances such as D_1 in Figure 16.4(b).

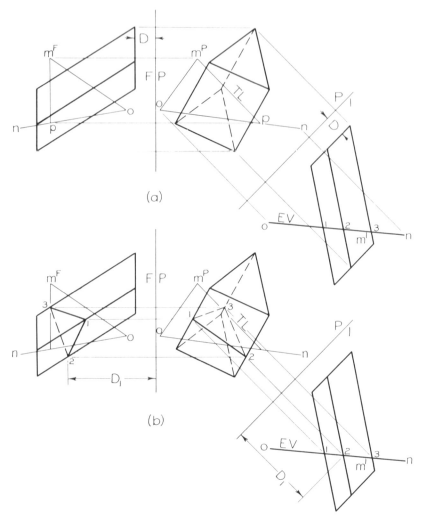

FIGURE 16.4   Intersection of plane and prism

16.5   Intersection of Plane and Cylinder

A convenient solution of this problem is obtained by the auxiliary-view method, as in the preceding example. An auxiliary view projected from the top view, Figure 16.5, is convenient because the bases of the cylinder in this case are horizontal and appear in edge view in any auxiliary elevation. Since there are no "edges" of the cylinder as there were in the prism, it is necessary to add corresponding elements in each view of the cylinder. Piercing points of these elements with plane MON appear in the auxiliary view, which shows the plane in edge view. These

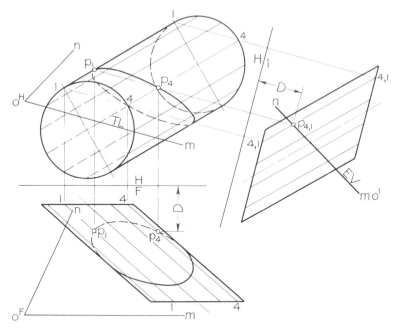

FIGURE 16.5   Intersection of plane and cylinder—auxiliary-view method

points, of which P_1 and P_4 are examples, are then projected back to the given views. The transfer distance D may be used as a check on the accuracy of the location of p_1 and p_4 in the front view. Drawing the resulting ellipses in correct visibility in the given top and front views completes the problem.

This problem also may be solved, Figure 16.6, using only the given views, by the two-view, piercing-point method of Chapter 6.

16.6   Intersection of Plane and Torus

Figure 16.7 shows the use of horizontal cutting planes to obtain the intersection of the plane ABC and the half-torus. These planes cut straight lines in the plane ABC and form semicircles on the surface of the torus. The construction for plane 3 is labeled in detail to illustrate the method. Plane 3 cuts line 3 in plane ABC and circles 3 on the torus. The intersections of line 3 and the corresponding circles in the top view produce points p_3 on the required line of intersection. The front views of these points are obtained by projection back to cutting plane 3. After an adequate number of points is secured by repetition of this procedure, the line of intersection is completed as shown.

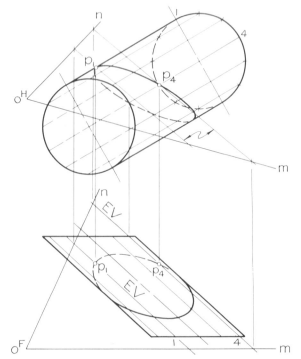

FIGURE 16.6   Intersection of plane and cylinder—two-view method

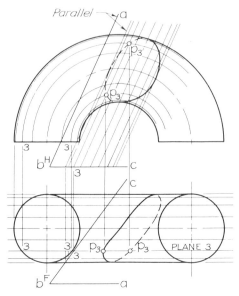

FIGURE 16.7   Intersection of plane and torus

16.7   Representation of Fair Surfaces

Double-curved surfaces such as a ship hull, an aircraft fuselage, or the various portions of an automobile body, which do not fall under standard geometrical classifications such as sphere, torus, and ellipsoid, are called *fair* surfaces. They are represented on a drawing by curves of intersection of various planes with the surfaces. For instance, the surface of the hull of a boat is described as illustrated in Figure 16.8. The *frame*

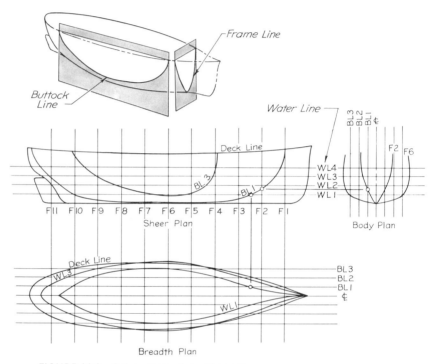

FIGURE 16.8   Fair surface—lines of a boat

lines are the intersections with the hull of vertical planes perpendicular to the longitudinal axis of the boat. The *buttock lines* are the intersections of vertical planes parallel to the boat's axis. The *water lines* are the intersections of horizontal planes with the hull. These various lines serve as elements of the curved surfaces in intersection problems such as the following (§16.8).

16.8   Intersection of Windshield and Aircraft Fuselage

In Figure 16.9 it is required to find the line of intersection of the plane of the windshield with the fuselage, given the buttock lines and the straight sides of the windshield. Since the buttock lines appear as straight lines in the plan view, vertical cutting planes containing the buttock lines appear in edge view coinciding with the buttock lines in the plan view. The parallel lines of intersection of these planes with the plane of the windshield are then located by projecting points 1, 2, 3, and 4 to the side view as shown, and their intersections in turn with buttock lines 10,

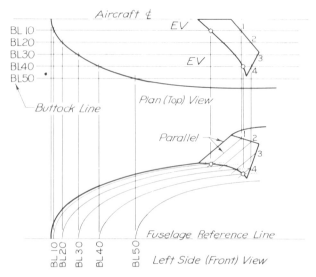

FIGURE 16.9   Intersection of windshield and fuselage of aircraft

20, and 30, and 40 are points on the required curve. The plan views of the points on the curve are then located by projection and the curves are drawn as shown. Note that the term *side view* as applied in the aeronautical industry refers to the view showing the left side of the aircraft, which conventionally is used as the principal view.

16.9   Pictorial Intersection of Plane and Cylinder

Some previously introduced concepts are applied to the problem of Figure 16.10, in which the intersection of unlimited plane ABC with the cylindrical form is desired. In the following solution only two of at least ten or twelve necessary points on the curve of intersection are provided for illustration clarity.

An element MN of the cylinder is drawn; and containing this element a vertical cutting plane MRST is introduced. Next, a horizontal line CK of plane ABC is added parallel to the given horizontal line AB. Since lines AB and MR both lie in the same upper horizontal plane of the pictorial outline, these lines intersect at point 2, a point on the line of

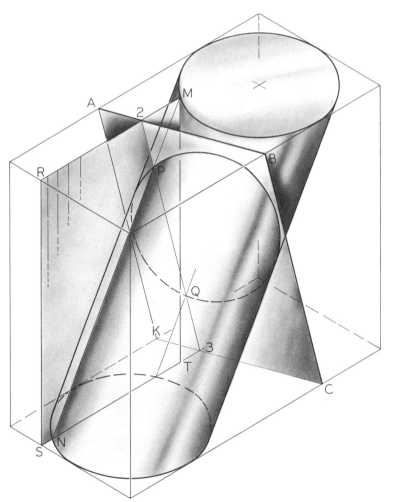

FIGURE 16.10 Pictorial intersection

intersection of planes ABC and MRST. A second point 3 on this line of intersection is at the intersection of CK and ST extended. Note that CK and ST both exist in the same horizontal base plane of the pictorial outline. Line of intersection 2-3 is then drawn.

Since element MN and line 2-3 each lie in cutting plane MRST, they intersect at point P, which establishes one of many needed points common to plane ABC and the surface of the given cylinder. Point Q is a second point established by the same cutting plane MRST. Additional cylinder elements and cutting planes are then added until an adequate number of points are secured to permit the representation of the curve of intersection as shown.

PROBLEMS

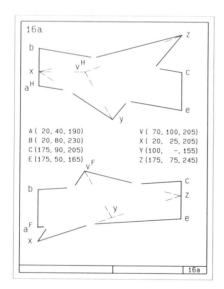

16a.

A (20, 40, 190)
B (20, 80, 230)
C (175, 90, 205)
E (175, 50, 165)

V (70, 100, 205)
X (20, 25, 205)
Y (100, -, 155)
Z (175, 75, 245)

16a

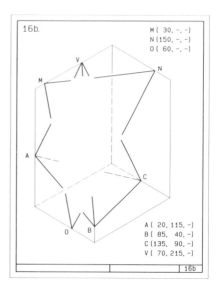

16b.

M (30, -, -)
N (150, -, -)
O (60, -, -)

A (20, 115, -)
B (85, 40, -)
C (135, 90, -)
V (70, 215, -)

16b

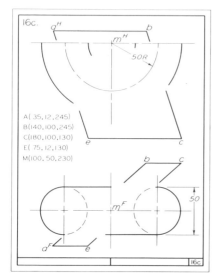

16c.

50R

A (35, 12, 245)
B (140, 100, 245)
C (180, 100, 130)
E (75, 12, 130)
M (100, 50, 230)

50

16c

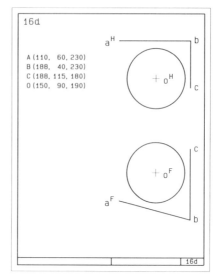

16d.

A (110, 60, 230)
B (188, 40, 230)
C (188, 115, 180)
O (150, 90, 190)

16d

16a. Establish the intersection of plane ABCE and the pyramid. Show appropriate visibility.

16b. In the isometric pictorial find the line of intersection of plane NMO and tetrahedron VABC. Complete the visibility.

16c. Find the intersection of the plane ABCE and the torus. Show complete visibility.

16d. Show the appropriate visibility of the section cut from the 64 mm diameter sphere by the unlimited transparent plane ABC.

237

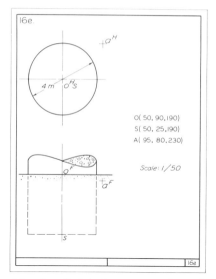

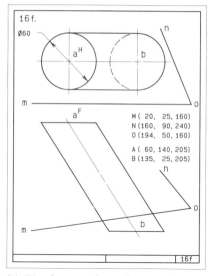

16e. A cylindrical support column is anchored into a rock ledge, the upper surface of which is a plane that passes through point A and has a strike of N 300° and a dip of 30°SW. Determine the volume in cubic meters of rock removed, and plot in the front view the intersection of the column and the ledge surface.

16f. Using only the given views, obtain the intersection of the unlimited transparent plane NOM and the cylinder.

16g. Indicate whether the following statements are true or false. If assigned, prepare written explanations or sketches to justify the answers.

(1) The intersection of an unlimited plane and a solid consists of a continuous closed path.

(2) A plane can cut the lateral surface of a cylinder in two straight lines.

(3) A plane intersecting a sphere always produces a circle.

(4) A plane intersecting a triangular prism produces lines of intersection that always take the form of a triangle.

(5) On a curved line of intersection it is desirable to obtain points more closely spaced, where the radius of curvature is greatest.

(6) A plane can cut a conical solid such that the lines of intersection form a triangle.

SELF-TESTING PROBLEMS

16A. Show the correct visibility of the intersecting plane and pyramid.

16B. Determine the intersection of the unlimited transparent plane ABC and the right cone.

16A.

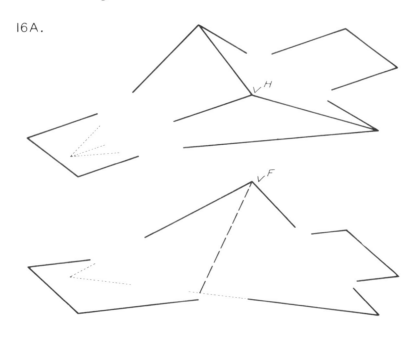

16B.

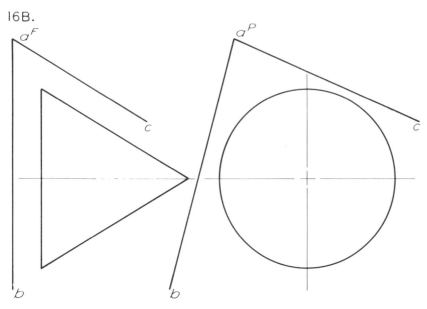

16C. Establish the line of intersection of the unlimited plane MNK and the pyramid VABC.

16D. Locate the line of intersection of the unlimited plane MNK and the oblique cone.

16C.

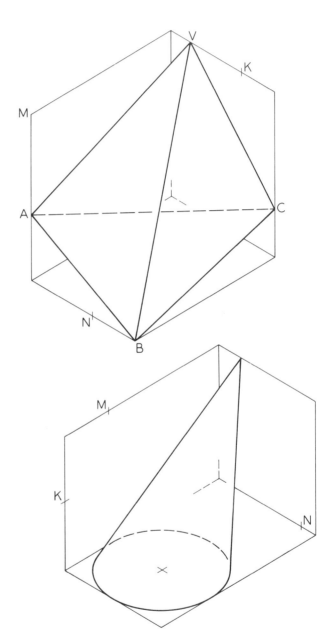

16D.

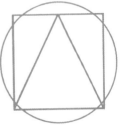

17 Developments

The development of a surface is the construction of a pattern that represents the unfolded or unrolled surface of the form as in Figure 17.1. The resulting plane figure gives the true size of each area of the form, so connected that when it is folded, or *fabricated*, the desired form is obtained. Containers and other products made of flat stock or sheet materials are first laid out on the flat sheets before the folding or

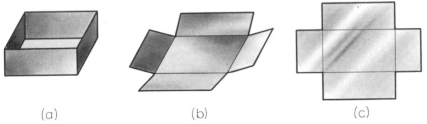

(a)   (b)   (c)

FIGURE 17.1   A simple development

bending operations bring them into the desired form. In case of quantity production, this process is used for the original layout necessary for setting up the blanking, stamping, or pressing operations. The fabrication of a commercial product from a sheet-metal layout is shown in Figure 17.2.

In the aircraft industry and in light-gage metal work such as is used in heating and ventilating, it is common practice to make the layout "outside up" so that when the form is completed the fold lines are on the outside. In heavy-gage and plate shops, it is the practice to make the layout so that the fold lines will be on the inside when the form is completed. However, the dividing line between "inside up" and "outside up" developments is not critically dependent upon the material. It is determined largely by the nature of the shop equipment used to fabricate the product and the material involved, such as sheet metal finished on one side only. In cases of doubt, therefore, a development

241

should be clearly labeled "inside up" or "outside up" as the case may be. For sake of uniformity, in this text all developments are "inside up."

The edges that must be joined are usually connected by seaming, riveting, welding, or soldering, and the allowance made for joining depends on the type of joint used. Usually the *length* of joint is kept to a minimum for reasons of economy and ease of handling.

FIGURE 17.2   Power leaf brake bending of square-to-round sheet-metal transition piece (courtesy Dreis & Krump Mfg. Co., Chicago)

On sheet metal thicker than 0.65 mm a *bend allowance* must be taken into account according to the practice of the particular trade because of the fact that the thicker sheets cannot be readily formed into absolutely sharp corners. However, for purposes of simplicity, this aspect is omitted from the developments in this text.

In general, the procedure in development is to select convenient lines on a given surface and then to find the true, or approximate, surface relationship of these lines. Reproduction of the relationships on a plane surface produces the development. For a pyramid or a cone the convenient lines are usually those passing through the vertex. For a prism or a cylinder the lines parallel to the axis are convenient.

Three general groups of developments, classified according to the type of surface involved or the method employed to construct the development, are as follows: (1) radial-line developments, the method used for pyramids and cones; (2) parallel-line developments, the method used for prisms and cylinders; (3) triangulation, the method used to break surfaces into a series of triangular areas, resulting in an exact development for a plane surface and an approximate development for other surfaces.

Warped surfaces and double-curved surfaces are theoretically not developable, but they may be approximated by division into units that are developable by one of these three methods.

17.1   Radial-Line Developments

The lateral surfaces of a pyramid are triangular areas, Appendix C.1, and the development consists of these triangles so arranged as to give the desired form when folded. The surface of a cone may be divided into narrow segments approaching the triangular form, Appendix C.2, and these "triangles" in proper sequence approximate the desired form when rolled to form the cone.

The true size of any triangle can be found by determining the true lengths of the three sides and assembling these true lengths to form the triangle (see §24.4).

17.2   Radial-Line Development—Pyramid

To develop the surface of the pyramid given in Figure 17.3(a), it is necessary to find the true lengths of each of the edges bounding the four triangular surfaces. In this instance edges OE and OA are revolved into a frontal plane, and their true lengths appear in the front view as $o^F e_r$ and $o^F a_r$, respectively. Since this pyramid is symmetrical about a frontal plane through the vertex O, the true lengths of edges OC and OB are identical to edges OE and OA, respectively. The base plane, being horizontal, shows in true size in the given top view, and consequently its four sides AE, EC, CB, and BA are shown true length in the top view. With the true lengths of all the edges known, the development now becomes a matter of assembly, Figure 17.3(b).

Notice that the development starts with edge OA, which is one of the shorter edges. (Edge OB could have been selected for a starting edge.) Point O is arbitrarily set and an arc with radius X equal to the true length of edge OA and edge OB is drawn. The true length of edge AB is then added by radius R. This now completes the surface OAB with

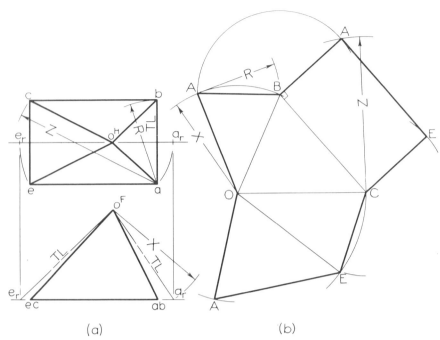

(a)   (b)

FiGURE 17.3   Radial line development—pyramid

the inside up. Surfaces OBC, OCE, and OEA are added in turn in a similar manner. The base plane ABCE is added to the development at a longer base edge such as edge BC. In this case it is easy to construct the rectangular base by utilizing 90° angles. As a check, or where the base angles are other than 90°, a diagonal such as radius Z may be used to construct the base as two triangles in a fashion similar to Figure 17.12 (or see Appendix 3.4).

17.3   Radial-Line Development—Truncated Pyramid

The development of a truncated pyramid with vertex accessible as given in Figure 17.4(a) is best constructed by first developing the whole pyramid and then eliminating the portion containing the vertex.

The true lengths of the lateral edges of the pyramid are found by revolution and are shown in the front view as $o^F c_r$ and $o^F e_r$. Since the pyramid is symmetrical about a frontal plane through the vertex O, the true lengths of edges OB and OA are identical to those of edges OC and OE, respectively. Since edges CE and BA of the base are frontal lines, their true length is obtained from their front view. The base edges CB and EA are horizontal lines and consequently are true length in the top view. The true lengths of all these edges now enable the designer to construct the layout of the pyramid as shown in Figure 17.4(b).

The use of the true length of edges OB and OC (radius X) together

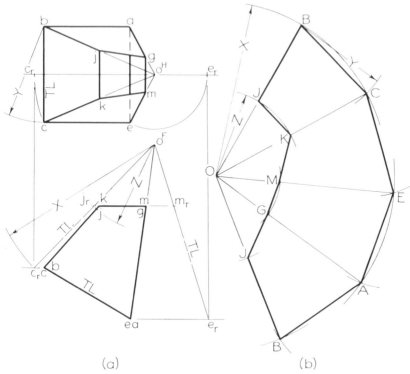

(a)   (b)

FIGURE 17.4   Radial-line development—truncated pyramid

with the true length of edge BC (radius Y) makes possible the construction of the panel OBC in true size and with the inside up in the layout. In similar fashion panels OCE, OEA, and OAB are added in turn. Next, points j_r and k_r are located as shown on the true-length view of edges OB and OC. These points are then transferred to the lateral edges in the layout as indicated by the radius Z. Similarly, points M and G are established, and the layout for the truncated pyramid is completed as shown.

17.4   Radial-Line Development—Right Circular Cone

To develop the surface of a truncated right circular cone as given in Figure 17.5(a), the conical surface may be broken into a series of small segments approximating triangles. These surfaces are then laid out in true size and assembled in proper sequence to approximate the surface of the original cone, Figure 17.5(b). It should be observed that arc O-1 on the cone is not exactly equal in length to arc O-1 on the development. This discrepancy will accumulate with the use of each succeeding segment, so that the resulting developed cone will be theoretically somewhat smaller than the original. If greater accuracy is necessary, the dimensions of the development may be calculated as follows: The right

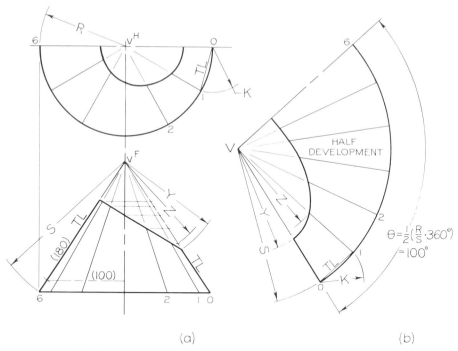

(a)

(b)

FIGURE 17.5   Radial-line development—right circular cone

circular cone development is a sector of a circle with a radius of *slant height* S, since all the elements are of the same length. The distance around the sector $2\pi S\theta/360°$, must be the same as the circumference of the base of the cone, $2\pi R$. When these two terms are equated and simplified,

$$\theta = \frac{R}{S}(360°).$$

Since in this case only a half-cone is shown, the angle used is one-half that required for a full development. The calculation may be performed graphically by setting off a "new" slant-height equal to 180 divisions (one-half of 360) at any convenient scale and measuring the corresponding radial distance at the same scale. This is shown by the dimensions in Figure 17.5(a).

To complete the layout of the truncated cone, the intersection points of the elements and the upper base are projected horizontally to the true-length extreme elements. The true distance of each intersection point from the vertex is now available and is transferred to the appropriate element in the layout. If the angle method is used to develop the cone, the intermediate elements are located by dividing the sector into the same number of equal parts as the base of the cone. A smooth curve is next faired through the points, and the resulting layout is a *half-development* of the cone divided along line V6, the line of symmetry represented by a center line. Half-developments are used frequently in

practice to save time. Obviously, a half-development may be used only
when the development is symmetrical, and moreover, the division *must*
be made along an axis of symmetry. In such cases, the caption "half-
development" should be added as shown in Figure 17.5(b).

17.5   Radial-Line Development—Oblique Cone

To develop the surface of the truncated cone given in Figure 17.6, the
cone is first divided along a line of symmetry and equally spaced elements are established. These elements are then revolved into a frontal
plane to establish their true lengths. The distance between them at the
base, radius Z in the top view, is used for the development as shown,
much as if the cone were a multisided pyramid. Since chordal distances
are employed rather than arc lengths, the development is somewhat inaccurate. The error is, however, usually minor and can be compensated
for in practice in the seam allowance together with slight deformation
of the material to secure a satisfactory fit.

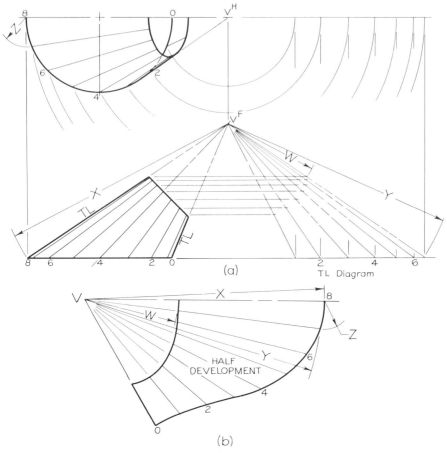

FIGURE 17.6   Radial-line development—oblique cone

The points of intersection of the elements and the upper-base are projected to positions on their respective true-length elements in the TL diagram. These points are in turn located in the layout, the resulting fair curve through the points completing the half-development.

17.6   Parallel-Line Developments

The forms involving parallel lateral edges or elements are the prism and cylinder, Appendices C.1 and C.2. In contrast to the radiating lateral edges or elements in the developments or the pyramid or cone, the lateral edges or elements or the prism or cylinder are parallel in the developments of these surfaces.

17.7   Parallel-Line Development—Truncated Right Prism

To develop the surface of a prism, three conditions must be known:

1. The true lengths of the lateral edges.
2. The relative positions of the parallel edges with respect to a *right section* (a plane figure formed by the intersection of a cutting plane perpendicular to the axis of a form).
3. The perimeter of the right section of the prism or perpendicular distances between the lateral edges.

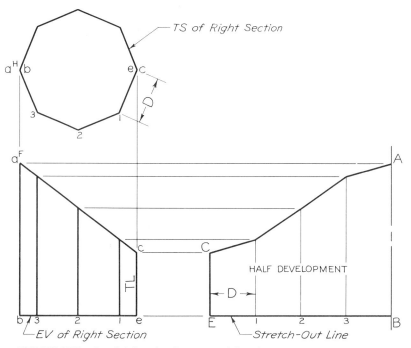

FIGURE 17.7   Parallel-line development—right prism

The two given views of the prism of Figure 17.7 conveniently give the true lengths of the lateral edges and the perimeter or distances between the edges. Since a development must show lateral surfaces in true size, angles as well as distances on the surface are retained in the development. In the given example the lower base plane is perpendicular to the lateral edges, and hence its perimeter unrolls into a straight line perpendicular to the parallel lateral edges in the development. This line is called the *stretch-out line*. The half-development shown is thus constructed by transfer of distances D between the lateral edges and by the transfer of the true lengths of the lateral edges. Note that the development is started on a line of symmetry and that for economy the seam is established at the shortest lateral edge CE.

17.8 Parallel-Line Development—Oblique Prism

In engineering practice the information necessary for the development of a surface is not always so conveniently given as in the preceding case.

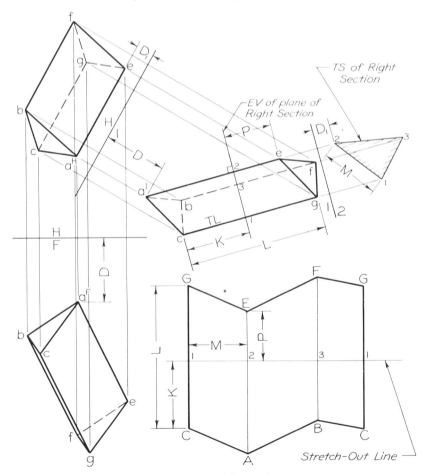

FIGURE 17.8   Parallel-line development—oblique prism

In order to find the true length of the lateral edges of the prism in Figure 17.8, the auxiliary elevation 1 is drawn. In this view the edge view of a right section such as 1-2-3 is added at any convenient location. Section 1-2-3, being perpendicular to the lateral edges, must unroll into a straight line (the stretch-out line). Auxiliary view 2 is now constructed to secure the true size of section 1-2-3, which gives the perimeter of the prism or the length of the stretch-out line.

Since in this example all lateral edges of the prism are the same length, the development may be started at any desired edge such as the edge CG. The stretch-out line is drawn at any convenient location, and edge CG is drawn perpendicular to it at one end. The length of edge CG is established in the development by transfer dimensions K and L.

The true distance between edges CG and AE, distance 1-2, is transferred from view 2 to the stretch-out line as indicated by dimension M, edge AE being selected as the second lateral edge in order to produce an inside-up development. The remaining dimensions of the right section are transferred in similar manner and in the proper order. The lengths of the lateral edges are then transferred to complete the development.

17.9   Parallel-Line Development—Right Circular Cylinder

If the curved surface of a cylinder is divided into segments by the addition of an appropriate number of elements, usually equally spaced for convenience, the surface may be developed in a manner similar to that used for a prism.

In Figure 17.9 the length of the development may be approximated by laying out the chordal distances 0-1, 1-2, 2-3, and so on, along the

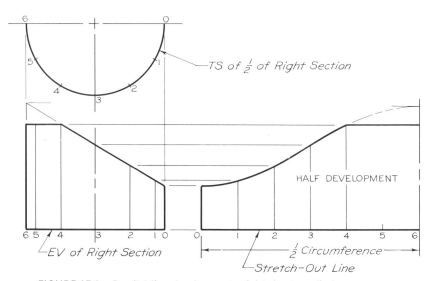

FIGURE 17.9   Parallel-line development—right circular cylinder

stretch-out line. For a more accurate development this length is made equal to the calculated semicircumference of the base circle as indicated. This distance is then subdivided into the same number of parts as those selected for the given views, and the elements are drawn perpendicular to the stretch-out line at these division points. The true length of each element is transferred from the front view. A smooth curve is faired through the endpoints of elements 0, 1, 2, 3, 4 as shown. The remaining outlines of the developed cylindrical surface are added to complete the half-development.

17.10   Parallel-Line Development—Oblique Cylinder

To construct the development of an oblique cylinder such as that given in Figure 17.10, an auxiliary view showing the axis of the cylinder in true length is drawn, and subsequently a secondary auxiliary view showing the true size of the right section of the cylinder is constructed. Note that the auxiliary view for the true length of the axis of the cylinder is projected from the *top* view, thereby retaining the convenience of the

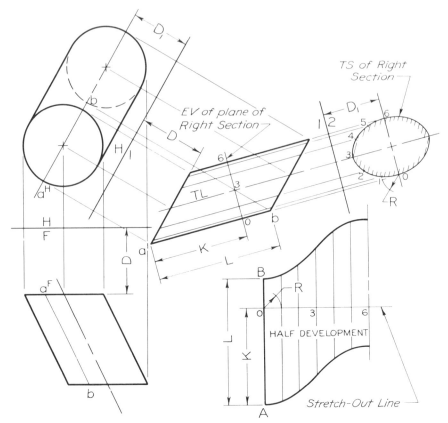

FIGURE 17.10   Parallel-line development—oblique cylinder

base planes in edge view rather than as plotted ellipses. In this view the circumference is divided into segments by points at equal chordal distances, thus establishing the elements of the cylinder that are used in the development. Any desired method may be used to construct the ellipse (Appendix B), but for convenience in development, the elements should be equally spaced. The formula for the circumference of an ellipse, while it would theoretically give a more accurate length for the development, is rather unwieldy and is therefore rarely used in practice. The circumference may be approximated by the expression $2\pi\sqrt{(a^2 + b^2)/2}$, where a is one-half the major axis and b is one-half the minor axis of the ellipse.

The elements of the cylinder are projected to the first auxiliary view. The development of the cylinder is then constructed by spacing the elements along the stretch-out line as indicated by the radius R and by transferring the lengths of the elements to the layout as shown. A smooth curve is faired through the endpoints of the elements to complete the half-development.

17.11   Special Application of Cylinder Development

Figure 17.11 exemplifies a practical application in which it is necessary to construct a development before the given views can be completed. The views as first given show the cylindrical screw-conveyor tube, the conical hopper, and their line of intersection MKN. For details of construction of the intersection, see §18.7. The problem is to add a reinforcing collar 130 mm wide about the intersection.

The first step is to construct the pattern (development) for the opening in the cylinder. Since the elements of the cylinder appear in true length in the front view, the pattern may be most simply constructed by transferring these lengths vertically downward to the available space below. The true right section of the cylinder appears in the profile view, from which the distances between the elements, such as D and D_1, are transferred to the layout as shown. A smooth curve faired through the points on elements 1, 2, . . ., 6 completes the development of the opening in the cylinder.

Since the opening corresponds to the inner edge of the required 130 mm reinforcing collar (neglecting thickness), the collar is now added to the layout by drawing a series of arcs of 130 mm radius centered at successive points on the layout of the opening and by fairing the curve of the outer edge tangent to these arcs.

The points of intersection of the elements 1, 2, . . ., 6 with the outer curve are now transferred to these elements in the front view. Additional elements 7, . . ., 10 are introduced in the development to establish additional points, and these are located first in the profile view and then in the front view. The curve representing the outer edge is now drawn in the front view to complete the solution.

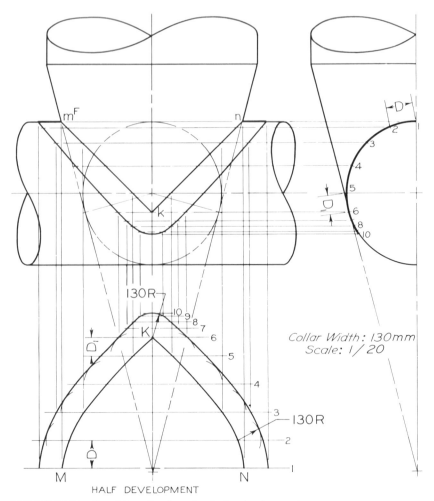

FIGURE 17.11   Special application of cylinder development

17.12 Triangulation

It is often necessary to construct patterns for surfaces that cannot be developed by the radial-line or parallel-line methods. Many such surfaces may be developed or approximately developed by *triangulation*.

EXAMPLE 1: OBJECT COMPOSED OF PLANE SURFACES

The hopper shown in Figure 17.12 (which is *not* pyramidal) is developed by breaking each of the four plane surfaces into two triangular surfaces. In this case the true lengths of the sides of the triangles are found by the construction of a true length diagram, a procedure frequently used in the sheet metal industry. The construction is drawn

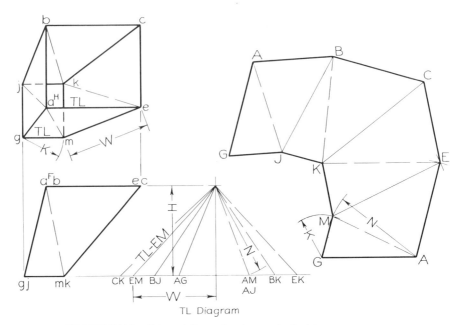

FIGURE 17.12   Triangulation—development of a hopper

to one side to avoid confusion. To exemplify the method, line em in the top view is assumed to be the edge view of a vertical plane. Edge EM of the hopper is then the hypotenuse of a right triangle lying in this vertical plane. The base of the triangle is equal in length to the top view em, distance W. The altitude of the triangle is the vertical distance from E to M, distance H, measured in the front view. These two lengths, W and H, are transferred to the TL Diagram as shown, and the true length of line EM is thus established (a comparison with §13.3 shows this procedure to be equivalent to finding the true length of a line by revolution). As a further aid to clarity, the solid and dashed true lengths are drawn on opposite sides of the vertical line in the diagram.

These lengths are then combined successively, starting with the shortest edge AG, to form the development as shown.

EXAMPLE 2: A TRANSITION PIECE

A piece that connects two differently shaped and/or sized conductors is known as a *transition piece*. This type of connector is used frequently in the heating and ventilating fields as well as in other industries.

As can be seen in Figure 17.13, the given transition piece could be employed to connect a vertical cylindrical conductor to an inclined rectangular duct. Further examination of the transition piece reveals that it is composed of four plane triangular surfaces, such as plane

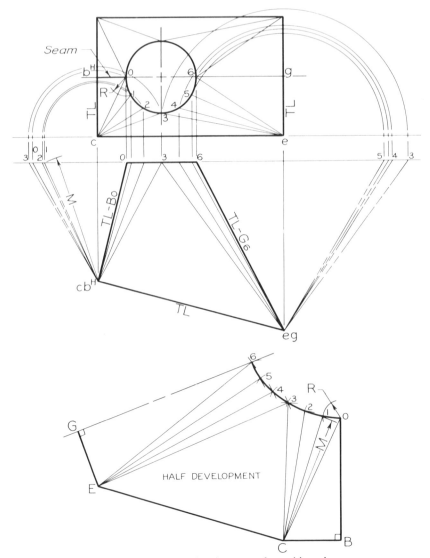

FIGURE 17.13 Triangulation—development of transition piece

C3E, and four portions of oblique cones, such as the conical surface O3C.

To produce the development, the true lengths of the sides of the plane surfaces are required plus the true lengths of the sides of the small approximate triangles created by breaking down the conical surfaces into several plane surfaces.

These true lengths are then assembled in proper relationship to reconstruct each triangular area, with the areas arranged in sequence to give the symmetrical development as shown. The seam is placed at the center of one of the flat surfaces for convenience in fabrication.

17.13   Warped and Double-Curved Surfaces— Approximate Development

Occasionally warped or double-curved surfaces (see Appendix C.3), which are theoretically nondevelopable, occur in technological practice. In such a case the surface is approximated with a series of small developable surfaces.

EXAMPLE 1: WARPED TRANSITION PIECE

In Figure 17.14 the surface of the transition piece is broken into a series of small triangles that approximate the original form. Note that the same number of equally spaced divisions is used on both the upper and lower openings. To avoid later confusion, the sides of the triangles are drawn alternately solid and dashed. The true lengths of the lines that make up each triangle are found in the manner discussed in Example 1 of §17.12 and are assembled to give the true size of each triangle in turn.

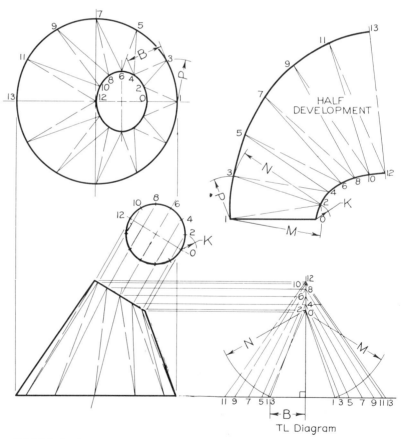

FIGURE 17.14   Approximate development—warped transition piece

EXAMPLE 2: SPHERE—GORE METHOD

Double-curved surfaces may be approximated by the use of small portions of cylinders or cones. The spherical surface in Figure 17.15(a) may be cut into a series of small *gores* by a number of cutting planes passed through the same diameter. Each gore is then approximated as a

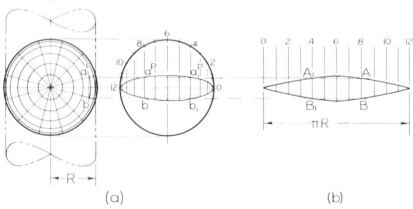

FIGURE 17.15 Approximate development of sphere—gore method

portion of a cylinder. The length of the gore around the half-cylinder is equal to the distance πR. This length is divided into segments corresponding to elements such as AB on the cylindrical surface. The length of element AB is transferred to its respective elements 4 and 8 in the development. The remaining points are similarly determined, with the resulting gore being one of twelve needed to approximate the development of the spherical surface. The method is also called the *polycylindric* method.

EXAMPLE 3: SPHERE—ZONE METHOD

The surface of a sphere may also be approximately developed by the division of the surface into a series of zones that may be approximated by cone frusta—the zone or *polyconic* method. Each zone, Figure 17.16, is a frustum of a right-circular cone. If the cones are circumscribed about the sphere, the resulting development will be slightly oversized. If the cones are inscribed in the sphere, the development will be slightly undersized. In practice either of these approximations is usually satisfactory.

If, however, it is necessary to secure a closer approximation of the surface of the sphere, the cones are so constructed that part of the sphere is outside and part is inside. This may be accomplished by dividing each arc, such as arc EC in Figure 17.16, into four parts and passing the extreme element of the cone through points 1 and 3, with

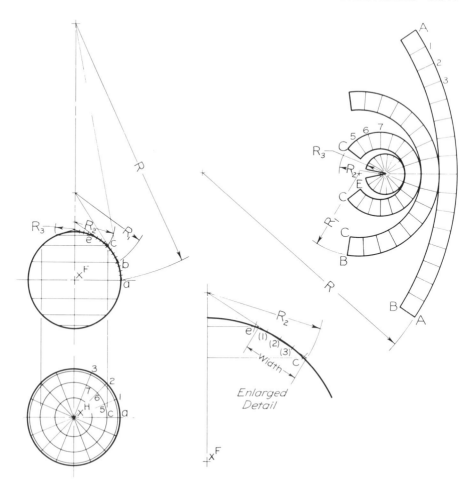

FIGURE 17.16   Approximate development of sphere—zone method

distance R_2 as the slant height of the complete cone of which the zone from point E to point C is considered a portion.

The remaining zones are treated similarly, resulting in the distances R, R_1, R_2, and R_3 for laying out the required cones. The size of the segment of the circle required for each frustum is determined by transferring the circumference of the base of each cone by chordal distances or by calculation as in §17.4. The widths of the conical surfaces needed for each frustum are obtained as indicated in the enlarged detail.

EXAMPLE 4: RIGHT HELICOID

A right helicoid is a helicoid (see Appendix C.3) in which the elements are perpendicular to the axis. In practice, as exemplified by the screw conveyor of Figure 17.17, the elements usually intersect the axis when extended and are all of the same length. For simplicity the views in

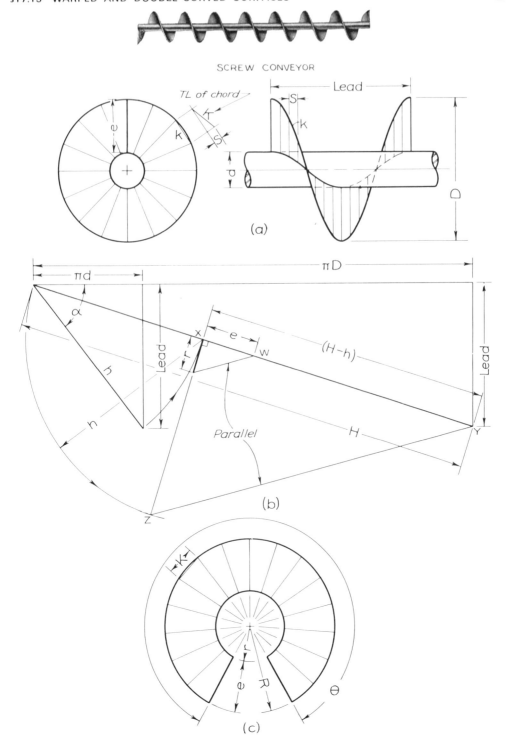

FIGURE 17.17 Approximate development—right helicoid

Figure 17.17(a) show only one complete turn or *flight* of the helicoid.

As is true of all warped surfaces, the helicoid is theoretically not developable. Its surface may, however, be closely approximated by division into small segments as indicated in Figure 17.17(a). Since in this example the elements e of the helicoid are all of the same length, the approximate development, Figure 17.17(c), is a sector of a circle with a circular central portion omitted to allow for the solid core of the conveyor.

The inner circular arc should have the same length (h) as the inner helix (on the cylindrical core) of the helicoid. This length is expressed by

$$h = \sqrt{(\pi d)^2 + (\text{lead})^2}$$

which may be calculated or obtained mathematically and graphically as shown at the left of Figure 17.17(b). In similar fashion the length of the outer helix, distance H, may be calculated or obtained graphically as shown.

In the approximate development the inner and outer circular arcs extend through the same total angle θ. This condition makes it possible to derive mathematically a formula for radius r of the development, the result being

$$r = \frac{h}{H - h}\,(e)$$

where distance e is the element length obtained from the circular view of Figure 17.17(a). Again the formula may be evaluated by calculation, or it may be solved graphically as follows, Figure 17.17(b).

Distance h is set off from one end of distance H to establish graphically the quantity (H − h). Distance h is then established at right angles to (H − h) and triangle XYZ is completed. Distance e is set off from point X along side XY to locate point W, and a line is drawn from point W parallel to hypotenuse YZ. The remaining leg of this smaller triangle is then equal to radius r, since by similar triangles,

$$\frac{r}{e} = \frac{h}{H - h} \qquad \text{or} \qquad r = \frac{h}{H - h}\,(e)$$

Using this radius and the radius R = r + e, the approximate development is now constructed. The distance around the circumference of the outer circle may be closely approximated by finding the true length of a chord of one section, distance k, as shown in Figure 17.17(a), and by setting off this chord successively around the outer circle of the development. The distance around the circle may also be established by calculating the angle θ from one of the following:

$$\theta = \frac{h}{2\pi r}\,(360°) \qquad \text{or} \qquad \theta = \frac{d}{2r\cos\alpha}\,(360°)$$

where α is the helix angle of the inner helix, Figure 17.17(b).

PROBLEMS

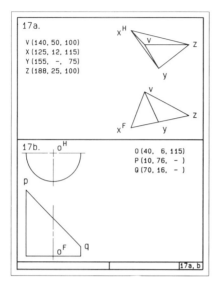

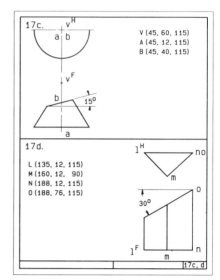

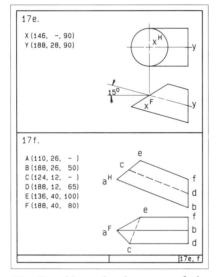

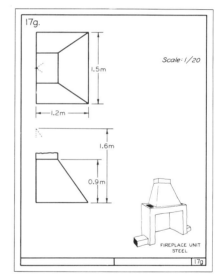

17a. Provide a development of the four surfaces of the pyramid.

17b. Provide a one-half development of the 60 mm diameter cylinder.

17c. Provide a one-half development of the cone, which has a 76 mm diameter base.

17d. Provide a development of the lateral surfaces of the triangular prism.

17e. Provide a development of the 40 mm diameter cylindrical duct.

17f. Provide a development of the lateral surfaces of the triangular prism.

17g. Develop the given pyramidal portion of the steel fireplace unit.

262

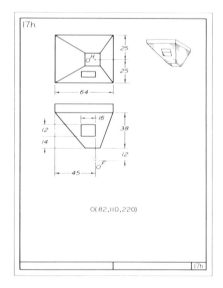

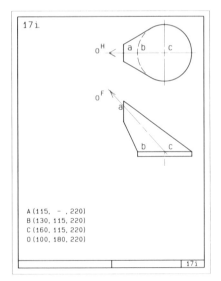

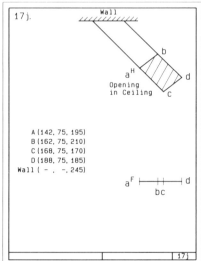

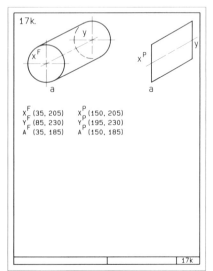

17h. Develop the pyramidal portion of the sheet-metal hopper.

17i. Lay out one-half the development of the oblique conical elbow section.

17j. The sheet-metal package chute makes an angle of 30° with horizon-tal. Show the true size of a right section and complete the front view. Lay out the development.

17k. Show the true cross section of the cylindrical duct. Lay out the development.

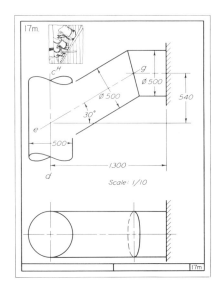

17m.

$\emptyset$ 500

$\emptyset$ 500

30°

540

500

1300

Scale: 1/10

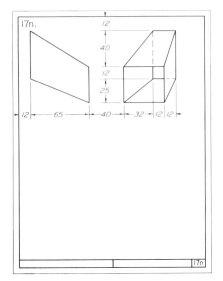

17n.

12

40

12

25

12   65   40   32 12 12

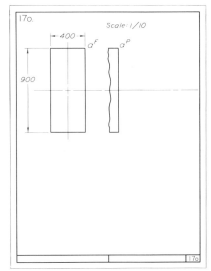

17o.

Scale: 1/10

400

900

a^F a^P

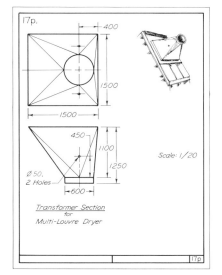

17p.

400

1500

1500

450

1100

1250

$\emptyset$ 50,
2 Holes

600

Scale: 1/20

Transformer Section
for
Multi-Louvre Dryer

17m. Show the intersection of the circular ducts about center lines CD and EG (see §18.7). Lay out one-half of the development of the duct about center line EG.

17n. Lay out the development of the transition piece.

17o. Design and complete the views

of a transition unit from the given rectangular cross section to a round of equal area. Then prepare a development of this unit.

17p. Develop the front half of the _transformer section_. Show the 50 mm hole on the development.

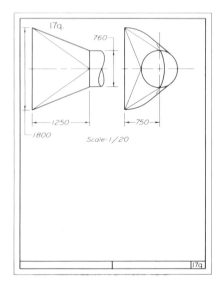

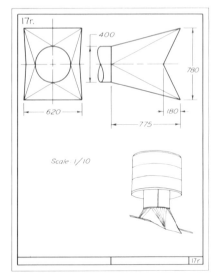

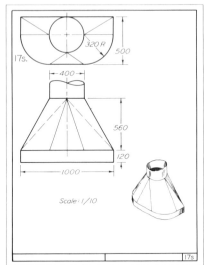

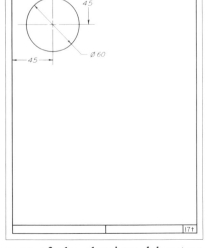

17q. Lay out one-half the development for the *dryer transition unit*.

17r. Lay out one-half the development of the *ventilator transition unit*.

17s. Lay out one-half of the development of the chemistry laboratory *hood*.

17t. Develop one-half the given sphere by the gore or zone method as assigned.

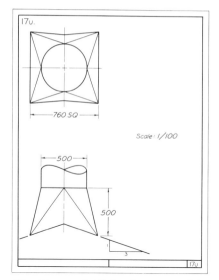

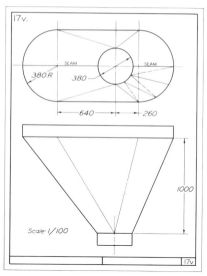

17u. Draw a half-development of the *ventilator transition unit*.

17v. On a separate sheet prepare a half-development of the transition piece which was designed for the processing of pharmaceutical products.

17w. Indicate whether the following statements are true or false. If assigned, prepare written explanations or sketches to justify the answers.

(1) The development of the lateral surface of a right cylinder is a rectangle.

(2) The development of a right cone is a triangle.

(3) Triangulation is the recommended method used in the development of a prism.

(4) The elements of an elliptical cone are equal in length.

(5) A right section implies that a right-side view shows the true perimeter of a prism or cylinder.

(6) A stretch-out line has an elastic quality such that it can be lengthened to provide extra material for a welded seam.

SELF-TESTING PROBLEMS

17A. Show the development of the prismatic sheet-metal duct.

17A.

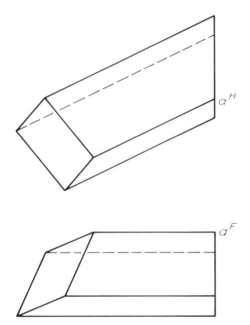

17B. Develop the given portion of the roof-mounted ventilator unit.

17B.

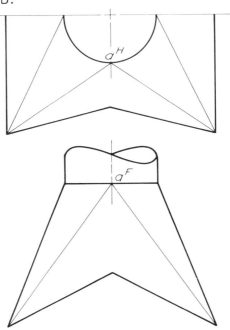

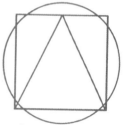

18 Intersections of Surfaces

A machine part or structure of any kind normally may be assumed to consist of a number of geometric shapes arranged to produce the desired form. Geometric shapes may sometimes be combined or interlocked in a pattern that is easily represented; but frequently the adjoining surfaces of these basic shapes meet in lines of intersection that require considerable effort to produce in multiview projection. When a form is constructed of sheet metal, other fabricated parts, or bonded material, an accurate representation of the intersecting surfaces becomes important, since the component parts must fit accurately and smoothly for optimum

FIGURE 18.1   Dustkop Cyclone Separator (courtesy Aget Mfg. Co., Adrian, Mich.)

functioning and appearance. Figure 18.1 shows a typical intersection pattern formed by the adjoining cone, cylinders, and prisms of the Dustkop Cyclone Separator.

The principles of intersections of planes and surfaces were discussed in Chapter 16. Here these principles will be applied to the more intricate problems resulting from the intersections of prisms, pyramids, cones, and cylinders. Selection of the particular procedure to be employed in a given problem is a matter of judgment based upon experience.

18.1   Intersection of Prisms—Edge Views Given

The intersection of two prisms involves the determination of the lines of intersection of the limited surfaces of the solids. These lines of intersection are obtained by finding the existing piercing points of the edges of one prism with the surfaces of the second prism, and then the existing

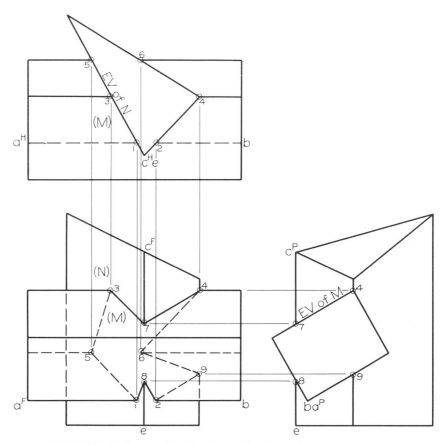

FIGURE 18.2   Intersection of prisms—edge views given

piercing points of the edges in the second prism with the surfaces of the first.

In Figure 18.2 the problem is somewhat simplified, since the top view shows edge views of the lateral surfaces of the triangular prism, and the side view shows edge views of the lateral surfaces of the rectangular prism. Thus the piercing points 1 and 2 of edge AB with two of the surfaces of the triangular prism are apparent in the top view and are projected to the front view as indicated. Points 3, 4, 5, and 6 are obtained similarly. The piercing points 7 and 8 of edge CE with two of the surfaces of the rectangular prism are observed first in the side view, which shows these surfaces in edge view, and are then projected to the front view. In completing the intersection of the prisms, care must be exercised to add only those lines which represent lines of intersection of the adjacent surfaces of the two prisms. For instance, a line of intersection between points 3 and 7 is drawn, since these two points are common to both planes M and N. Line 3-7 thus represents the line of intersection of these two surfaces. On the other hand, a line connecting points 1 and 3 would be incorrect, since it would lie *inside* the rectangular prism and thus could not be common to the *surfaces* of the prisms. The completed solution includes the correct visibility as shown.

Note that the edges of each prism are terminated at the points of intersection with the surfaces of the other prism. In other words, the completed problem is treated as if the intersecting forms consisted of a one-piece casting. In most practical applications such a form would either be solid or would be fabricated from sheet or plate, in which latter case it would need to be hollow for the passage of liquids, gases, or finely divided materials.

18.2 Intersection of Prism and Pyramid—Auxiliary-View Method

In contrast to the preceding problem the lateral surfaces of the prism of Figure 18.3 are not shown edgewise in either of the principal views. In this illustration an auxiliary view is drawn which does include the edge views of these planes. Next, piercing points 1, 2, 3, and 4 of two edges of the pyramid with two surfaces of the prism are located in this auxiliary view and are then projected to the top and front views. Piercing points 5, 6, 7, and 8 are obtained by the two-view piercing-point method, since the addition of the auxiliary views needed to show each surface of the pyramid in edge view would be cumbersome. An edgewise cutting plane, containing edge AB of the prism and intersecting the surfaces of the pyramid in lines YX and YZ, is introduced in the auxiliary view to locate piercing points 5 and 6. A similar construction is employed to secure points 7 and 8. The solution is completed by connecting the various piercing points as shown.

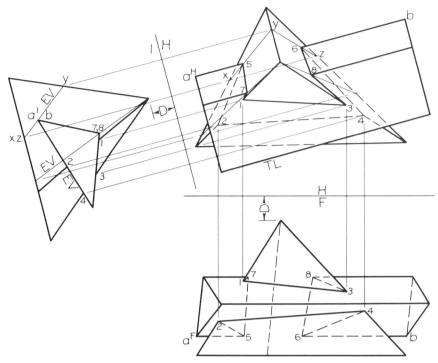

FIGURE 18.3   Intersection of prism and pyramid—auxiliary-view method

18.3   Intersection of Two Prisms—Two-View Method

By the use of the two-view, piercing-point method, the construction for the intersections of prisms and pyramids may be confined to any two views.

For simplicity the construction needed for the location of only three of the several piercing points is shown in Figure 18.4. An edge-view cutting plane passed through edge AB in the front view intersects one surface of the triangular prism in line 1-2, which, when projected to the top view, locates piercing point M of edge AB in this surface. This point is then projected back to the front view. Piercing point N is obtained by the use of a cutting plane through edge CE, intersecting a surface of the quadrangular prism in line 5-6. A similar construction is shown to locate point S. Although some cutting planes may be tried that do not produce piercing points within the areas of the limited surfaces of the solids, this trial approach can be reduced to some extent by a careful preliminary study of the problem to locate edges that obviously do not pierce the other solid. After all the existing piercing points are secured, the figure of intersection forms one or more closed paths.

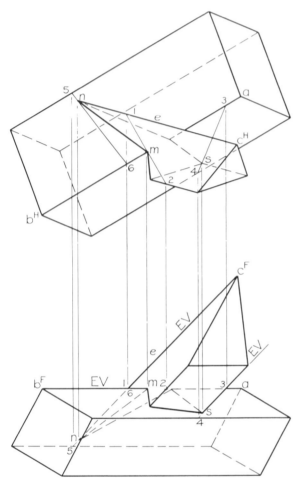

FIGURE 18.4 Intersection of two prisms—two-view method

18.4 Intersection of Prism and Cone

The intersection of the prism and cone of Figure 18.5 represents a problem occurring frequently in practice. For this solution profile cutting planes 1, 2, 3, and 4 are used. Since these cuts are parallel to the circular base of the right cone, they intersect the cone in four easily drawn circles. The intersection of circle 2 in the side view with the front edge-view surface of the prism results in the two piercing points X and Y, which are then projected to the front view. Also found by this procedure are the remaining points on the curved figure of intersection. This curve is a portion of a hyperbola (§21.3). Note that only those cutting planes are used that produce circles on the surface of the cone which will intersect the lateral surfaces of the prism. With this in mind, the observing student would probably first add the concentric circles in the side view and then project to the front view to secure the corresponding cutting plane lines.

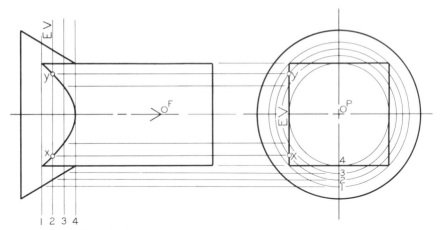

FIGURE 18.5   Intersection of prism and cone

18.5   Intersection of Right Circular Cylinders

In sheet-metal duct work, in piping, and in steel-plate tank work, the representation of the figure of intersection of cylinders is a frequently occurring problem. In Figure 18.6 frontal cutting planes are used, since these cuts intersect both cylinders in elements, as illustrated in the pictorial. In the top view the intersection of cutting plane 2 with the circular view of the vertical cylinder locates elements 2 of this cylinder, which are then projected to the front view. Since the front view of the corresponding element 2 of the horizontal cylinder cannot be projected directly from the top to the front view, use is made of the partial side view. The element is first projected to the side view and then transferred to the front view by means of transfer distance D_1. The intersections of the correspondingly numbered elements in the front view locate points X and Y on the required figure of intersection. Other cutting planes are added, spaced appropriately, to locate points on the intersection where they are most needed to produce an accurate curve.

If intersecting cylinders, such as those illustrated in Figure 18.7(a), have a greater relative difference in their diameters, the figure of intersection more nearly approaches the form of a circular arc. Thus for many practical problems of this type, where representation is desired and development is *not* necessary, a circular arc may be a satisfactory substitute for the actual curve. The radius of the larger cylinder is used, since the resulting arc when properly centered on the axis of the smaller cylinder passes through the critical points a[F] and b of the true curve.

When the relative diameters differ even more, as illustrated in Figure 18.7(b), the conventional practice is to ignore entirely the slight curvature of the intersection unless a development is needed.

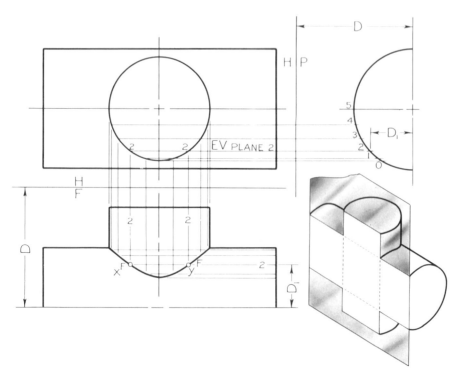

FIGURE 18.6 Intersection of right circular cylinders

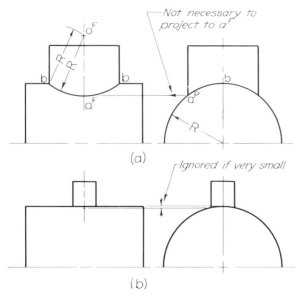

FIGURE 18.7 Conventional intersections—used where intersection is not prominent and accuracy is unimportant

18.6 Intersection—Right Cylinder and Cone

For the intersection problem of Figure 18.8 horizontal cutting planes are used, since they cut circles on the cone and elements on the cylinder. The pictorial shows the cone and cylinder intersected by a single horizontal plane. Cutting plane number 4, as an example, cuts the cone in circle 4, the top view of which is obtained by direct projection. This same cutting plane cuts the cylinder in element 4, which is located in the top view by the use of the transfer distance D_1 secured in the partial side view. The intersection of circle 4 and element 4 in the top view is x^H, a point on the required figure of intersection. Point x^F is obtained by projection to the front view. Additional cutting planes are added until a sufficient number of points are located to produce an accurate curve. The complete curve will be symmetrical in the top view. One half is shown here.

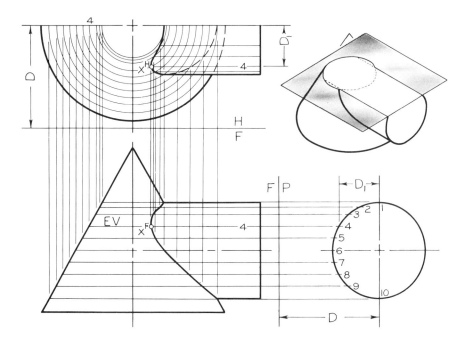

FIGURE 18.8   Intersection of right cylinder and cone

Alternative Method. Under similar circumstances the figure of intersection of a right cone and cylinder may also be plotted by assuming a series of cutting planes containing the vertex of the cone and which appear in edge view in the side view, Figure 18.9. In this case appropriately spaced elements of the cone are introduced in the side view, including the important elements 1 and 11 tangent to the cylinder. These elements are then located in the top view, as suggested by transfer distance D_1 for element 3, and are projected to the front view. Edge-view cutting planes containing the cone elements are then introduced in the side view. These planes also cut elements from the cylinder which, when projected to the front and top views, locate points of the intersection such as A_1 and A_2 on element 3.

Still another method for plotting the figure of intersection of a right circular cone and cylinder is discussed in §18.7.

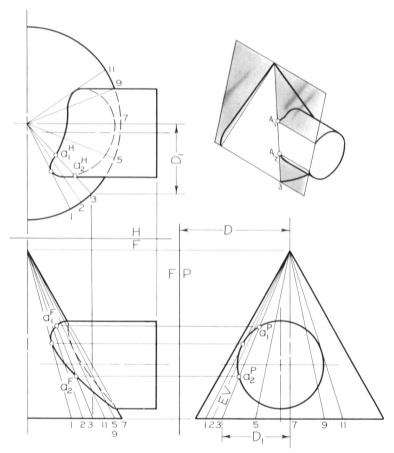

FIGURE 18.9   Intersection of right cylinder and cone—alternative method

18.7   Intersection of Surfaces—Sphere Method

A unique one-view method may be used for finding the figure of intersection of two curved surfaces when the following conditions exist:

1. The surfaces have circular right sections.
2. The axes intersect.
3. The axes appear true length in the same view.

This solution is based on the fact that a sphere of sufficient diameter, centered on the axis of a surface of revolution such as the circular cylindrical surface of Figure 18.10(a), intersects the surface in circles which appear edgewise in a view that shows the true length of the axis of the cylinder. A sphere of the same diameter as the cylinder would be tangent to the surface and contact the surface in only one circle.

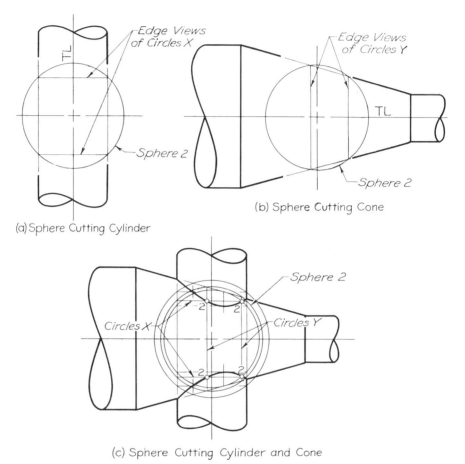

(a) Sphere Cutting Cylinder

(b) Sphere Cutting Cone

(c) Sphere Cutting Cylinder and Cone

FIGURE 18.10   Intersection of surfaces of revolution—sphere method

Figure 18.10(b) shows a similar intersection for a sphere and a conical surface.

When the cylindrical and conical surfaces intersect as in Figure 18.10(c), an imaginary sphere centered at the intersection of their axes cuts edgewise circles from both these surfaces. These four circles intersect to determine points common to both surfaces, since the circles lie on the same imaginary spherical surface. As an example, sphere 2 cuts edgewise circles X on the cylinder and circles Y on the cone. The crossing points 2 of X and Y are points on the required intersection. Other concentric cutting spheres are used as shown to complete the solution.

A special case of this spherical cutting plane method materializes when a sphere centered on the intersecting axes of circular cones or cylinders is tangent to both surfaces, Figure 18.11. For these conditions the resulting figures of intersection are ellipses appearing edgewise in any view showing both axes true length.

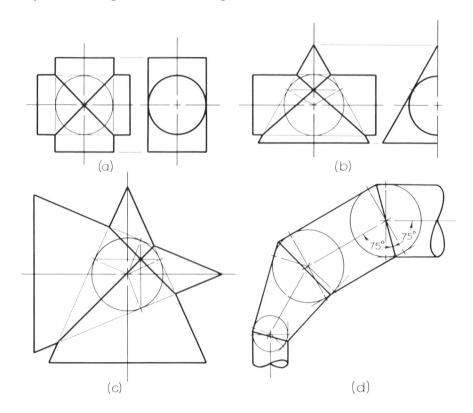

FIGURE 18.11 Intersection of surfaces of revolution—special cases

The sphere enveloped by the two cylinders of Figure 18.11(a) is tangent to the surfaces of both cylinders only when they have equal diameters. For this situation the resulting intersection *appears* as straight lines as shown. Figures 18.11(b) shows the resulting apparent straight-line figure of intersection for a cone and cylinder that tangentially envelop the same sphere. In Figure 18.11(c) is shown the figure of intersection for the two cones fulfilling the same conditions. For simplicity of design and fabrication, sheet-metal elbows in circular ducts are usually designed in this manner, Figure 18.11(d).

18.8 Planes Cutting Elements from Oblique Cones and Cylinders

In previous illustrations of intersecting cylinders and cones, the given forms have been in normal positions—horizontal or vertical in space.

In these cases planes that cut elements from the surfaces appear in edge view in one or more of the given views, and their representation is relatively simple. When the given forms are oblique, however, planes containing elements are also oblique, and their representation is correspondingly more involved. This section is concerned with the requirements that must be met by planes that cut elements from oblique cones and cylinders and with the representation of such planes. The subsequent examples and explanations apply these principles to actual problems.

Elements are cut from a cone by any planes that pass through the vertex of the cone and intersect or are tangent to the base of the cone. Figure 18.12(a) shows in pictorial form the cutting plane OXY which intersects the cone in the elements 1 and 2. Note that line XY of the cutting plane lies in the base plane of the cone. Figure 18.12(b) shows this same relationship in two views. Also shown is a plane OWZ, which being tangent to the cone (§15.2), contains the single element 3.

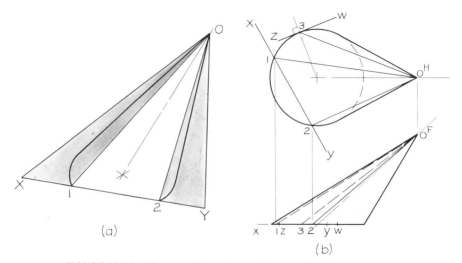

FIGURE 18.12   Planes cutting elements from an oblique cone

Elements are cut from a cylinder by any plane that passes parallel to the axis of the cylinder and intersects or is tangent to the cylinder. Figure 18.13 shows first in pictorial form and then in multiview projection the cutting plane XYZW, which intersects the cylinder in the elements 1 and 2.

The introduction of a plane that cuts elements from a cone or a cylinder is a convenient method for finding the points in which a line pierces a conical or cylindrical surface, Figure 18.14. If lines are drawn from the vertex of the cone, Figure 18.14(a), and intersecting given line AB at any two convenient points such as P and B, a cutting plane is formed. The intersection XY of this plane with the base of the cone, line XY, establishes elements 1 and 2 of the cone, which in turn intersect line AB at piercing points M and N.

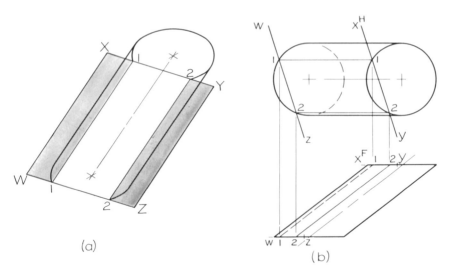

FIGURE 18.13   Plane cutting elements from an oblique cylinder

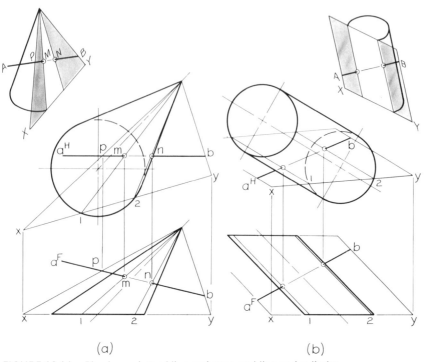

(a)   (b)

FIGURE 18.14   Piercing points of line and cone and line and cylinder

A similar procedure is employed in Figure 18.14(b) to find the points in which line AB pierces the cylindrical surface. In this case the cutting plane is established by introducing two lines parallel to the axis of the cylinder and intersecting line AB.

18.9   Intersection of Oblique Cone and Cylinder

To obtain the figure of intersection of the oblique cone and cylinder of Figure 18.15, oblique cutting planes are used that cut elements from both the cone and cylinder. The cutting planes must pass through the vertex of the cone to cut elements from the cone; the same cutting planes must also pass parallel to the axis of the cylinder in order to cut elements from the cylinder.

To secure the type of cutting plane that will accomplish these objectives, a line VS, Figure 18.15(a) and (b), is drawn through the vertex of the cone and parallel to the axis of the cylinder. Any plane containing this line and intersecting the cone and cylinder will cut elements from each. These elements, being in the same cutting plane, must intersect in points common to the cylindrical and conical surfaces. Thus the points are points on the figure of intersection.

The next step in the construction is therefore to establish the cutting planes and the corresponding elements. Point P on line VS is the intersection of this line with the plane of the bases. Line VP together with any horizontal line emanating from point P and intersecting the bases represent a plane that cuts elements from the cone and cylinder. The pictorial, Figure 18.15(b), shows one cutting plane and the four points secured by the intersecting elements thus obtained.

Planes VPN and VPJ of Figure 18.15(a) are called *limiting* planes. Each is tangent to one surface and cuts through (is *secant* to) the other surface. Any plane outside the narrow segment of space between these limiting planes is of no value to the construction since it cannot contact both surfaces.

A numbering system for the elements and the resulting points on the curve has been developed for any combination of cylinders and cones when the type of cutting plane shown is used to find the figure of intersection. In Figure 18.15(a) the limiting planes and an appropriate number of intermediate planes are established.

The system and its properties follow:

1. Point number 1 is assigned to any point where the horizontal line of any one of the cutting planes crosses either base outline. Other points on the same base are numbered consecutively, either clockwise or counterclockwise, except
2. When a secant limiting plane is reached, the order is reversed, and the numbers are continued around in the other direction.

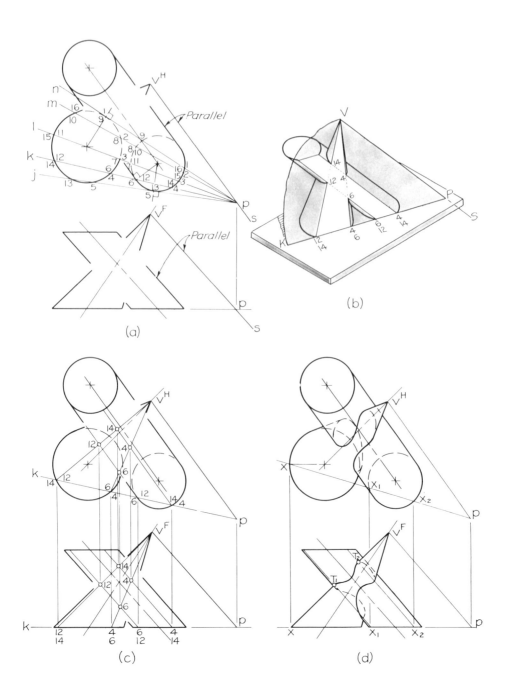

FIGURE 18.15 Intersection of oblique cone and cylinder

3. The numbering is continued until every point on the first base has two numbers, except the secant points, which have only one.

4. The second base is numbered, following the same rules; it is necessary that numbers 1 and 2 are assigned to lines of the same planes as on the first base. This procedure results in like numbers appearing on the bases for each cutting plane, this fact serving as a check on proper numbering.

5. The elements of each surface are then drawn, and the intersections of the like-numbered elements noted. Each point is identified by a single number corresponding to the intersecting elements. As an example, Figure 18.15(c) shows the single cutting plane given in the pictorial together with the resulting elements and points.

6. After all points are located, a preliminary curve is sketched consecutively from point 1 to point 2, then through 3, 4, and so on.

7. Intermediate cutting planes are added where needed to assure an accurate curve. In particular, the exact locations of points on extreme elements are found in each view. At each such point the curve is tangent to the extreme element. The locations of some of these tangent points may be obtained by the strategic location of the original cutting planes as shown in Figure 18.15(a), or by a subsequent addition of one or more intermediate cutting planes as illustrated at (d).

8. The visibility of the portion of the curve extending between any two successive tangent points is determined by a test of the visibility of any point on that segment of the curve. *A point on the curve is visible only if that point is established by two elements which are each visible in their respective surfaces.* The four points on the curve in Figure 18.15(c) are all visible points in both views since these points lie on visible elements of both forms in the top view and again in the front view. Points 6 and 12 in the top view, being located on the extreme elements of the cylinder in that view, represent the tangent points at which the curve in this problem changes visibility. The final curve for the figure of intersection is drawn as shown at (d).

The figure of intersection will always be a single continuous curve where one limiting plane is tangent to the first form and the other limiting plane is tangent to the second. Two separate curves or figures of intersection result when two limiting planes are tangent to the same form, as shown in Figure 18.16. This illustration is discussed in §18.10, following. If a limiting plane is tangent to both bases the result according to the numbering system is two curves with a point in common.

18.10 Intersection of Two Oblique Cones—Bases in Same Plane

To cut elements from each of two intersecting cones, the cutting planes must pass through the vertex of each cone. If in Figure 18.16 a line V_1V_2 is drawn, each of the cutting planes must contain this line. The given pictorial shows one of these planes cutting from the cones elements 5 and 13, the intersections of which locate the correspondingly numbered points on the required intersection. Point P, the intersection of line V_1V_2 with the plane of the bases of the cones, is used in Figure

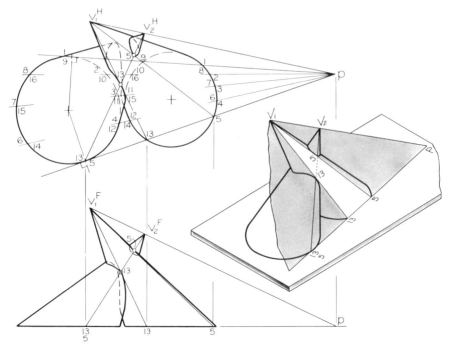

FIGURE 18.16 Intersection of oblique cones—bases in same plane

18.16 in establishing the cutting planes. The remaining construction and numbering is performed as explained in §18.9. In Figure 18.16 the two limiting planes are tangent to the same surface. The result is therefore two separate figures of intersection, one being established by points 1 through 8 and the other by points 9 through 16.

For the special case in which the vertices of the two cones are at the same elevation, the line through the vertices may be regarded as intersecting the base plane at infinity. The lines of intersection of the cutting planes with the base plane are then parallel to the line through the vertices.

18.11   Intersection of Two Cones—Bases in Nonparallel Planes

To cut elements from both the given cones, the cutting planes must pass through the vertices of the given cones. Thus in Figure 18.17, line V_1V_2 is drawn as in the preceding example. In this case, however, line

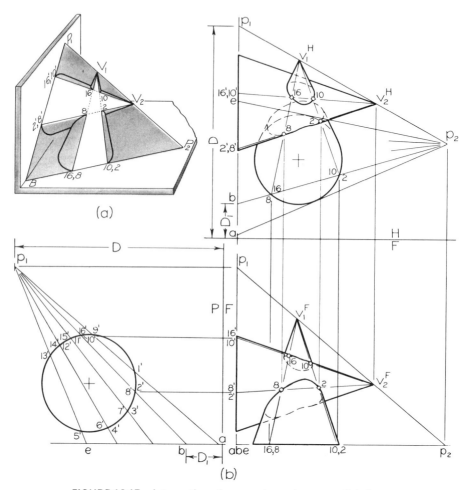

FIGURE 18.17   Intersection of cones—bases in nonparallel planes

V_1V_2 intersects the base planes in separate points P_1 and P_2. A partial left-side view showing the circular shape of the base of one cone is added to provide a means of transferring points on this base from one view to another.

In the multiview illustration, cutting plane P_1P_2B is shown in detail. This plane is first drawn in the top view locating elements 2, 10 and 8, 16 on cone V_1. Line P_1B of this same plane is added to the side view by use of transfer distances D and D_1. The intersection of this line with the circular base of cone V_2 locates the elements 2′, 8′ and 16′, 10′ for this cone. After these elements of both cones are located in the top and front views, the intersections of the correspondingly numbered elements are noted. Other cutting planes are added as shown. As is evident in the top view, limiting planes $P_1 P_2 A$ and $P_1 P_2 E$ are both tangent to cone V_1. Hence there are two separate figures of intersection as shown.

18.12 Intersections of Oblique Cylinders—Bases in Same Plane

In order to cut elements from both of two intersecting cylinders, the cutting planes must pass parallel to the axis of each cylinder. Since each

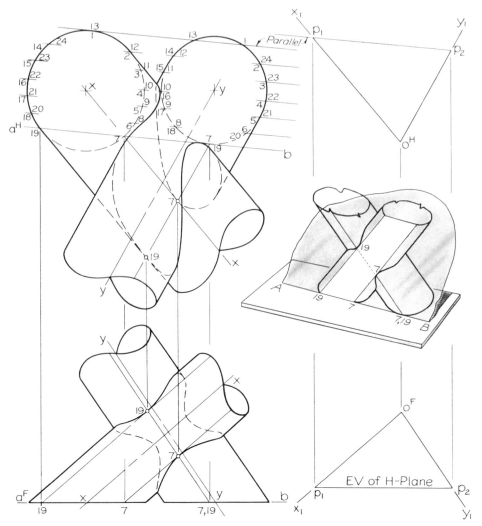

FIGURE 18.18   Intersection of oblique cylinders

cutting plane must be parallel to these two axes, it follows that the cutting planes will be parallel to each other. If in Figure 18.18 a plane OX_1Y_1 is passed through any point O and parallel to the axes of the two given cylinders (see §9.3), a plane is thus established to which the cutting planes must be parallel. The lines of intersection of the cutting planes with the horizontal base plane must therefore be parallel to a line of intersection such as P_1P_2 of a horizontal plane and plane OX_1Y_1. Thus the cutting planes are represented by drawing lines in the top view which intersect the bases and are parallel to line p_1p_2. Their intersections with the bases of the cylinders locate the resulting cut elements. Cutting plane AB is shown in detail in the pictorial and in both views of the multiview drawing, together with the resulting elements and intersection points 7 and 19. This plane, being a limiting plane, produces only two main points on the intersection. Other points are obtained similarly to produce the complete figure of intersection as shown.

18.13   Line Tangent to Curve of Intersection

A line tangent to a curved surface lies in a plane tangent to the surface. A line tangent to the figure of intersection of two curved surfaces is tangent to both surfaces. Therefore, a line tangent to a figure of inter-

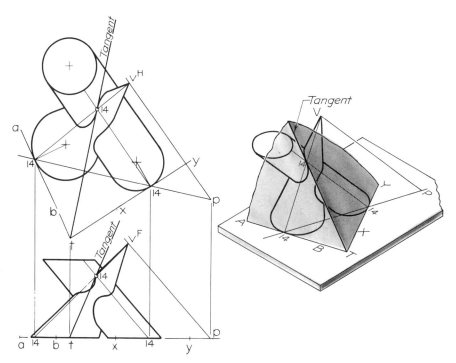

FIGURE 18.19   Line tangent to curve of intersection

section at a chosen point is the line of intersection of the two tangent planes containing the particular point.

Figure 18.19 is a reproduction of Figure 18.15(d). Let it be required to construct a line tangent to the curve of intersection at point 14. The plane tangent to the cone and containing point 14 is represented by element 14 of the cone and line AB tangent to the base of the cone. The plane tangent to the cylinder and containing point 14 is represented by element 14 of the cylinder and line XY tangent to the base of the cylinder (see §§15.2 and 15.3). Lines AB and XY are both in the base plane and hence intersect at point T. This point and point 14 of the curve are common to the two tangent planes, and thus the line joining them is tangent to both surfaces at point 14 of the figure of intersection. As a rough visual check, when such constructions are properly performed, the resulting line will appear tangent to the curve. If the line seems to cross rather than lie along the local portion of the curve, the construction is probably incorrect.

PROBLEMS

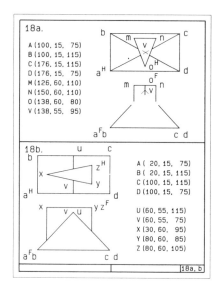

18a.

A (100, 15, 75)
B (100, 15, 115)
C (176, 15, 115)
D (176, 15, 75)
M (126, 60, 110)
N (150, 60, 110)
O (138, 60, 80)
V (138, 55, 95)

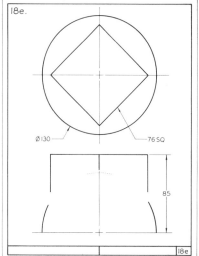

18b.

A (20, 15, 75)
B (20, 15, 115)
C (100, 15, 115)
D (100, 15, 75)

U (60, 55, 115)
V (60, 55, 75)
X (30, 60, 95)
Y (80, 60, 85)
Z (80, 60, 105)

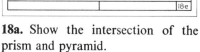

18a, b

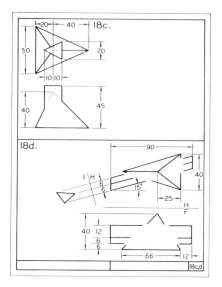

18c.

18d.

18c,d

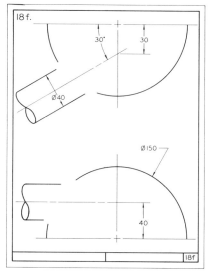

18e.

Ø130   76 SQ

85

18e

18f.

30°   30

Ø40

Ø150

40

18f

18a. Show the intersection of the prism and pyramid.

18b. Add a right-side view showing the intersection of the two prisms.

18c. Complete the front view of the intersecting forms and then add a right-side view.

18d. Complete the views of the intersecting pyramid and prism.

18e. Plot the intersection of the prism and sphere.

18f. Plot the intersection of the cylinder and sphere.

290

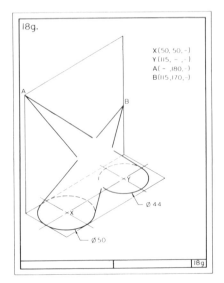

18g.

X(50, 50, -)
Y(115, -, -)
A(-, 180, -)
B(115, 170, -)

Ø 44

Ø 50

18g

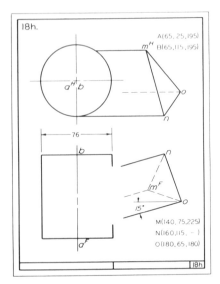

18h.

A(65, 25, 195)
B(65, 115, 195)

76

15°

M(140, 75, 225)
N(160, 115, -)
O(180, 65, 180)

18h

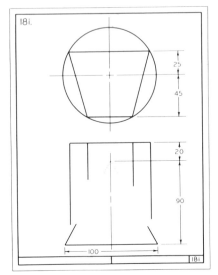

18i.

25

45

20

90

100

18i

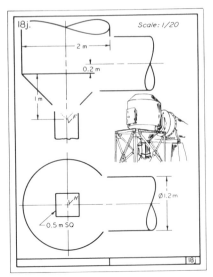

18j.

Scale: 1/20

2 m

0.2 m

1 m

Ø 1.2 m

0.5 m SQ

18j

18g. Complete the isometric pictorial including the intersection of the two cones.

18h. Show the intersection of the prism and the cylinder.

18i. Show the intersection of the prism and the cone.

18j. Find the intersection of the cone and the square outlet tube, and the intersection of the horizontal cylinder and the vertical cylinder and cone.

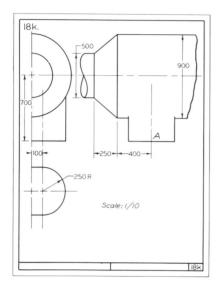

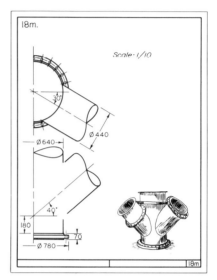

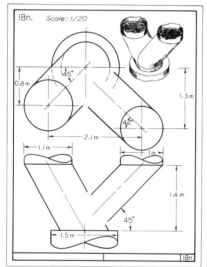

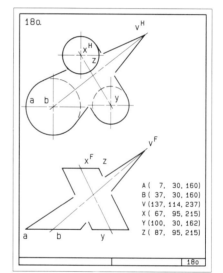

18k. Lay out the development for a 150 mm wide reinforcing collar for the intersection of outlet A with the 900 mm diameter main. Complete the views. Neglect thickness.

18m. Show the line of intersection between the vertical and lateral pipes. Small details for which dimensions

are not given may be approximated or omitted.

18n. Complete the front and top views of the oblique cylindrical ducts, showing their line of intersection. The axes do *not* intersect.

18o. Complete the views of the intersecting forms.

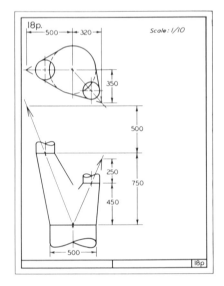

18p. Scale: 1/10

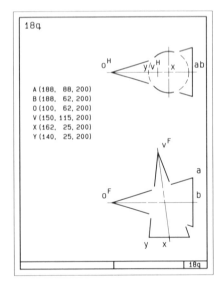

18q

A (188, 88, 200)
B (188, 62, 200)
O (100, 62, 200)
V (150, 115, 200)
X (162, 25, 200)
Y (140, 25, 200)

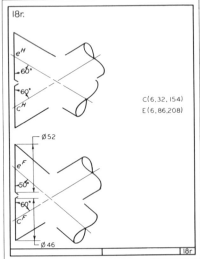

18r.

C(6, 32, 154)
E(6, 86, 208)

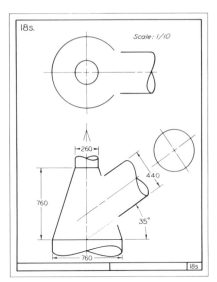

18s. Scale: 1/10

18p. Complete the views of the branches of the Y dust-collector fitting.

18q. Complete the front and top views of the intersecting cones.

18r. Complete the front and top views of the intersecting oblique cylinders.

18s. Find the intersection of the inverted funnel and cylindrical duct, using only the front and top views.

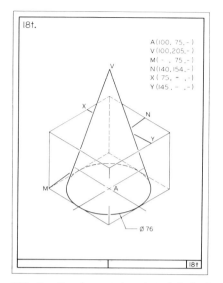

18t.

A(100, 75, -)
V(100,205,-)
M(- , 75,-)
N(140,154,-)
X(75, - , -)
Y(145, - , -)

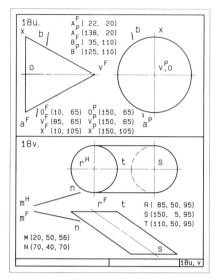

18u.

A_P^F (22, 20)
A_P^F (138, 20)
B_P^F (35, 110)
B_P^F (125, 110)

0_F^F (10, 65) 0_P^P (150, 65)
V_F^F (85, 65) V_P^P (150, 65)
X_F (10, 105) X^P (150, 105)

18v.

R (85, 50, 95)
S (150, 5, 95)
T (110, 50, 95)

M (20, 50, 56)
N (70, 40, 70)

18t. In the isometric pictorial find the points at which lines MN and XY pierce the cone.

18u. Using only the given views, determine the intersection of line AB and the cone.

18v. Using only the given views, determine the intersection of line MN and the cylinder.

18w. Indicate whether the following statements are true or false. If assigned, prepare written explanations or sketches to justify the answers.

(1) The lines of intersection of two prisms form either one or two closed figures.

(2) If two spheres intersect, the intersection is a circle.

(3) A plane through the vertex of a cone may intersect the cone in two elements.

(4) A single plane cannot cut elements from each of two cones unless the cones have a common vertex.

(5) In determining intersection by the use of cutting planes, it is essential that the cutting planes appear edgewise in one of the given views.

SELF-TESTING PROBLEMS

18A. Determine the intersection of line AB and the right cone.

18B. Show clearly the preliminary construction for establishing the intersection of the cone and cylinder. Locate only those points on the curve of intersection secured by the use of the two limiting planes.

18A.

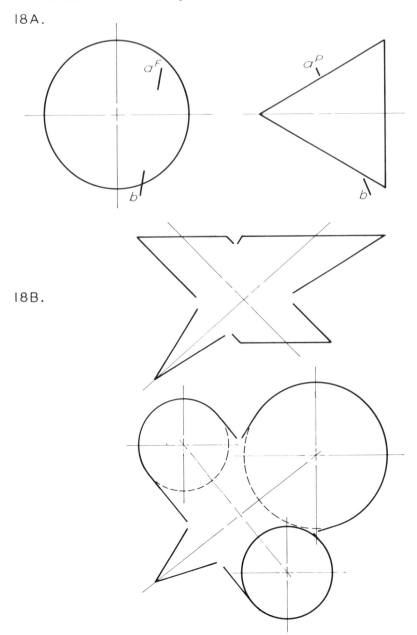

18B.

18C. and **18D.** Locate the two piercing points of line AB and the surfaces of the given forms.

18C.

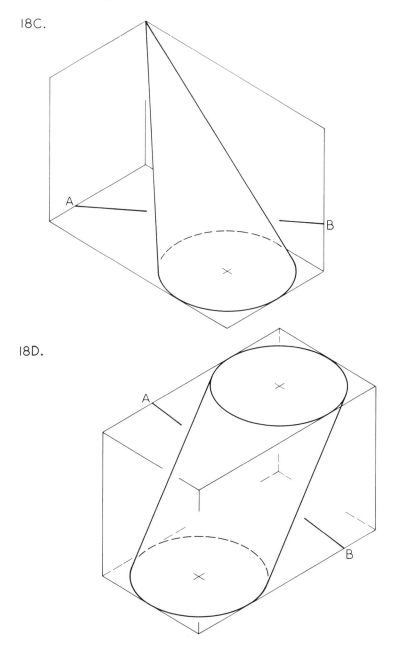

18D.

19 Shades and Shadows

The architect and engineer frequently employ shades and shadows to help produce an illusion of depth in a drawing that of necessity is made on a plane surface. In the field of production illustration in particular, the technique of shading is used to promote realism and advertising appeal. This technique applied to an architect's *rendering* presents to the

FIGURE 19.1 Shadow study—C-5 jet transport (courtesy Lockheed-Georgia Co., Marietta, Ga.)

prospective client a picture emulating and often surpassing the effect of a photograph. The United States Patent Office in its *Guide for Patent Draftsmen* specifies that patent drawings should include a shading technique for edges and surfaces. Although several methods exist for finding exact or approximate shades and shadows, the discussion in this text will be confined to a single approach utilizing the principles of descriptive geometry.

297

19.1   Shade, Shadow, and Umbra

Shade is that surface area of a geometric form from which direct light is excluded by the form itself, Figure 19.2.

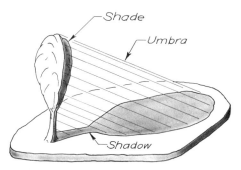

FIGURE 19.2   Shade, shadow, and umbra

Shadow is that surface area of a geometric form from which direct light is excluded by a second intervening form. Many physical objects are composed of several geometric forms. For purposes of distinction between shade and shadow, the above definitions are applied to the various forms independently.

Umbra is that portion of space from which light is excluded.

As the late Professor Eliot Tozer frequently cautioned his students, the phrase, "Standing in the shade of the old apple tree," should technically be stated, "Standing in the *umbra* of the old apple tree" or "Standing *on* the *shadow* of the old apple tree."

19.2   Light Source and Direction

Although any source of light may be used as a basis for establishing shade and shadow, it is conventional to employ a distant source such as the sun, since the rays may then be considered parallel. In addition, these parallel rays are assumed to have the direction of the diagonal of a cube extending from the upper left to the lower right, Figure 19.3(a). This direction is particularly convenient because the resulting principal views of this diagonal are at 45° with horizontal on the paper, Figure 19.3(b).

19.3   Shadow of a Prism

The shadow cast by the prism of Figure 19.4 is determined by locating the shadows of the corners of the prism. The shadow of a point is the

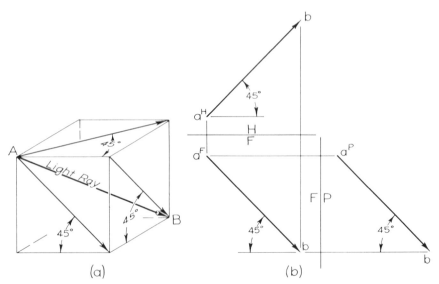

FIGURE 19.3   Conventional light-ray direction

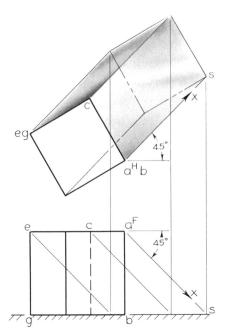

FIGURE 19.4   Shadow of a prism

intersection of the light ray from the point with the surface on which the shadow is cast.

As an example, the shadow of point A is found by introducing the conventional light ray AX, which is then extended to point S, its intersection with the horizontal plane on which the prism rests. The shadows of points C and E are found by the same method. Since points B and G are in the given horizontal plane, they coincide with their shadows. The shadow of the edges of the prism are then represented by the lines connecting the shadows of the corners. Note that the shadow on a horizontal plane of any vertical edge falls along the 45° ray in the top view.

While in this illustration the theoretical shadows of the undesignated edges are also indicated (in phantom form), they are of course not necessary, since only the outline of a shadow is actually observed. A shadow obtained by this method is an oblique projection.

19.4   Shade and Shadow of a Pyramid and Wall

In Figure 19.5 the shadow of the wall is similar to that of the prism in the preceding example. The shadow of the pyramid falls partly on the ground and partly on the front face of the wall. To secure the shadow on the ground, the wall is temporarily considered removed. The light ray through apex A is introduced, and its intersection with the ground, point K, is established. Since the base of the pyramid rests on the ground,

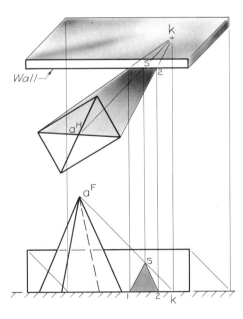

FIGURE 19.5   Shade and shadow of pyramid and wall

the required shadow extends from point K to the base corners. With the wall considered replaced, this ground shadow terminates at the front face of the wall as shown in the top view, leaving the shadow of the pyramid incomplete. The remaining portion of the shadow falls on the frontal wall face. To secure this portion, the intersection point S of the ray from point A with the front face of the wall is found. Points S, 1, and 2 are projected to the front view to complete the outline of this portion of the shadow. The two back faces of the pyramid are in shade as indicated in the top view.

19.5   Shade and Shadow of Object Composed of Cylinders

The shadow of a circle cast on a plane to which it is parallel is a true circle of the same diameter. Thus the shadows of the various bases of the cylinders of Figure 19.6 appears as circles in the top view. Point 5 is the

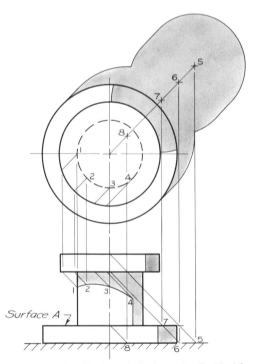

FIGURE 19.6   Shade and shadow of cylindrical forms

center of the circular shadow of the upper base of the cap, and point 6 is the center of the shadow on the ground of the lower base of the cap. Point 7 is the center of that portion of the shadow of this same lower

base that falls on surface A. Point 8 is the center of the shadow of surface A on the ground. The corresponding circles together with the
tangent lines representing the shadows of extreme elements are drawn as
shown to complete this portion of the shadow.

The lower base of the cap casts a portion of its shadow on the cylindrical support column. The curved outline of this shadow is established
by finding the shadows of points 1, 2, 3, and 4 of appropriate points of
the lower base.

The shade areas appearing in the front view are established by locating
the elements along the light rays that are tangent to the cylinders.

19.6   Shade and Shadow of Cone and Prism

The shadow of the prism of Figure 19.7 is cast on the ground as
shown. Part of the shadow of the cone falls on the ground and part on
the upper surface B of the prism. To obtain the portion that falls on the

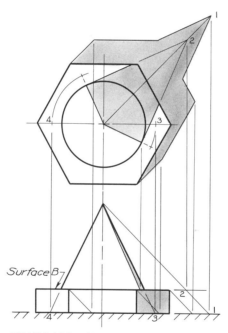

FIGURE 19.7   Shade and shadow of cone and prism

ground, the prism is temporarily considered removed, and the lateral
surface of the cone is extended to the ground as shown. To find the
portion that falls on surface B, surface B is extended to secure the
imaginary shadow, point 2, of the cone apex. The shadow together with
the shade area as shown complete the solution.

19.7 Shade and Shadow of Shelf and Support

The shelf and support of Figure 19.8 is an illustration of a shadow problem involving a front and left-side view. Although no new fundamentals are presented in this solution, the direction of the light rays in

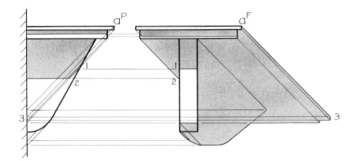

FIGURE 19.8 Shade and shadow of shelf and support

the left-side view should be noted. Of interest in this problem is the fact that the shadow falls partly on different surfaces of the object itself and partly on the frontal wall.

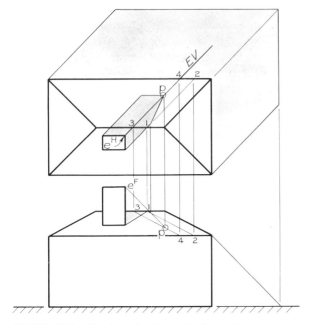

FIGURE 19.9 Shadow of prism on inclined plane

19.8   Shadow of Chimney on Inclined Roof Planes

To find a shadow on an oblique or inclined plane in which the edge view is not given, the two-view, piercing-point method of §6.2 is used to secure the intersection of the light rays with this type of surface. As an example, the shadow of point E, Figure 19.9, is obtained by passing a vertical cutting plane through the top view of the ray from point E. This cutting plane intersects the rear roof plane in the line 3-4. The intersection of the light ray and the front view of line 3-4 locates the required shadow, point P. Other points are obtained as indicated to complete the solution. Note in the front view that only the visible shadow is shown.

19.9   Pictorial Shade and Shadow

The extension of the principles of pictorial intersections to pictorial shade and shadow provides an interesting application of these descriptive geometry principles. More importantly, it is expected that the visual perception demanded in the solution of such problems will serve to expand one's ability to observe and learn by more effective utilization of that most dominant learning sense, sight.

Some introductory concepts of pictorial shade and shadow are presented in the oblique pictorial of Figure 19.10, which utilizes the selected light ray direction of line AB. Direction line NB represents the projection of line AB on a horizontal plane and line MB provides the projection on a frontal plane.

The shadow of point 1 on the horizontal base plane is secured by the introduction of pictorial ray 1-11 parallel to line AB above and its horizontal projection 0-11 parallel to line AB. Since lines 0-11 and 1-11 both exist in a single plane, they intersect at point 11, the shadow of point 1 on the horizontal base. As point 0 already lies in the same base plane, line 0-11 is the shadow of vertical line 0-1 on the plane. Note that *the shadow of a vertical line on a horizontal plane takes a direction parallel to* line NB.

The shadow of point 2 on the base plane is established at the intersection of pictorial ray 2-12 and its horizontal projection. Observe that the shadow of horizontal line 1-2 on the horizontal base plane is a parallel line 11-12. Note carefully the similar construction needed to obtain points 13 and 14, which represent the shadows of points 3 and 4.

A portion of the shadow of line 5-6 falls on horizontal plane H of the pictorial itself. The shadow of point 5 on plane H is located at the intersection of pictorial ray 5-15 and its horizontal projection. Although this shadow point 15 lies beyond the limits of plane H, line 15-6 does establish the appropriate initial direction of the shadow for line 6-5. Since the actual shadow of point 5 does not fall on plane H, it should be apparent that it must fall on the frontal plane 7-3-2. This shadow point

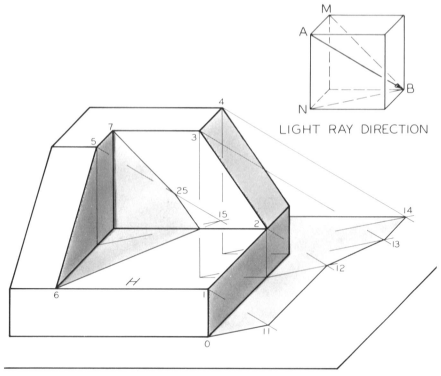

FIGURE 19.10 Oblique shade and shadow

25 is established at the intersection of pictorial ray 5-25 and its frontal projection that is added parallel to line MB through point 7.

Other shadow lines are added as shown and shaded areas are designated to complete the solution.

19.10 Isometric Shade and Shadow

The solution of the shade and shadow project of Figure 19.11 involves some extension of the concepts presented in the previous section. Shadow point 11 is located at the intersection of the pictorial ray from point 1 with its projection on the horizontal base plane. Other shadow points, such as point 12, on the base plane are similarly established.

The elliptical shadow segment of the arc 3-5 falls primarily on horizontal plane H of the pictorial itself. The procedure for obtaining the shadows of the three points 3, 4, and 5 is demonstrated. Note that shadow point 13 is desirable even though this point exists beyond the limits of plane H. Similarly, shadow point 24 is needed to help orient that elliptical shadow of arc 3-5 that falls on the base plane.

The shadow of point 6, which falls on the inclined plane of the pictorial, requires the use of the cutting-plane technique for locating the intersection of pictorial ray 6-16 with this inclined plane. The cutting

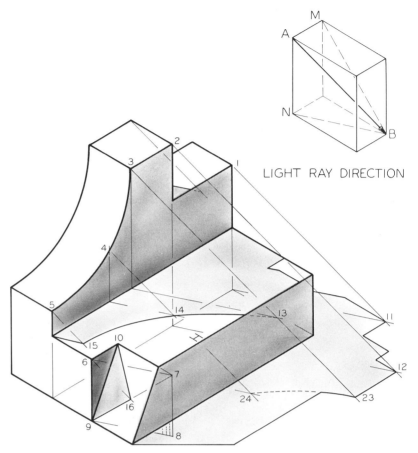

LIGHT RAY DIRECTION

FIGURE 19.11   Isometric shade and shadow

plane consists of the horizontal lines 6-7 and 9-8 drawn parallel to line
NB above. This cutting plane intersects the inclined plane in the line 9-7.
Point 7 is a point on this line of intersection, since line 10-7 of the inclined plane and line 6-7 of the cutting plane both lie in the same
horizontal plane. By a comparable analysis, it can be observed that
point 9 is another point on this line of intersection. Now, since line 9-7
and ray 6-16 both exist in the same cutting plane, point 16 represents
the piercing point, or shadow, of the pictorial ray with the desired
pictorial plane. Additional shadow lines are added to designate the
shaded areas to complete the solution.

19.11   Shadow on a Cylindrical Surface

The shadow of a line on a cylindrical surface involves the use of
inclined cutting planes that will intersect the cylindrical surface in
elements of the cylinder.

A typical construction is demonstrated for point 2 of Figure 19.12. Pictorial light ray 2-3 is introduced parallel to light direction line AB above, and its horizontal projection 2-4 is added parallel to line NB. The intersection of ray 2-3 with the end plane 5-6-10 of the cylinder is now needed. This intersection point 3 is established by first locating point 4 at the crossing of lines 2-4 and 6-7 and then transferring this point vertically downward to ray 2-3.

The required cutting plane that contains ray 2-3 and that intersects the cylindrical surface in an element is provided by the introduction of lines 3-9 and 2-7, both parallel to cylindrical elements. The enclosed inclined cutting plane is the parallelogram 2-7-3-9, which intersects the cylindrical surface in the element 0-0. Since ray 2-3 and element 0-0 both lie in this same cutting plane, they intersect at point 12, the desired shadow of point 2 on the cylinder. Additional points are secured similarly to complete the elliptical shadow boundaries.

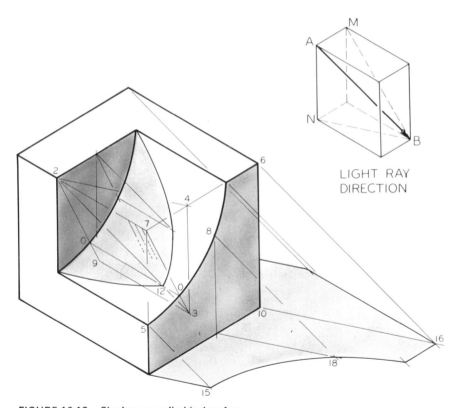

FIGURE 19.12   Shadow on cylindrical surface

PROBLEMS

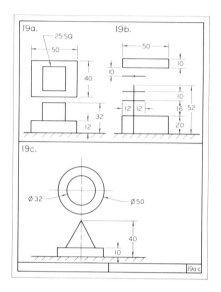

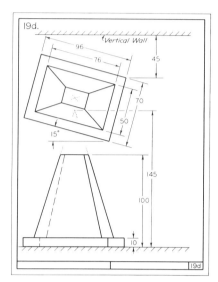

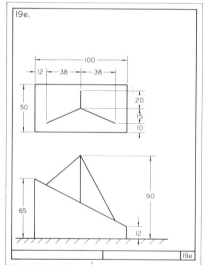

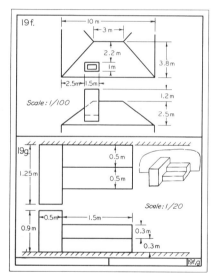

19a. Establish the shadow of the two prisms.

19b. Locate the shadow of the wall and cross pole.

19c. Determine the shade and shadow of the cone and the cylindrical base.

19d. Determine the complete shade and shadow.

19e. Establish the shadow of the inclined base, vertical pole, and three guy wires.

19f. Determine the shadow of the chimney on the inclined roof planes.

19g. Show the shadow of the wall and steps.

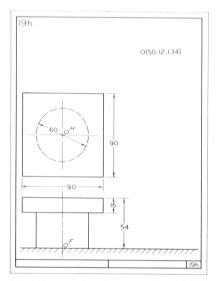

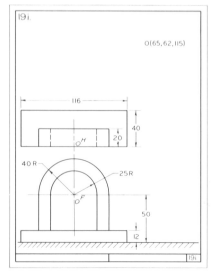

19h. Determine the complete visible shade and shadow of the prism and cylinder.

19i. Determine the shade and shadow of the archway and base.

19j. Indicate whether the following statements are true or false. If assigned, prepare written explanations or sketches to justify the answers.

(1) The shadow of a vertical line on a frontal wall is a vertical line.

(2) The shadow of a horizontal circle on a frontal wall is a circle.

(3) The shadow of a line on two parallel planes is two parallel lines.

(4) A top view always shows the true area of a shadow.

(5) A roof sloping at 30° and with no overhang does not by itself cast a shadow.

(6) The shadow of a frontal circle on a frontal wall is a circle.

(7) The shadow of a line is always longer than the line itself.

(8) The shadow of a circle may be a straight line.

SELF-TESTING PROBLEMS

19A. Show the complete shade and shadow of the given combination of geometric forms.

19A.

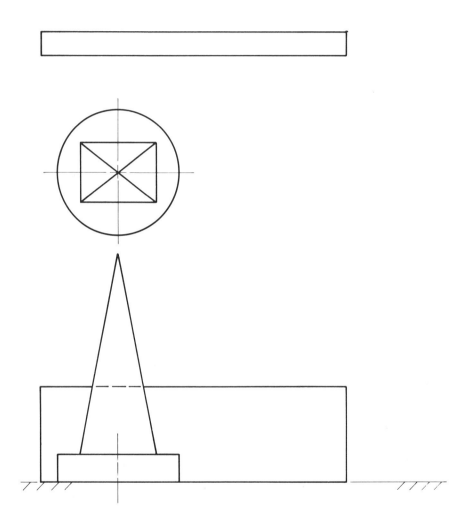

19B. Using the light-ray direction AB, establish the shade and shadow of the given oblique pictorial.

19C. Using the light-ray direction MN, obtain the shade and shadow of the oblique pictorial.

19B.

19C.

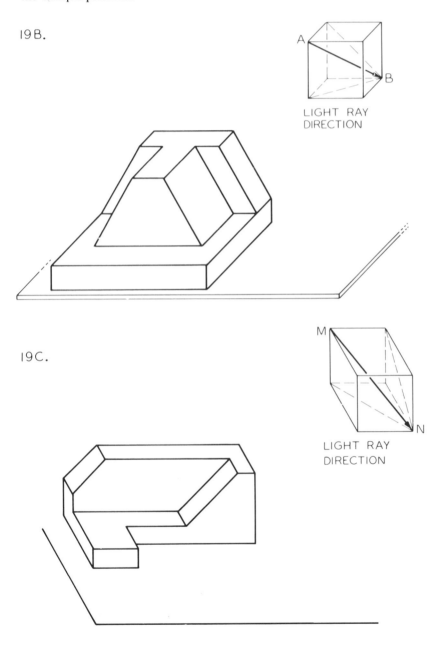

LIGHT RAY
DIRECTION

LIGHT RAY
DIRECTION

19D. and **19E.** Using the light-ray direction AB, provide the shade and shadow of the isometric pictorial.

19D.

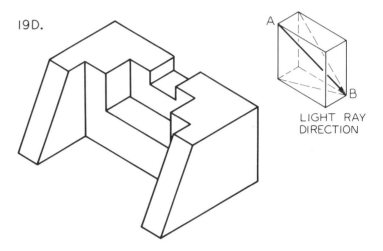

LIGHT RAY
DIRECTION

19E.

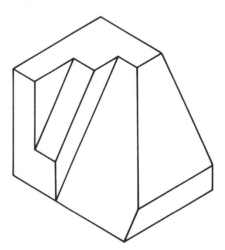

20 Perspective Projection

In Chapter 1 it was pointed out that multiview projection is an exact method of representing complex space relations, that is, a direct method for laying out and measuring on paper all the true distances and angles involved in a space or three-dimensional problem. If the engineer and others in technology are to communicate by this method, the recipients of the information must of course understand or be able to "read" multiview drawings. Since, however, the general public and some technicians are not trained in this subject, the various types of pictorial drawings occupy an important sector of the field of graphical communication. As was mentioned, it is frequently difficult to supply in a pictorial drawing the complete technical information necessary for the manufacture of an article, but lay people do not normally need all the details. For them the general appearance and method of operation of the product are sufficient, and a pictorial drawing is usually satisfactory. The technical illustrator should therefore be familiar with the more widely used types of pictorial drawing and the principles of projection used in their construction. Since an extensive treatment of this subject is beyond the scope of this textbook, only a few examples of perspective construction employing descriptive geometry principles are discussed here.

20.1 Perspective or Central Projection

Axonometric and oblique projections are widely used in pictorial work because of their relative ease of construction. They do not and cannot, however, represent an object in a completely natural manner. In both methods, as well as in multiview projection, the projection lines from the object to the picture plane are parallel. In nature, projection lines are actually *visual rays* extending from an object to the eye of the observer. In order for these rays to be parallel, the observer must be at an infinite distance from the object. Since this is not the situation for

313

normal viewing, such projections can at best only roughly simulate a true picture of an object. *Perspective* projection takes into consideration the fact that an observer in nature is at some finite or measurable distance from any object seen. While the methods of construction employ descriptive geometry or multiview methods, the resulting projection closely approximates an actual optical image.

Figure 20.1 shows the basic arrangement in space for perspective projection. The T-square is the object viewed. The eye of the observer is at an established point SP, the *station point*. The visual rays are represented by lines drawn from the station point through the various points

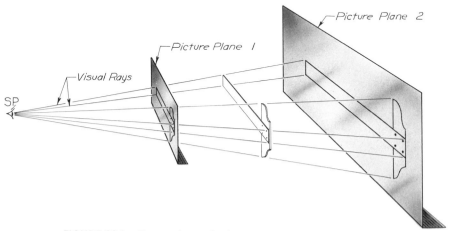

FIGURE 20.1   Perspective projection

on the T-square. If a picture plane is introduced, the visual rays pierce it at points that are then the perspective projections of the respective points on the T-square. Lines properly joining the piercing points complete the perspective projection on the particular picture plane. It is evident that on picture plane 1, which is between the observer and the object, the perspective projection is smaller than the object. On picture plane 2, which is placed beyond the object, the perspective projection is larger than the object. On a picture plane placed behind the observer (to the left of SP) the perspective projection would be inverted as in a real image formed by the lens of a camera.

20.2   Perspective Construction by Multiview Projection of Piercing Points

The basic construction for perspective projection is shown in multiview form in Figure 20.2. A frontal picture plane is employed and appears in edge view in the plan and right-side views. The plan view of the object is turned so that an observer at the given station point may see clearly two of the vertical faces of the object. It should be noted that the

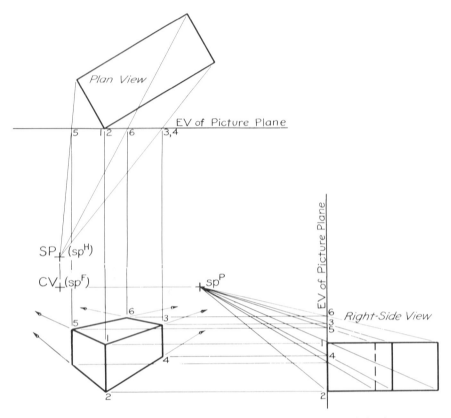

FIGURE 20.2 Construction of a perspective by multiview projection of piercing points

plan view is an ordinary top view placed or fastened in a revolved position, and that the resulting right-side view is a special view of the revolved object, an objectionable feature of the construction that will be eliminated later.

The visual rays are now drawn from station point SP to the points of the object in the plan and side views. Their piercing points with the picture plane (for example, points 1, 2, 3, and 4) are projected to the front view. When all the visible piercing points are located, they are connected with lines representing the edges of the object to complete the perspective projection.

Several important fundamentals of perspective projection may be observed by a careful study of Figure 20.2:

1. *Lines parallel to the picture plane remain parallel to their original positions in the perspective projection.* Note, for example, lines 1-2 and 3-4.

2. *Lines in the picture plane* (such as 1-2) *appear true length in the perspective projection*, as well as parallel to their original positions.

3. *All lines parallel to each other but not parallel to the picture plane converge to a common point in the perspective projection.* If extended, lines 1-3 and 2-4 intersect at some point to the right of the perspective in Figure 20.2. As will be seen later, line 5-6 intersects 1-3 and 2-4 *at the same point* in the perspective projection.

20.3   Two-Point and One-Point Perspectives

The geometry of the construction in Figure 20.2 is such that the intersection or *vanishing point* (VP; §20.4) of lines in the perspective is at *eye level* (level with sp^F) if the edges they represent are horizontal. In the terminology of perspective projection, sp^F is called the *center of vision*, CV, since sp^F coincides with the central point toward which the observer is looking. The picture plane is perpendicular to the visual ray through the center of vision.

Figure 20.3 illustrates perspectives in which the vanishing points are actually shown. To produce the perspective in Figure 20.3(a), the *steps* were placed with the vertical edges parallel to, and the horizontal edges inclined to, the picture plane. In the perspective the vertical edges remain vertical and the horizontal edges converge to two vanishing points. Thus the perspective is called *two-point perspective*. If there were inclined or oblique lines present, such lines would have still other vanishing points. The term *two-point* refers to the number of vanishing points of *principal edges* only.

For Figure 20.3(b) the steps were placed with two sets of principal edges, the vertical edges and one set of horizontal edges, parallel to the picture plane. These sets of parallel lines remain parallel in the perspective, and only the second or receding set of horizontal edges have their perspective projections converging to a vanishing point. This is therefore called *one-point perspective*. The vanishing point coincides with the center of vision.

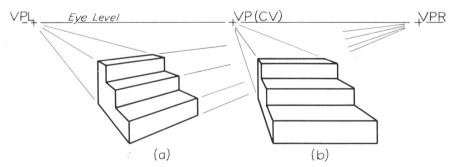

FIGURE 20.3   Two-point and one-point perspectives

20.4 Construction for Vanishing Points

As was pointed out in §20.1, the perspective of a point is that point at which the visual ray from point SP to the point pierces the picture plane. This is illustrated again in Figure 20.4. If a construction line

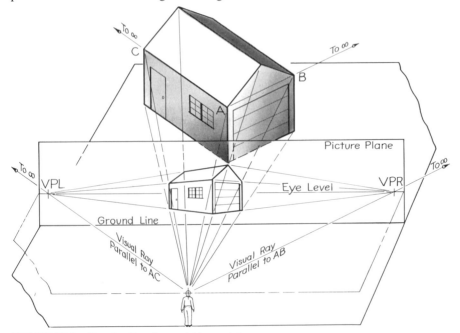

FIGURE 20.4 Construction for vanishing points

through points A and B of the building is extended an infinite distance, the visual ray to the infinitely distant point on line AB is parallel to line AB. The piercing point of this visual ray in the picture plane, point VPR (vanishing point to right), is thus the perspective of the infinitely distant point. Since the same visual ray is also parallel to several other lines on the building, the single point VPR is the perspective of infinitely distant points on all these parallel lines and is thus the vanishing point of these lines. In similar fashion, the visual ray parallel to the other set of horizontal principal lines of the building, of which line AC is a member, pierces the picture plane at VPL (vanishing point to left), which is the vanishing point of this second set of principal edges. It is evident that, since the two visual rays are horizontal in space, point VPL and point VPR are at the same height as point SP, which is at eye level (as described previously).

A third visual ray, parallel to the vertical lines of the building, would not pierce the picture plane at all, proving that these lines have no vanishing point and thus must be parallel in the perspective. This is then another two-point perspective, as expected from the arrangement of object and picture plane.

20.5 Construction of a Two-Point Perspective

In Figure 20.5(a) the construction for vanishing points is performed in multiview form. The point labeled SP, it must be remembered, is the top or plan view of the station point. The visual rays parallel to the horizontal edges of the block pierce the picture plane in points that appear in the plan view as points vpl^H and vpr^H. The points labeled VPL and VPR are the *front views* of these piercing points which are at eye level, the *horizon*. The location of the horizon therefore must be known before the perspective can be constructed. Since the perspective itself is merely the front view of the pattern formed by the piercing points of the visual rays to the object, points VPL and VPR are the two vanishing points to be used in the construction of the perspective. The beginner must take special care to avoid being confused by the overlapping top and perspective (front) views. This arrangement is customary because of the space saved.

With the vanishing points established, construction of the perspective can be started, Figure 20.5(b). The forward or leading vertical edge 1-2 of the object has been placed in the picture plane and thus is its own perspective (like edge 1-2 in Figure 20.2). This line is therefore true length in the perspective and is drawn directly below the corresponding corner of the plan view at its true height with respect to the horizon. It is convenient to place an available elevation view, in this case the right elevation (right-side view), at this given height with respect to the horizon so that height dimensions may be transferred by horizontal projection as shown for edge 1-2.

Construction lines containing horizontal edges extending from point 1 are now drawn to the corresponding vanishing points. The piercing points 3 and 4 of visual rays to the other ends of these horizontal edges are determined in the top view and projected to the front view as shown to establish points 3 and 4 of the perspective. Point 5 is now located by a visual ray and projection to edge 1-3 in the perspective. From point 5 a line is drawn toward point VPR and point 6 located on this line by a vertical projection line from the piercing point of the visual ray for the corresponding point in the plan view. These procedures are repeated for other points until the perspective of the base of the object is completed.

When an object is so placed that its leading edge is not in the picture plane, the perspective of the leading edge is neither in true length nor at true height. This is the situation for the prismatic upper portion of the given object of Figure 20.5. To locate the perspective of the leading edge of this prism, its vertical front surface is imagined to be extended to intersect the picture plane in line 7'-8', Figure 20.5(c). Since points 7' and 8' are in the picture plane, and since the lines extended to locate them are horizontal in space, points 7' and 8' must be at true height in the perspective. Thus they are located as shown by projection from the given side view. Construction lines representing the infinitely long extensions of the corresponding edges of the object may now be drawn in

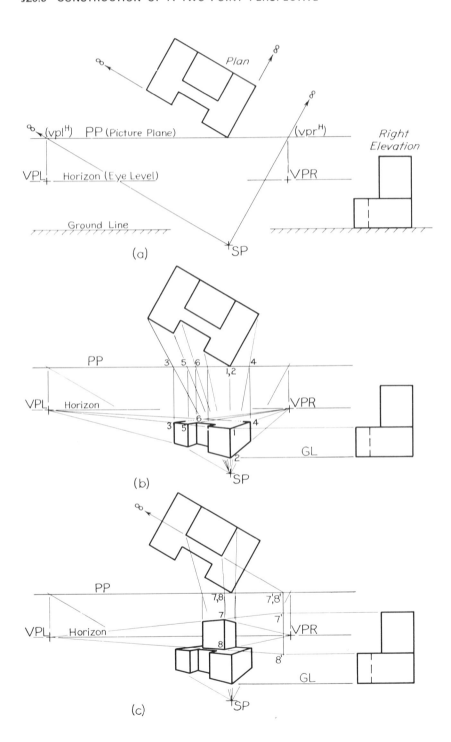

FIGURE 20.5   Construction of a two-point perspective

the perspective from points 7′ and 8′ to point VPL. Points 7 and 8 are then located on these construction lines by projection from the piercing points of the corresponding visual rays, obtained in the plan view. With the leading edge 7-8 of the prism located, the remainder of the perspective is completed in the same manner as the first portion.

20.6   Construction of a One-Point Perspective

As stated in §20.3, if an object is placed with two sets of principal edges parallel to the picture plane, only the third set of principal edges has a vanishing point. When a visual ray is drawn parallel to this third set of edges, Figure 20.6, the vanishing point turns out to be the center

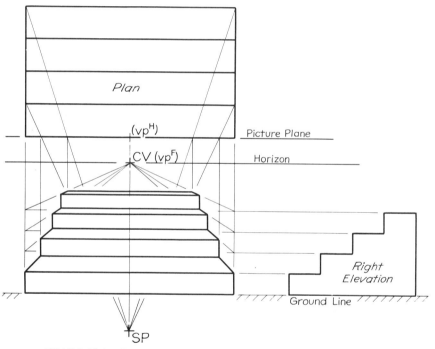

FIGURE 20.6   Construction of a one-point perspective

of vision, point CV. From this stage the perspective construction is the same as in two-point perspective.

One-point perspectives are popular for showing interiors and long, receding vistas. For instance, see Figure 20.7, which is a one-point perspective except for minor discrepancies introduced by the optics of photography.

FIGURE 20.7 One-point perspective (courtesy The Partlow Corp., New Hartford, N.Y.)

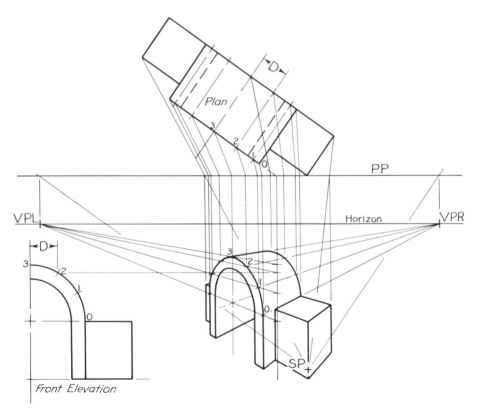

FIGURE 20.8 Curves in perspective

20.7 Curves in Perspective

By the methods explained in §20.5, the perspective of any given point can be located. The construction of the perspective of a curve is merely repetition of this fundamental construction to locate a sufficient number of desirably spaced points so that a smooth curve may be drawn through them.

In Figure 20.8 the straight-line portions of the perspective are located as before. A series of points is then introduced on the circular view of the larger arc, and their corresponding plan views are located by transfer distances such as D. Now, if horizontal construction lines are drawn through these points on the front surface of the object and extended to pierce the picture plane, the piercing points are at true height in the perspective. The perspectives of the construction lines then extend from the piercing points to point VPL. The points themselves are located on the construction lines through the use of visual rays in the usual way. Points on the smaller arc are established in similar fashion.

The foregoing procedure could be repeated to locate points on the visible portion of the rear arc, but it is simpler to draw the elements of the cylindrical surface, in the plan and in the perspective, through the numbered points previously established. The points on the rear arc are then points on these elements and may be found directly by drawing visual rays to the points in the plan view and projecting from the piercing points of the rays.

20.8 Three-Point Perspective

If an object is placed so that none of its principal edges are parallel to the picture plane, there will be three principal vanishing points, and the construction becomes more involved. Therefore, three-point perspective is not widely used. It is effective, however, for realistic illustration of tall structures and for bird's-eye views where the heights of the objects are to be emphasized.

As an illustration, Figure 20.9, a simple rectangular prism is drawn in three-point perspective. Given the front and top views of the prism and station point, Figure 20.9(a), and given that the observer is looking toward the center point C of the prism, an auxiliary view showing ray SP-C in true length also shows the picture plane in edge view and perpendicular to sp^{1}c. Since the vanishing point of a line is the perspective of an infinitely distant point on the line, visual rays are drawn parallel to the edges of the prism in the top and auxiliary views. Their piercing points with the picture plane are located in the auxiliary and top views and are transferred to Figure 20.9(b) by means of distances X, Y, and Z. Figure 20.9(b) is to be a true-size view of the picture plane (and perspective), but its position on the paper is more natural if it is constructed

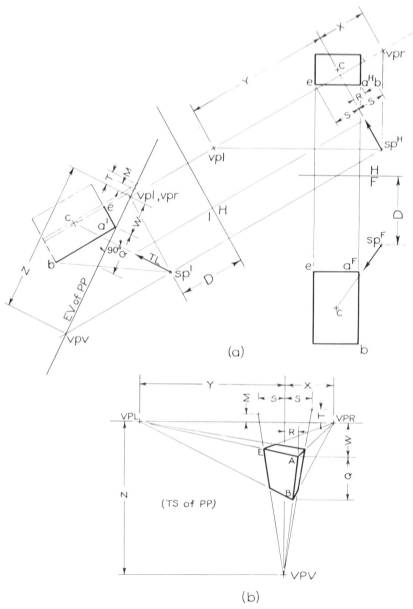

FIGURE 20.9 Three-point perspective

with line VPL-VPR horizontal than if it is projected from view 1 as a secondary auxiliary view.

Point A of the prism is in the picture plane and is located in the perspective by transfer of distances R and W. The three edges meeting at point A then lie along the construction lines from point A to the three vanishing points. The ray from sp^1 to b pierces the picture plane at a distance Q below a^1. This distance then locates point B in the perspective.

As an illustration of another method of locating the perspective of a point not in the picture plane, the prism edge through e is extended in view 1 and pierces the picture plane at a distance M from vpl, vpr. This piercing point is transferred to the perspective by means of distance M, and distance S obtained in the top view. The extended edge of the prism then appears in the perspective as the line from this piercing point to point VPV. Its intersections with the lines from points A and B to point VPL determine the endpoints of the edge of the prism through point E. Repetition of these methods completes the perspective.

Of course, if the entire auxiliary view of the object is drawn as indicated in phantom, the method is objectionable for representing a complicated structure. If the object is composed principally of horizontal and vertical lines, however, the entire auxiliary view need not be completed. The piercing points of extended vertical edges may be located instead by projection from the top view to the picture plane in view 1.

20.9   Perspective Intersection

Some concepts of isometric and oblique intersections can be applied to perspective pictorials as illustrated in Figure 20.10, where it is desired to obtain the intersection of unlimited plane MNK with the given two-point perspective.

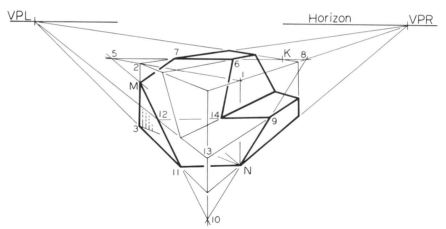

FIGURE 20.10   **Perspective intersection**

Initially, a cutting plane 1-2-3-N is established containing line MN. Since lines MN and 1-2 both lie in this cutting plane, they intersect at point 5. Similarly, since point K and point 5 both lie in the same top surface of the perspective pictorial, line 6-7 provides an initial portion of the desired lines of intersection. Line 5-6 extended to meet line 1-VPR secures point 8, which lies in the same perspective plane as point N, thus establishing line N-9 as an additional segment of the required intersection.

Line 8-N is then extended to location point 10; and on line 10-M, point 11 is identified, providing two additional line segments, M-11 and 11-N. Perspective lines 13-VPL and M-10 intersect at point 12, which lies at the same perspective height as point 9, thus securing line 9-14. The establishment of line M-7 completes the required intersection.

20.10   Perspective Shade and Shadow

In the perspective shade-and-shadow project of Figure 20.11, a single light source point S is located somewhat to the left and to the rear of the pictorial. The projection of point S on the horizontal base plane of the pictorial is designated as point H.

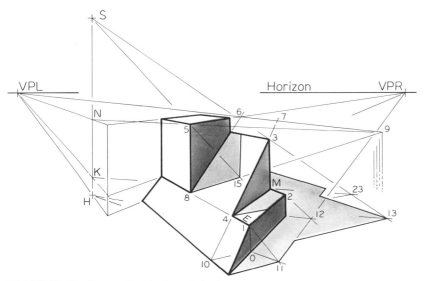

FIGURE 20.11   Perspective shade and shadow

The shadow of point 1 on the base plane is found by the introduction of light ray S-1 and the projection of this ray H-0 on the horizontal base. Since S-1 and H-0 both lie in the same plane, they intersect at 11, the shadow of point 1. The shadow of point 2 is similarly obtained. Line 11-12, which vanishes to VPR, provides the shadow of line 1-2, a portion of which is later found to be covered by other shadows.

The shadow of point 3 is located at point 13, and 10-13 provides the direction of that portion of the shadow of inclined line 3-4 that falls on the pictorial base plane. The remaining portion of the shadow of line 3-4 falls on plane E.

For this construction, the shadow of point 3 on plane E is needed. First the height of horizontal plane E is transferred to location K. The desired shadow, point 23, is located at the intersection of ray S-3 and

its projection on plane E, which is represented by line K-M, extended. Line 4-23 provides that portion of line 3-4 that falls on plane E.

The shadow of point 5 falls on the inclined plane of the pictorial. For this construction, the slanted face is extended to line 6-7, the same height as the top surface of the pictorial. Next, height 5 is transferred to location N. Then a vertical cutting plane N-5-8 is introduced containing ray S-5. This cutting plane intersects the pictorial slanted surface in the line 8-9. Since lines S-5 and 8-9 both lie in the cutting plane, they intersect at point 15, the shadow of point 5. Line 8-15 is the shadow of line 5-8, and line 5-6 establishes the direction of that portion of shadow 5-6 that falls on the slanted surface. Additional shadow lines, as required, are found in a similar manner.

PROBLEMS

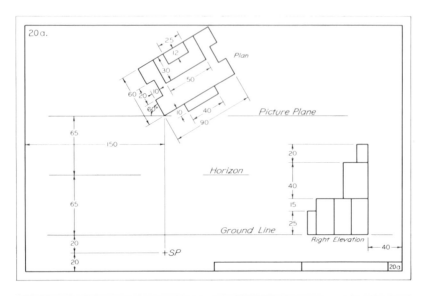

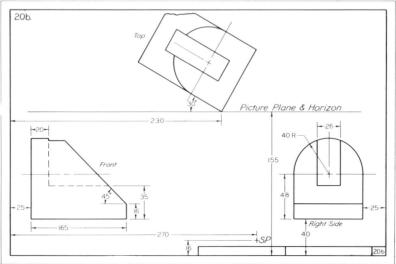

20a. Show the perspective of the *office building* outline.

20b. Draw the perspective of the *slide block*. The front view may be deleted.

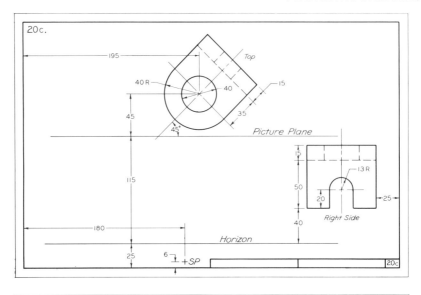

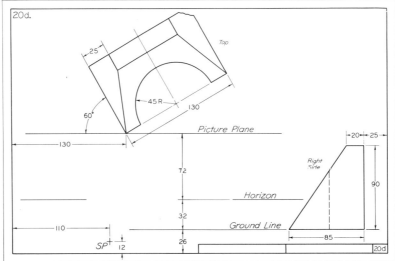

20c. Construct the perspective of the *guide collar*.

20d. Draw the perspective of the *slide bushing*.

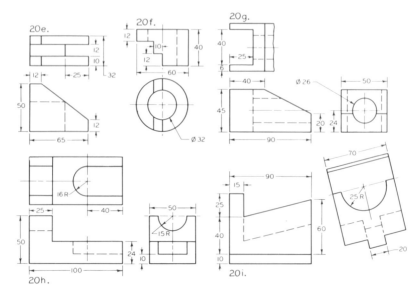

20e–i. Draw axonometric, oblique, or perspective projections as assigned.

20j. Indicate whether the following statements are true or false. If assigned, prepare written explanations or sketches to justify the answers.

(1) Hidden lines are usually omitted from pictorial drawings.

(2) A circle can appear in its true shape in a perspective pictorial.

(3) In two-point perspectives not more than two vanishing points are ever desirable.

(4) On a perspective drawing a line can project longer or shorter than its true length.

(5) The perspective projection of a straight line is always a single straight line.

(6) In three-point perspective all vertical lines converge to a single vanishing point.

SELF-TESTING PROBLEMS

20A. Prepare a perspective pictorial utilizing the given layout.

20A.

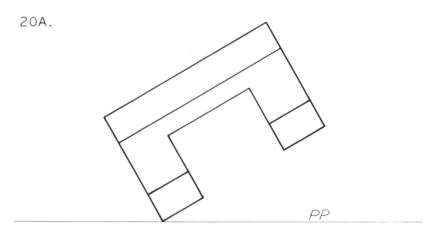

PP

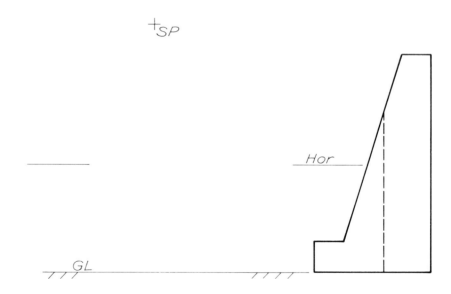

+SP

Hor

GL

20B. and **20C.** Establish the intersection of the unlimited plane MNK and the perspective pictorial.

20B.

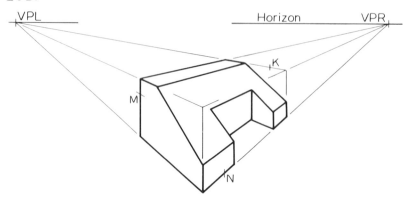

20C.

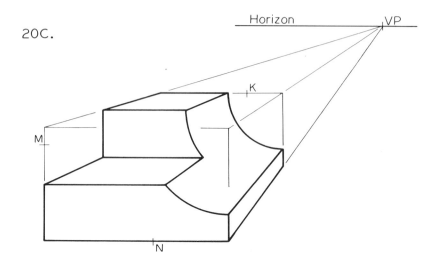

20D. and **20E.** For the perspective pictorial, provide the shade and shadow using the light source S.

20D.

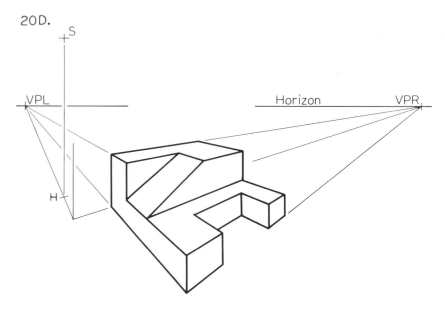

20E.

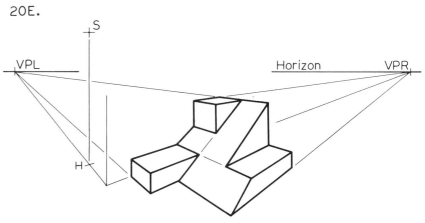

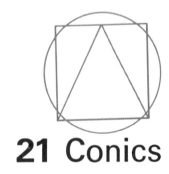

21 Conics

The intersections of planes and right circular cones produce the more common plane curves or *conic sections*, namely, the *circle, ellipse, parabola,* and *hyperbola*. When a plane is perpendicular to the axis of a right circular cone, the section is a circle. An ellipse is formed when the plane intersects every element of the cone and is not perpendicular to the axis, Figure 21.1. If the cutting plane is passed parallel to an element of the cone, the section is a parabola, Figure 21.2. The hyperbola is formed when the plane cuts both *nappes* (Appendix C.2) of a conical surface, Figure 21.3.

21.1   The Ellipse

In Figure 21.1, the intersection of a plane and a right circular cone is shown. The plane is placed so as to cut each and every element of the cone. The introduction of a series of horizontal cutting planes permits the establishment of points on the ellipse in the auxiliary view. The *foci* are determined by the projections of the centers of two spheres inscribed in the cone and tangent to the cutting plane. The *directrices* are projected to the auxiliary view as the lines of intersection between the given cutting plane and the horizontal planes through the tangent circles of the spheres and cone.

The ellipse may be defined as the locus of points in a plane whose distances from a fixed point (either of the foci) are in constant ratio, less than unity, to their distances from a fixed line (the corresponding directrix). Note the procedure used to locate point K on the ellipse.

Another definition of the ellipse describes it as the path of a point moving in a plane in such a manner that the sum of its distances from two fixed points (the foci) is constant and equal to the longest dimension of the ellipse. This definition is utilized in the pin-and-string method given in most general mechanical drawing texts; for still other methods of constructing an ellipse, see Appendix B.

333

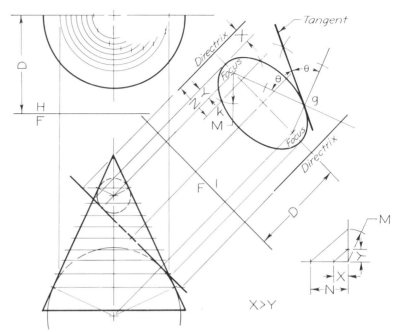

FIGURE 21.1   The ellipse

A line tangent to an ellipse at a given point such as point G on the ellipse, Figure 21.1, bisects the angle formed by the two *focal radii* through the given point.

21.2   The Parabola

A parabola is formed when the cutting plane is *parallel* to an element of the right-circular cone, Figure 21.2. The line of intersection between the cone and cutting plane is determined by a series of horizontal cutting planes and is shown in true size in the auxiliary view. The focus is the projection of the center of a sphere inscribed in the cone and tangent to the cutting plane. The directrix is the line of intersection between the cutting plane and a horizontal plane passed through the circle of tangency of the inscribed sphere and cone. The axis of the parabola is the projection of the axis of the cone.

The parabola may also be described as the path of a point moving in a plane so that its distance from a given point (the focus) is always equal to its distance from a given straight line (the directrix). Point K in the auxiliary view illustrates application of this principle in constructing the parabola.

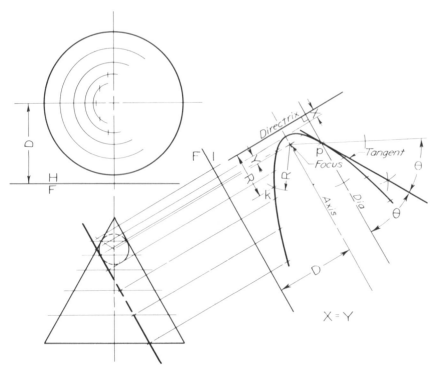

FIGURE 21.2   The parabola

A line tangent to a parabola through a given point bisects the angle formed by a focal radius and a diameter (a perpendicular to the directrix) through the given point. Figure 21.2 shows the tangent through a point P.

21.3   The Hyperbola

Figure 21.3 shows a hyperbola, which is composed of the intersections of the cutting plane with both nappes of the conical surface. For convenience, the cutting plane has been passed parallel to the axis. The focus for each nappe is the projected center of the sphere inscribed in the corresponding nappe of the cone and tangent to the cutting plane. The directrices are the lines of intersection between the cutting plane and planes passed through the circles of tangency of the spheres and the nappes of the cone. The *asymptotes* are the projections of the extreme elements of the nappes. They are the lines to which the curves theoretically would be tangent at an infinite distance.

The hyperbola also may be described as the path of a point moving in a plane in such manner that the difference of its distances from two fixed

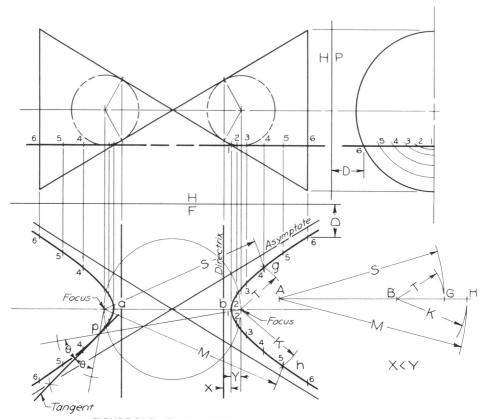

FIGURE 21.3   The hyperbola

points (the foci) is constant and equal to distance AB. This principle, used in Figure 21.3 to locate points G and H, is applied for as many points as desired to draw the complete hyperbola.

A line tangent at a given point on a hyperbola bisects the angle formed by the two focal radii, which intersect at the given point. The construction for the tangent at point P illustrates this procedure.

PROBLEMS

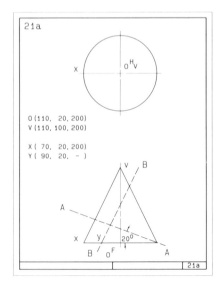

21a.

O (110, 20, 200)
V (110, 100, 200)

X (70, 20, 200)
Y (90, 20, -)

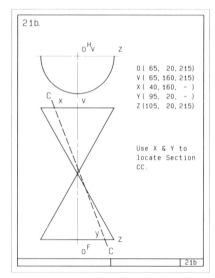

21b.

O (65, 20, 215)
V (65, 160, 215)
X (40, 160, -)
Y (95, 20, -)
Z (105, 20, 215)

Use X & Y to
locate Section
CC.

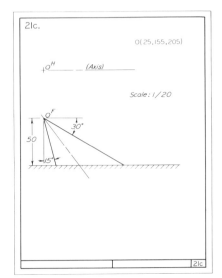

21c.

O (25, 155, 205)

(Axis)

Scale: 1/20

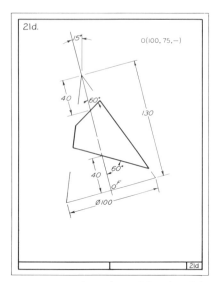

21d.

O(100, 75, -)

21a. Plot and name the curves produced by the cutting planes A-A and B-B. Show the foci and directrices of the curves.

21b. Plot and identify the curves of intersection of cutting plane C-C with both nappes of the conical surface. Locate the foci and asymptotes.

21c. Show the true-size view of the ground area illuminated by the right circular conical beam of a floodlight at point O. Calculate the area illuminated. [Area of ellipse $= \dfrac{\pi}{4}$ (major axis $\times$ minor axis).]

21d. Draw the views that show the true sizes of the bases of the given portion of a right circular cone.

337

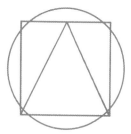

22 Map Projection

Although the earth more closely approaches the shape of an ellipsoid, for an introductory study of map-making its shape will be considered spherical. The characteristics of the surface of the earth are best portrayed on the familiar spherical globe found in geography classrooms. But since the three-dimensional globe is impractical for many uses, other methods are necessary for representing the earth's surface. These methods all involve laying out the earth's surface on a single plane. This, then, is a development problem; and if a sphere were theoretically developable, an accurate representation would be possible. Since a sphere is not developable, approximate methods are used, §17.13. Although some accuracy is lost in these approximations, there are unique inherent advantages for some approximations that make them more useful than the globe.

The most fundamental principle of cartography (map making) is the establishment of a coordinate system on the earth's surface to which any point can be related. The most frequently used set of coordinates

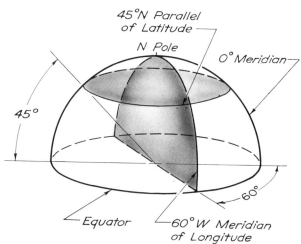

FIGURE 22.1 Latitude and longitude

is the *parallel-meridian* system. In this system there are 90 parallels of *latitude* between the equator and each pole. Figure 22.1 shows the 45° North parallel of latitude. The other set of coordinates consists of 180 *great circles* radiating from the pole at equal angles and dividing the parallels into 360 degrees of *longitude*. Figure 22.1 shows the 60° West meridian of longitude. It is customary to reckon longitude east and west from a prime (zero) meridian up to 180 degrees. The meridian through Greenwich, England, is used by almost all countries as the prime meridian. This system of coordinates provides a framework upon which the position for each spot on the earth's surface may be located by its latitude and longitude.

22.1 Representation of Earth's Surface

The earth's surface may be approximately represented by projection on a plane or on a developable geometric shape such as a cylinder or cone, Figure 22.2. The projectors used may be either parallel or converging.

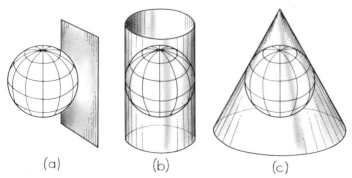

(a)   (b)   (c)

FIGURE 22.2   Types of developable surfaces on which the surface of a globe may be projected

Since the earth is essentially flat for limited areas, no problem exists in providing satisfactory maps for such areas. It is for relatively large areas that many types of maps have been devised in an attempt to minimize distortion.

22.2 Projection on a Tangent Plane

If the earth's axis is inclined and the surface is projected orthographically to a plane as shown in Figure 22.3, the resulting map has a

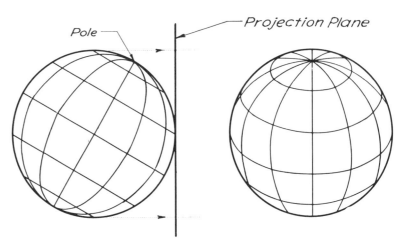

FIGURE 22.3   Orthographic projection

pictorial effect that makes it particularly useful for many elementary textbook illustrations.

In *gnomonic* projection the projection lines emanate from the center of the sphere. Figure 22.4 shows a *polar-gnomonic* projection with projection lines radiating from the center to a plane tangent at the North pole. The principal advantage of gnomonic projection is the fact that all great circles are projected as straight lines, a property of particular value in navigation since the shortest distance between two points on the earth's surface lies along a great circle.

Most maps in common use are not obtained by such geometrical means as previously introduced but are nongeometrical variations of these types. For instance, the map of Figure 22.5 is one in which the

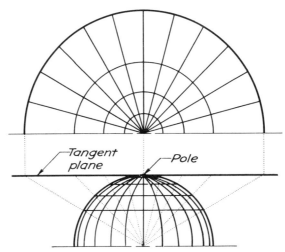

FIGURE 22.4   Polar-gnomonic projection

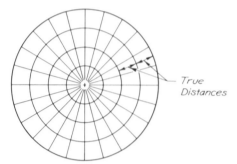

FIGURE 22.5   Nongeometrical polar map

gnomonic projection is altered so that the distances between parallels are equal. Thus measurements made radially from the pole are true distances on the earth's surface. Distances measured in other directions, however, are not true.

22.3   Projection on a Cone

The surface of the earth may be projected on a tangent cone, resulting in the geometrically produced map of Figure 22.6. Here the parallels are projected as concentric circles and the meridians as converging straight lines. The parallel in contact with the enveloping cone is projected true and is called the *standard parallel*. Other features of the earth are not

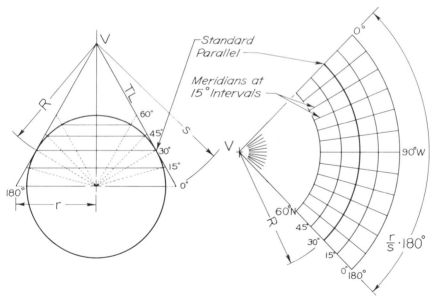

FIGURE 22.6   Conical projection—one standard parallel

projected true, with the distortions increasing in proportion to the distance from the standard parallel. To avoid undue distortion the area projected on the cone is limited to that included between the 0° and 60° parallels.

For practical use a nongeometrical variation of this conical projection has been developed, Figure 22.7. A standard parallel is used as before,

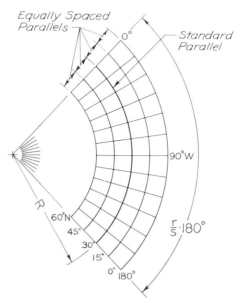

FIGURE 22.7   Nongeometrical conical map

along which true distances are laid off. In contrast to the theoretical projection, here the parallels are spaced at true distances. Again the number of parallels is limited to avoid increasing scale variations.

Projection on a cone passed through the earth as shown in Figure 22.8 results in two standard parallels. In this method a somewhat greater area may be projected without objectionable scale variation. For practical use this projection is altered so that the parallels are evenly spaced.

Another nongeometrical variation of conical projection is the *polyconic map*, Figure 22.9. This map is an alteration of the zone method of a sphere development (see §17.13, Example 3).

In the polyconic map the parallels appear as nonconcentric circles, the radii of which are the slant heights of the corresponding tangent cones of Figure 22.9(a). The spacing of the parallels is controlled by the straight central meridian, which is drawn to true scale, Figure 22.9(b). The points at which the meridians cross the parallels are located by setting off along each parallel the corresponding distances between the meridians. These distances may be approximated by transfer of the chordal distances such as X from the top view, or by the rectified-arc method of

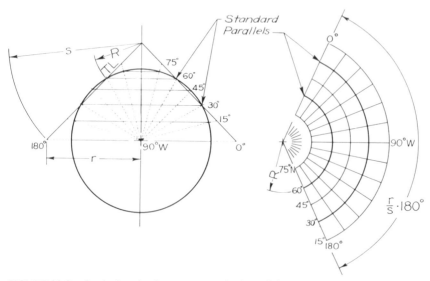

FIGURE 22.8   Conical projection—two standard parallels

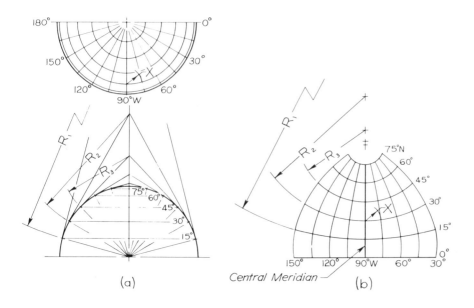

FIGURE 22.9   Nongeometrical polyconic map

Appendix B.11. Alternatively, the total angle through which each parallel extends may be calculated in the same way as in the development of a right circular cone, §17.4. The parallel is then subdivided to correspond to the number of meridians included. The meridians passing through these points are curved lines. The scale along the meridians increases as the distance from the central meridian increases. Thus the polyconic map has but little scale distortion near the central meridian but rapidly increasing distortion at the east and west extremes.

The polyconic map is well known in the United States since it has had considerable use by the U.S. Coast and Geodetic Survey.

22.4   Cylindrical Projection

A cylindrical surface is another developable form on which the earth's surface may be projected to secure a geometrically produced map. Figure 22.10 shows a cylindrical projection in which the projectors emanate from the center of the sphere. On the development of the cylindrical surface both the parallels and meridians are produced as straight lines. The equator is projected at its true scale, but other parallels have a scale distortion that increases in relation to the distance from the equator.

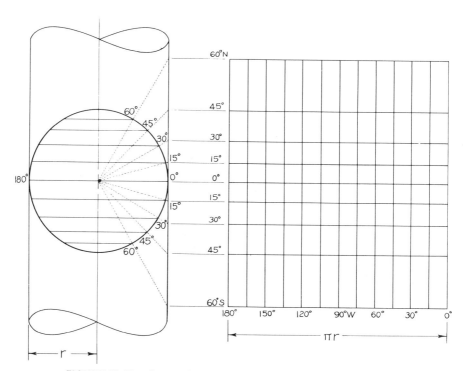

FIGURE 22.10   Geometrical cylindrical projection

Variations may be obtained in cylindrical projections by the use of orthographic projection or by the use of a cylindrical form that cuts through the sphere in about the same manner as the cone of Figure 22.8.

The most popular and useful nongeometric variation of cylindrical projection is the Mercator map, Figure 22.11. Here the scale of the

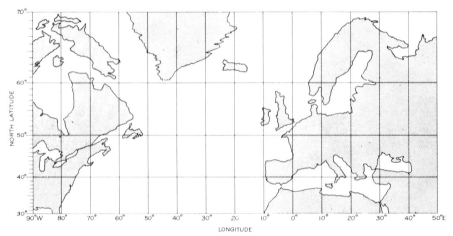

FIGURE 22.11   A Mercator map

meridians is increased in the same proportion as the scale of the parallels. Thus at any point on the map the scale is the same in every direction, but areas are considerably exaggerated for the extreme north and south latitudes. An advantage in navigation is the fact that a constant compass bearing appears as a straight line (a *rhumb* line) on this map.

PROBLEMS

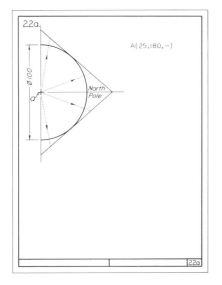

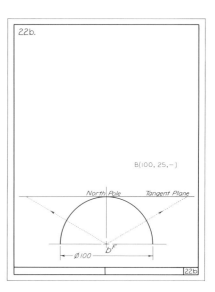

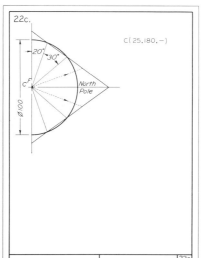

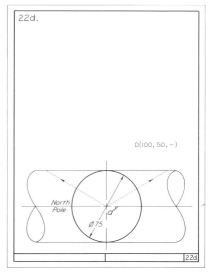

In problems 22a–22f locate points on the map as assigned by the instructor.

22a. Project as indicated the surface of the earth on a cone tangent at the 40° parallel and construct the map. Show the parallels at 10° intervals from 10° to 70° and the meridians at 10° intervals from 0° to 180°W.

22b. Draw the polar-gnomonic projection showing the parallels at 10° intervals from the pole to the 30°N parallel. Show meridians at 10° intervals. Show the great circle course between two assigned points.

22c. Project as indicated the surface of the earth on the cone. On the map show the parallels at 10° intervals from the equator to 70°N. Show the meridians at 10° intervals.

22d. Project as indicated the surface of the earth on the cylinder. On the map show the parallels at 10° intervals from 60°N to 60°S and the meridians at 10° intervals from 0° to 180°W.

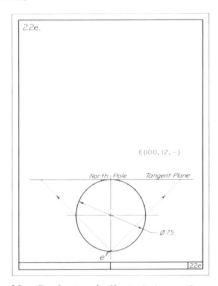

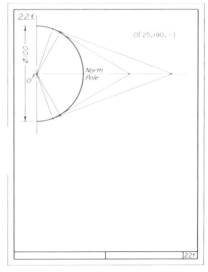

22e. Project as indicated the surface of the earth on the given tangent plane. Show the parallels at 15° intervals in the northern hemisphere. Show the meridians at 15° intervals.

22f. Using the given layout as stated, construct a polyconic map with the parallels at 10° intervals from 20°N to 70°N. Show the meridians at 10° intervals from 0° to 180°W.

22g. Indicate whether the following statements are true or false. If assigned, prepare written explanations or sketches to justify the answers.

(1) On the earth's surface all meridians pass through the poles.

(2) On the earth's surface the parallels of latitude are great circles.

(3) New York is east of the zero meridian.

(4) On a Mercator map the length of the equator is the only line that appears true length.

(5) Neither the North nor South pole can be represented on a Mercator map.

(6) A rhumb line is a constant bearing line.

(7) Any great circle appears as a straight line on a gnomonic projection.

(8) The polyconic projection provides a true representation of the earth's surface.

(9) In conic projection only one or two meridians are true length.

(10) The Cartesian coordinate system is normally used to identify the location of points on the earth's surface.

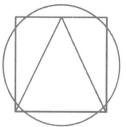

23 Spherical Triangles

Three great circles of the sphere of Figure 23.1(a) intersect to form the spherical triangle ABC. This triangle together with the radial line to the vertices form the *spherical pyramid* OABC, Figure 23.1(a) and (b). The subsequent discussion is concerned with the relationships that exist among the parts of this pyramid.

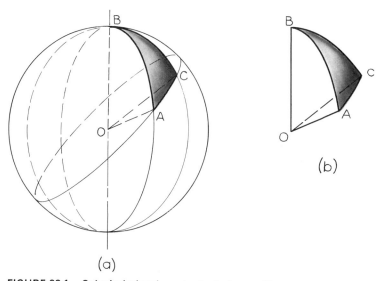

FIGURE 23.1   Spherical triangle and spherical pyramid

Side BC is measured by the plane angle BOC; side AC is measured by the plane angle COA; and side BA is measured by the plane angle BOA. *The sum of the sides of a spherical triangle is less than* 360°.

The angle at B of the spherical triangle is measured by the dihedral angle formed by the intersecting faces BOA and BOC, the angle at C by the dihedral angle between faces COB and COA, and the angle at A by the dihedral angle between faces AOB and AOC. *The sum of the dihedral angles of a spherical triangle is greater than* 180° *and less than* 540°.

An understanding of spherical triangles is basic to the solutions of problems in terrestrial and celestial navigation and inertial guidance.

23.1 Solution of Spherical Triangle, Three Sides Given

Let it be given that in a spherical pyramid OABC, side BC = 40°, side BA = 60°, and side CA = 35°. Let it be required to draw the front and top views of the pyramid and to solve for the angles of the spherical triangle.

The front and top views of any conveniently sized sphere are drawn, Figure 23.2. For simplicity of solution, face OBC is placed in a frontal plane. Being frontal, side BC is laid off at its given angle of 40° in the front view. In this position bc appears as a true circular arc. Point c in the top view is located by projection.

With line OB as an axis, face BOA is assumed revolved into this same frontal plane, and the given angle of 60° for side BA is set off as indi-

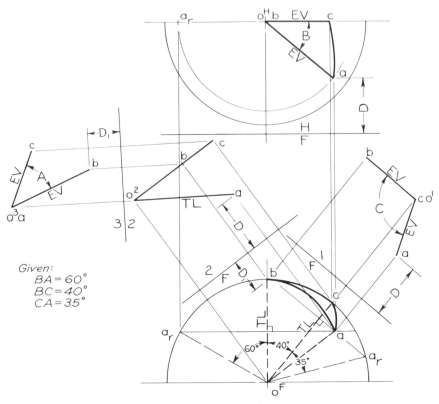

FIGURE 23.2 Solution of spherical triangle, three sides given

cated. About side OC as an axis, face COA is revolved into this same frontal plane, and the given angle of 35° for side CA is set off. The front view at this stage is thus a development of the faces of the spherical pyramid. The actual front views of points O, B, and C coincide with these same points of the development. In the top view these three points are also in their correct positions. To complete the views of the actual pyramid, only point A needs to be located. The front view of A is secured by counterrevolution about the true-length axes of revolution o^{F}b and o^{F}c. These paths of revolution, which appear perpendicular to the true axes, intersect to locate a. The path of revolution about the vertical axis OB appears as a true arc in the top view as shown. The intersection of this arc with a projector from a in the front view locates a in the top view. To aid visualization, the views of the spherical pyramid may be completed by drawing the views of the great circle arcs ac and ab. These elliptical curves are not, however, necessary for the solution.

Angle B of the spherical triangle, which is measured by the included dihedral angle at B is measured in the top view, since the intersecting planes of this angle appear edgewise in this view. Angle C is obtained in auxiliary view 1, for which the direction of sight is along the true-length line of intersection o^{F}c. Since the line of intersection of planes BOA and OAC is oblique, successive auxiliary views 2 and 3 are used to produce edge views of these planes. In view 3 angle A is measured as shown, completing the solution.

23.2   Solution of Spherical Triangle, Two Sides and Included Angle Given

In a spherical pyramid OABC let it i given that side BA = 42°, side BC = 60°, and angle B = 50°. Let it be required to draw the front and top views of the pyramid and solve for the remaining side and angles. A sphere of convenient size is drawn, Figure 23.3, and plane BOC is placed in a frontal plane as shown, using the given 60° angle for side BC. Plane OBA is revolved into this same frontal plane about line OB as an axis, utilizing the angle BA = 42° as given. In the top view the dihedral angle B is laid off at its given angle of 50°, since the line of intersection OB of the planes appears as a point in this view. Point A is then counterrevolved. This path of revolution appears as a circular arc in the top view, and its intersection with the edgewise view of plane OBA locates a. The path of revolution appears perpendicular to the axis o^{F}b in the front view. The intersection of this path in the front view with a vertical projector from the top view locates a in the front view. As in the previous example, angle C is obtained by use of the single auxiliary view 1; angle A is found by use of the successive auxiliary views 2 and 3. Side CA is obtained by the revolution of plane COA into the frontal plane about line CO as an axis.

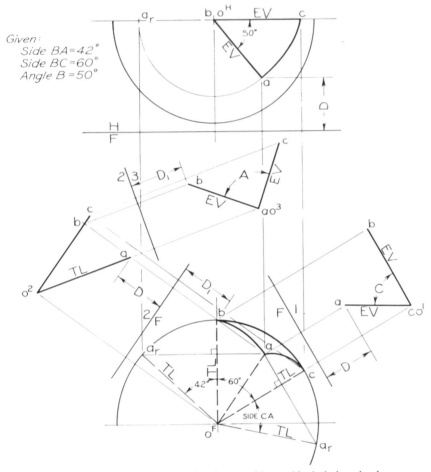

Given:
Side BA = 42°
Side BC = 60°
Angle B = 50°

FIGURE 23.3   Solution of spherical triangle, two sides and included angle given

23.3   Solution of Spherical Triangle, Three Angles Given— Polar-Triangle Method

The *poles* of a great circle are the two endpoints of that diameter of the sphere which is perpendicular to the plane of the great circle. Six poles exist for the three great circles of a spherical triangle ABC. These poles can be joined to form eight spherical triangles. The *polar triangle,* however, is that triangle for which: pole B_1 is that pole of side AC which falls on the sames side of side AC as does pole B; pole C_1 is that pole of side BA which falls on the same side of side BA as does pole C; and pole A_1 is that pole of side BC which falls on the same side of side BC as does pole A.

The polar triangle provides a means for solving a spherical triangle indirectly. For example, a spherical triangle with the three angles given can be transformed into a problem of the related polar triangle with the three sides known. The solution of this polar triangle problem also provides the solution of the given triangle.

The following relationships are of particular importance:

1. *If a second triangle is the polar triangle of a first, the first triangle is also the polar triangle of the second.*
2. *An angle of a spherical triangle is the supplement of the opposite side of its polar triangle.*

This latter relationship can be explained by the following: Figure 23.4 shows a given true dihedral angle X between two great circle planes of a

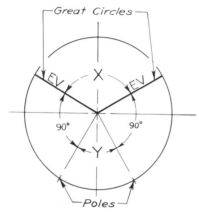

FIGURE 23.4   Poles of great circles

spherical triangle. The plane angle at Y is produced by the perpendiculars used to obtain the poles of these great circles. It can be seen that angle Y is the supplement of angle X. Angle Y is the side of the polar triangle opposite angle X of the given triangle.

Thus a problem in which the three angles of a spherical triangle are given can be solved by means of its polar triangle.

Problem
  Given:

$$\text{Angle } A = 120°$$
$$\text{Angle } B = 130°$$
$$\text{Angle } C = 125°$$

The sides of the related polar triangle are the supplements of the given angles:

$$\text{Side } B_1 C_1 = 180° - 120° = 60°$$
$$\text{Side } A_1 C_1 = 180° - 130° = 50°$$
$$\text{Side } A_1 B_1 = 180° - 125° = 55°$$

This polar triangle, having its three sides known, may be solved by the method of §23.1. Then, in turn, the sides of the given spherical triangle can be computed, since they are supplementary to the angles of the polar triangle.

Figure 23.5 shows the polar triangle $A_1 B_1 C_1$, the views of which are found as explained in the foregoing. Vertices A and C of the given triangle are formed by drawing perpendicular lines 2 and 3 to the great circle planes $OB_1 C_1$ and $OB_1 A_1$, which appear edgewise in the top view. Auxiliary view 1 is added to secure an edge view of the great circle plane $OC_1 A_1$. Vertex B is found by drawing the perpendicular line 4 to the edge-view plane $OC_1 A_1$ as shown. When the vertices A, B, and C are

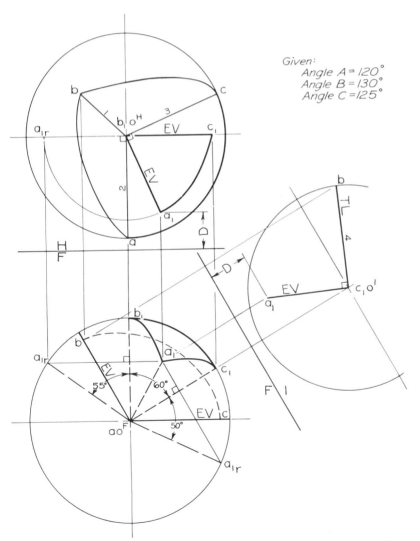

Given:
Angle A = 120°
Angle B = 130°
Angle C = 125°

FIGURE 23.5 Solution of spherical triangle, three angles given—polar triangle method

located in each of the given views, the views of the spherical triangle are completed as shown, the great circle lines being represented here for visualization purposes. The sides of the spherical triangle may now be solved as in previous examples (not shown here). As pointed out in the preceding paragraph, it is simpler to solve the polar triangle and then calculate the sides of the given spherical triangle.

23.4   Navigation Problem

On the surface of the earth, the shortest distance between two points is along the great circle containing the points. If the earth is assumed to be a perfect sphere, problems involving great circle distances may be solved by the use of spherical triangles.

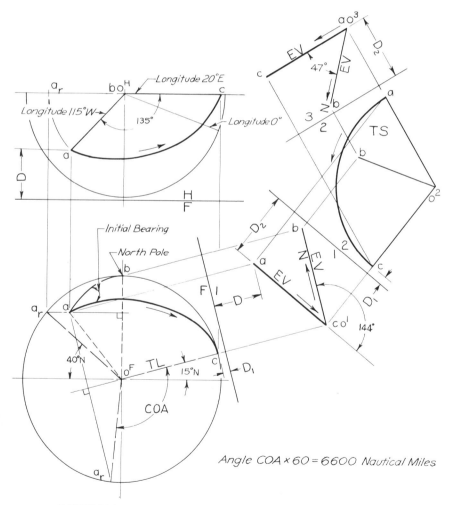

FIGURE 23.6   Navigation problem

Given point C, latitude 15°N, longitude 20°E, and point A, latitude 40°N, longitude 115°W (see Figure 22.1), let it be required to solve for the distance between them.

Point B in the solution, Figure 23.6, is regarded as the North pole for convenience. Plane OBC is placed in the frontal plane through the pole, so that point C is located directly by using the angle of latitude 15° as given. Plane OBA is assumed to be revolved into the frontal plane about line OB as an axis. For this revolved position, the given 40° angle of latitude is used to locate a_r.

The dihedral angle between planes OBC and OBA is equal to the difference in longitude of points A and C. Point C is 20° east of the prime meridian and point A is 115° west of the prime meridian; thus the included angle between the two locations is 115° + 20°, or 135°. This angle is laid off as shown on the top view. Point A_r is counterrevolved to locate point A as shown. Plane OCA is revolved about line OC as an axis to secure an angular measurement of the required distance of the side AC. Since 1° of arc is approximately equal to 60 nautical miles on the earth's surface, the great circle distance is obtained by multiplying the angle COA by 60 (1 nautical mile equals approximately 1.85 kilometers).

Since a great circle course would necessitate a constantly changing bearing, an exact great circle course is impractical for manual navigation. However, this shortest distance course can be approached by traveling a series of constant bearing courses, called *rhumb lines* (§22.4), which approximate the great circle. The initial rhumb-line bearing is equal to the dihedral angle between planes OBA and OCA. This angle is secured in auxiliary view 3 and is specified as N47°. The final bearing is equal to the supplement of the dihedral angle between planes OCA and OBC. This angle is obtained in auxiliary view 1 and is specified as N144°. Of interest is the fact that the course traveled is initially a northeasterly direction and is finally southeasterly.

PROBLEMS

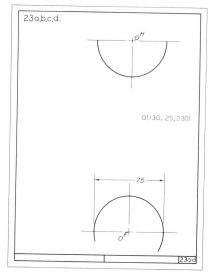

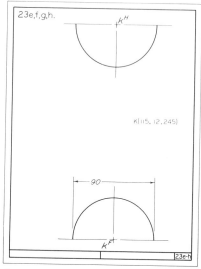

23a. Given that side BC = 30°, side CA = 40°, and side BA = 60°, solve for the angles A, B, and C of the spherical triangle ABC.

23b. Given that side BC = 35°, side BA = 45°, and angle B = 70°, solve for side CA and angles A and C.

23c. Given that side BC = 40°, angle B = 60°, and angle C = 110°, solve for angle A and sides BA and CA.

23d. Given that angle A = 120°, angle B = 105°, and angle C = 140°, solve for sides BC, CA, and BA.

23e. Given: point C, lat. 40°N, long. 20°W; and point A, lat. 30°N, long. 100°W. Determine the great-circle distance between the two points and plot the course in the given views.

23f. Given: point C, lat. 50°N, long. 30°E; point A, lat. 15°N, long. 90°W. Determine the great-circle distance between the two points and plot the course in the given views.

23g. Given: point C, lat. 60°N, long. 60°E; point A, lat. 0°, long. 120°W. Determine the great-circle distance from point A to point C and the initial and final bearings.

23h. Given: point C, lat. 50°N, long. 20°W; point A, lat. 30°N, long. 105° E. Determine the great-circle distance from point A to point C and plot the course in the given views. Find the initial and final bearings.

23i. Indicate whether the following statements are true or false. If assigned, prepare written explanations or sketches to justify the answers.

(1) The sides of a spherical triangle are portions of great circles.

(2) The sum of the sides of a spherical triangle is equal to 180°.

(3) The shortest distance between two points along the earth's surface lies along a great circle through the points.

(4) The center of a great circle is always the center of the sphere.

(5) One degree of arc of a great circle is approximately equal to 60 nautical miles on the earth's surface.

(6) Only one great circle can be passed through two points on the earth's surface.

SELF-TESTING PROBLEM

23A. Determine the distance in nautical miles and plot the path of the great-circle course between points A and C on the earth's surface.

A—Latitude 40°N, Longitude 80°W.
C—Latitude 20°N, Longitude 40°E.

23A.

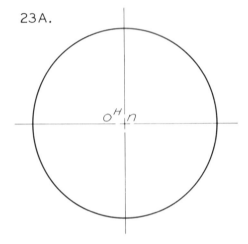

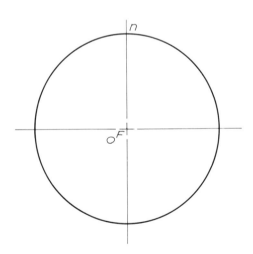

24 Review

In Chapters 1 through 23 the material was organized in a logical order for development of student understanding. Fundamental constructions were not introduced until needed. For example, primary auxiliary views were discussed in Chapter 2, but the introduction of successive auxiliary views was deferred until Chapter 5. Since there are a great many fundamental principles and constructions involving primary auxiliary views, it was felt that there was no need to burden the student with successive auxiliary views until after extensive application of the principles of primary auxiliary-view constructions to actual problems. Now that the student has almost completed a course in descriptive geometry, he or she should be aware of the fact that the really basic principles of the subject are few in number—that the majority of the topics covered are actually applications of these fundamentals to specific problems. This review chapter is *not* therefore an outline of the text but is an attempt to summarize and correlate in concise form the basic principles studied.

For ease in reading, back references are not given. It is expected that students will use the index in the back of this book, as needed, to locate the details of a particular construction.

24.1   Construction of Additional Views

In the early chapters several examples of the steps performed in the construction of additional views were discussed. These steps were worded in a general way because, as the student probably now realizes, the same procedure is used for the construction of *any* additional view of an object when two adjacent views are given.

It is perhaps the most important fundamental of descriptive geometry that if two points are established in adjacent views, their spatial relationship is thereby established. Hence information is available for the construction of any additional views of these two points. It may be that in a particular case the true spatial relationship of two points may not be so readily observed as in other cases. For example, if the front and top views of a profile line are given, the relative positions of the endpoints in space are more easily seen after a side view is added. Nevertheless, with only the front and top views *given*—implying that every point is shown and identified—all pertinent information about their relative positions is *available*.

Thus, with two adjacent views of an object given, any other views desired may be constructed. The procedure for constructing additional views adjacent to one of the given views is always the same. It is illustrated in Figure 24.1, which is discussed in the following four steps.

Views 1 and 2 are assumed to be given. It is required that a third view be constructed adjacent to view 2.

Step 1. Establish the line of sight for the desired information.
         In Figure 24.1 the line of sight indicated by the arrow

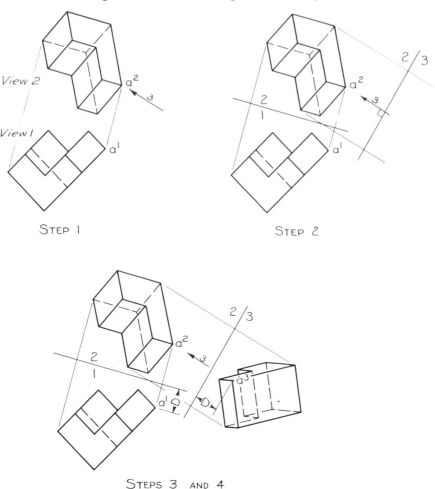

FIGURE 24.1   Constructing an additional view with two adjacent views given

marked 3 is employed to make this illustration general and does not produce a particularly useful view 3. The general principles of establishing lines of sight in particular problems are reviewed in §24.2.

It might be said that this statement encompasses a large part of the understanding of descriptive geometry. The remaining steps 2, 3, and 4 become routine after a short period of application, so that the solution of almost any problem in reality reduces to the selection of the proper line of sight.

Step 2.   Introduce the necessary folding lines.
           One folding line 1/2 is always perpendicular to the projection lines joining the given views. The second folding line 2/3 is always perpendicular to the line of sight for the desired additional view and is drawn at any convenient distance from the given view to which the required view is adjacent.

Step 3.   Transfer distances to the new view.
           In Figure 24.1 the additional view 3 is drawn adjacent to view 2. Hence a distance such as D for point a^1 is obtained in view 1 and is set off in view 3 along the projection line from a^2 to view 3.

Step 4.   Determine the visibility and complete the view.

TABLE 24.1   *Uses of Additional Views*

Use	Position of Line of Sight		Typical Applications
	In space	*On multiview drawing*	
1. True length of line (TL)	Perpendicular to line	Perpendicular to any view of the line or directed toward a point view of the line	(1) Prerequisite to use 2 (2) Slope or grade (3) Angle between line and principal plane
2. Point view of line	Parallel to line	Parallel to true-length view of line	(1) In connection with use 3 (2) Distance between line and point (3) Distance between line and another line
3. Edge view of plane (EV)	Parallel to plane	Parallel to true-length view of line in plane or directed toward a true-size view of the plane	(1) Prerequisite to use 4 (2) Angle between two planes (3) Distance from point to plane, line perpendicular to plane
4. Normal or true-size view of plane (TS)	Perpendicular to plane	Perpendicular to edge view of plane	(1) Area of plane figure (2) Angle between intersecting lines (3) Other plane geometry constructions

24.2 The Four Fundamental Uses of Views

Under Step 1 in §24.1 it was stated that selection of the proper lines of sight is of basic importance in the solution of descriptive geometry problems. While a seemingly infinite variety of problems may be encountered in practice, each problem can be reduced to one or a combination of only four projected views. These are listed in Table 24.1, together with the corresponding positions for the lines of sight and a few frequently encountered applications.

A drawing employing the four uses listed in Table 24.1 is shown in Figure 24.2. Examples of information obtained in the auxiliary views are:

> View 1. True length of edge AB (use 1) and angle between AB and frontal plane (θ_F).
>   True size of surface ABCE (use 4).
> View 2. Point view of line AB (use 2).
>   Angle between surfaces ABCE and ABGK (use 3).

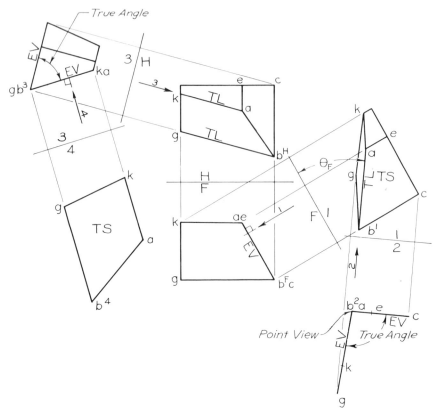

FIGURE 24.2   Applications of uses of additional views

View 3. Point views of lines AK and BG (use 2) and angle
        between surface ABGK and horizontal plane (use 3).
View 4. True size of surface ABGK (use 4).

24.3   Projective and Nonprojective Properties

Relationships between lines and planes that are evident in their
orthographic projections are called *projective properties.*

Parallelism. If two lines are parallel, they appear parallel in any view
(except in those views in which they coincide or appear as points).

If two lines are parallel in two views, the lines are parallel in space
(unless the views are parallel to the projection lines between them, in
which case the lines may or may not be parallel in space).

Perpendicularity. If two lines are perpendicular, they appear perpendicular in any view showing at least one of these lines in true length
(unless one of the lines shows as a point). In a view in which neither of
two perpendicular lines appears in true length, the lines will not appear
perpendicular in the view.

If a line is perpendicular to a plane, it is perpendicular to all lines in
the plane. Consequently, a line perpendicular to a plane appears perpendicular to any true-length view of a line in the plane. A line perpen-
picular to a plane also must appear perpendicular to any edge view of
the plane.

Points on Lines, Lines in Planes. If a point is on a line, the views of the
point must lie on the respective views of the line.

If two lines intersect, they have a point in common; and in any two
adjacent views, the two views of the point of intersection must lie on a
common projection line.

If a line is in a plane, it must either intersect or be parallel to any other
line in the plane, and these properties are retained as the lines are
projected from view to view.

Proportional Division. Views of a series of points along a line segment
divide the corresponding views of the line segment in the same propor-
tions as the points divide the line segment. Thus the midpoint of a line
appears as the midpoint of any view of the line.

On the other hand, the views of a series of lines dividing a plane angle
do *not* necessarily divide in the same proportion the corresponding views
of the angle except where the angle appears in true size. For example,
except in a true-size view, a view of the true bisector of an angle does *not*
necessarily bisect the corresponding view of the angle.

24.4 Example Problems

EXAMPLE 1: TRUE SIZE BY TRIANGULATION

Given

Triangle ABC represented by its front and top views, Figure 24.3(a).

Required

Construct the true size of the triangle without using auxiliary views.

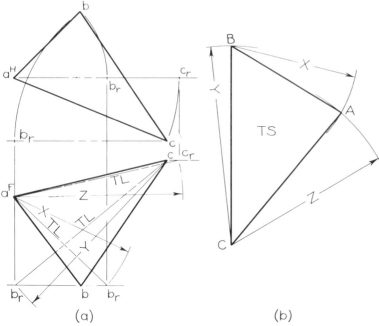

(a)   (b)

FIGURE 24.3   True size by triangulation

Analysis

From plane geometry any triangle can be reproduced if the lengths of the three sides are known.

Graphic Solution

Since no auxiliary views are permitted, the revolution method is used for finding the true lengths, Figure 24.3(a). At (b) one of the edges, say BC, is drawn to length. Intersecting arcs of radii equal to the true lengths of edges AB and AC are drawn with points B and C as respective centers to determine point A.

This method is used extensively in the development of surfaces by *triangulation* and by the *radial-line method*. It may also be used to reproduce polygons of more than three sides by dividing them into triangles by means of diagonals.

EXAMPLE 2: INTERSECTION AND DEVELOPMENT OF CYLINDERS

Given

Front and top views of two intersecting right circular cylinders, Figure 24.4.

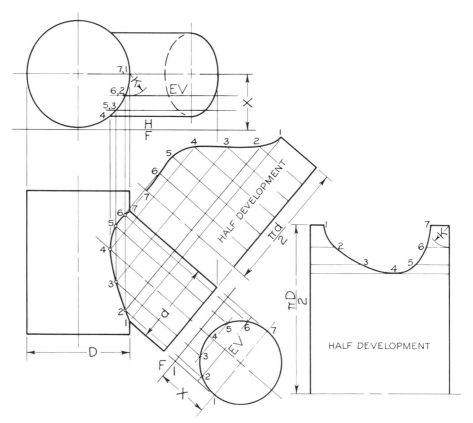

FIGURE 24.4   Intersection and development of cylinders

Required

Plot the line of intersection and develop the surfaces of the cylinders, showing the line of intersection.

Analysis

A series of frontal cutting planes will be parallel to the axes of both cylinders and will cut elements from their surfaces.

Graphic Solution

Partial auxiliary view 1 showing the inclined cylinder in its true circular form is added to facilitate the location of elements. The frontal cutting planes appear in this view as lines parallel to folding line F/1. They are drawn through equally spaced points 1 to 7 for convenience. Elements 1 to 7 are then located in the top view, in which their piercing points with the surface of the vertical cylinder are apparent. By projection, the front views of these piercing points are located and the front view of the intersection is completed.

Since the cylinders and their intersection are symmetrical about a center line, half-developments are sufficient. For accuracy in development the half-circumferences of the cylinders are calculated by $\pi D/2$ and $\pi d/2$. These distances are laid out to scale and are divided into parts corresponding to the divisions of the circular views. For the larger cylinder, the elements containing the intersection points are located by chordal distances such as K. While theoretically these distances are too short, the error is slight and does not accumulate to an objectionable extent when the number of successive divisions is small, as in this case. The endpoints and intersection points on the elements are now located in the developments as shown to complete the problem.

EXAMPLE 3: LOCATION OF REFLECTED LIGHT RAY

Given

Front and top views of a light ray GK and the plane ABCE of a mirror, Figure 24.5(a).

Required

Locate the reflected ray without using auxiliary views.

Analysis

The *image* of point G is an imaginary point located on the opposite side of the mirror from point G and the same distance from the mirror. The direction of the reflected ray is determined by the image of point G and the intersection or piercing point of ray GK and plane ABCE. See the diagram in Figure 24.5(b).

Graphic Solution

Line GG_1 is constructed perpendicular to plane ABCE by drawing its top view perpendicular to true-length view c^He and its front view perpendicular to true-length view c^Ff. (Line CF is a frontal line added to the plane for this purpose.) The piercing point P of line GG_1 and plane ABCE is then located by the two-view method. Length d of g^Fp is set off from p to locate g_1, thus making line segments GP and PG_1 equal. The top view g_1^H is located by projection, or it could be located by transfer of the distance between g^H and p. Point G_1 is thus the image of point G.

Piercing point Q of ray GK (extended) and plane ABCE is then located by the two-view method. The required reflected ray QS is the extension of the line G_1 to Q.

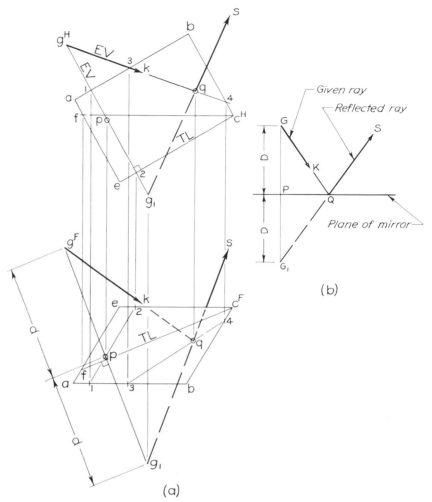

FIGURE 24.5   Location of reflected light ray

EXAMPLE 4. RESULTANT AND RESOLUTION OF VECTORS

Given

The front and top views of plane MNOP and of concurrent vectors AB and AC, point A being in plane MNOP. Figure 24.6.

Required

Find the resultant of the vectors and its magnitude. Locate the component (of the resultant) that acts normal to plane MNOP.

Analysis

The resultant may be found in the given views by application of the *parallelogram of forces*. Its magnitude is its true length measured at the given scale. If the given plane is shown in edge view, any vector acting perpendicular to the plane must appear perpendicular to this edge view and in true length. The resultant may then be resolved into a component

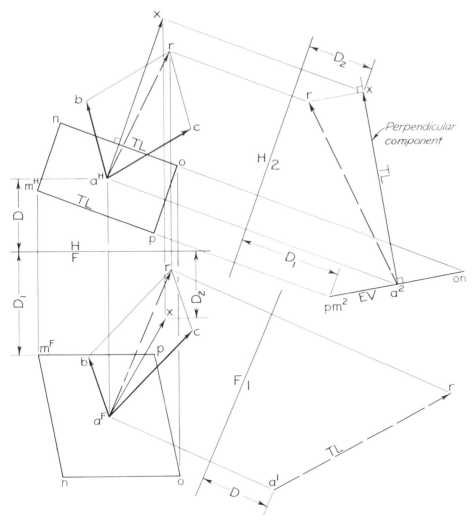

FIGURE 24.6 Resultant and resolution of vectors

along this perpendicular (plus a component parallel to the plane).

Graphic Solution

Lines are drawn through points B and C parallel respectively to vectors AC and AB. These intersect at point R and diagonal line AR is the required resultant. Its true length is found in auxiliary view 1.

Since lines MP and NO are in true length in the top view, it is convenient to obtain an edge view of plane MNOP by projection from the top view as shown. Resultant AR is also projected to this auxiliary view and is resolved as required into component a^2x acting perpendicular to plane MNOP. Since a^2x is true length, the top view a^Hx is parallel to folding line H/2, and x is located by projection from view 2. The front view of component AX is then established by projection from a^Hx and transfer of distance D_2 from the auxiliary view.

PROBLEMS

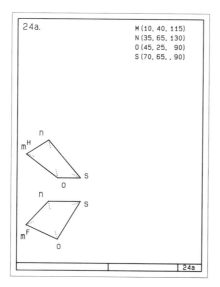

24a.

M (10, 40, 115)
N (35, 65, 130)
0 (45, 25, 90)
S (70, 65, , 90)

24a

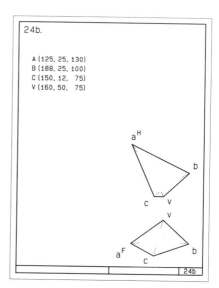

24b.

A (125, 25, 130)
B (188, 25, 100)
C (150, 12, 75)
V (160, 50, 75)

24b

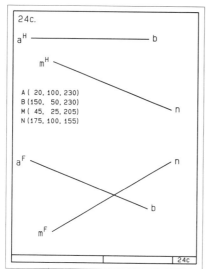

24c.

A (20, 100, 230)
B (150, 50, 230)
M (45, 25, 205)
N (175, 100, 155)

24c

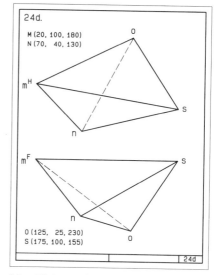

24d.

M (20, 100, 180)
N (70, 40, 130)

0 (125, 25, 230)
S (175, 100, 155)

24d

24a. Complete the visibility of the tetrahedron. Determine the true size of the angle formed by planes MNO and NOS. Show the true size of face MNO.

24b. Complete the visibility of the given pyramid. Find the true length and views of the altitude with point V as the apex. Provide a development of the four surfaces of the pyramid.

24c. Using only the given views, find the true length and the views of the shortest connector between the lines AB and MN.

24d. Locate the center of a sphere that passes through the four corners of the tetrahedron MNOS.

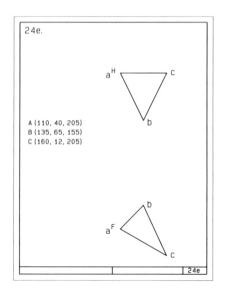

24e.

A (110, 40, 205)
B (135, 65, 155)
C (160, 12, 205)

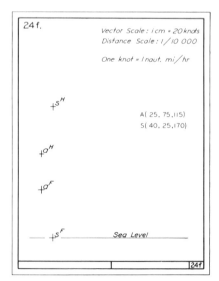

24f.

Vector Scale : 1cm = 20 knots
Distance Scale : 1/10 000

One knot = 1 naut. mi/hr

A(25, 75,115)
S(40, 25,170)

Sea Level

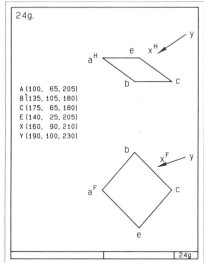

24g.

A (100, 65, 205)
B (135, 105, 180)
C (175, 65, 180)
E (140, 25, 205)
X (160, 90, 210)
Y (190, 100, 230)

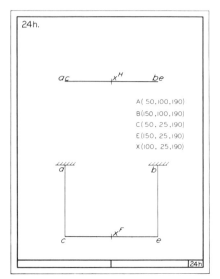

24h.

A(50,100,190)
B(150,100,190)
C(50, 25,190)
E(150, 25,190)
X(100, 25,190)

24e. Show the views of a cube, one face of which lies in plane ABC with line AB as a diagonal of this face of the cube.

24f. A ship at point S is bearing N70° at 32 knots. An aircraft at point A is bearing N40° at 100 knots and is diving at a rate of 360 m in 1000 m.|How close will the aircraft come to the ship?

24g. Determine the true angle that light ray XY makes with the mirror surface ABCE. Show the views of the reflected ray and the true size of the mirror.

24h. Using only the given views, locate the new positions of the cross bar CE and its supporting cables AC and BE if the bar is revolved 75° clockwise about a vertical axis through point X.

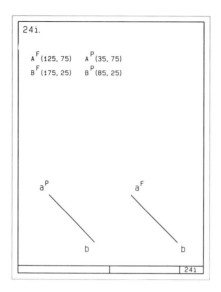

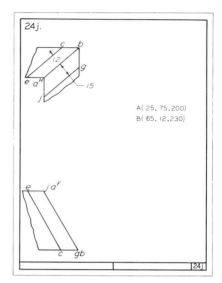

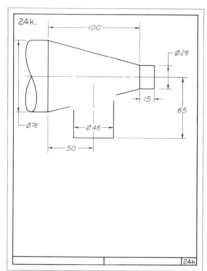

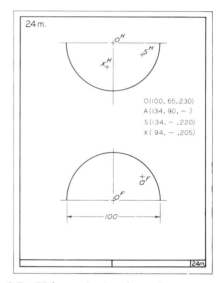

24i. Establish a plane that contains line AB and makes an angle of 65° with a profile plane. Determine the angle that this plane makes with a horizontal plane.

24j. Determine the bend angle for the reinforcing plate CEABJG. Show the development of this plate.

24k. Using only the given view, construct the intersection of the conical and cylindrical ducts (sphere method). Develop the surface of either or both as assigned, showing the intersection.

24m. Using only the given views, pass a plane tangent to the sphere at point A. Locate the gnomonic projections of points S and X on this tangent plane.

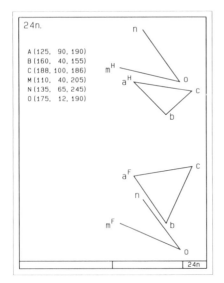

24n.

A (125, 90, 190)
B (160, 40, 155)
C (188, 100, 186)
M (110, 40, 205)
N (135, 65, 245)
O (175, 12, 190)

24n

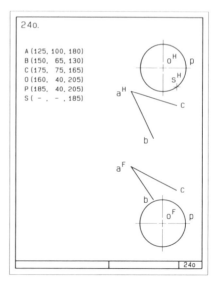

24o.

A (125, 100, 180)
B (150, 65, 130)
C (175, 75, 165)
O (160, 40, 205)
P (185, 40, 205)
S (– , – , 185)

24o

24n. Establish the views of the line in limited plane ABC that lies 65 mm from plane MON. Determine the bearing, grade, and true length of this line.

24o. (a) Determine the shortest distance from the surface of the sphere to plane ABC. (b) Point S lies on the surface of the sphere. Obtain the views of the orthographic projection of point S on the given plane.

24p. Prepare written statements and sketches (if assigned) explaining your answers to the following locus visualization questions:

(1) What is the locus of points at distance D_1 from a given plane and distance D_2 from a point?

(2) What is the locus of points equidistant from two parallel lines?

(3) What is the locus of points a fixed distance D from each of two skew lines?

(4) What is the locus of points equidistant from three points and lying in a given plane other than that of the three points?

(5) What is the locus of points equidistant from two points and from two intersecting lines?

(6) What is the locus of points 25 mm from a 75 mm diameter circle?

(7) What is the locus of points equidistant from two skew lines?

(8) What is the locus of points equidistant from four points no three of which lie in a straight line?

SELF-TESTING PROBLEMS

24A. Using only the given views, determine the true length and views of the shortest connector between the given lines. Add the views of a vertical connector.

24A.

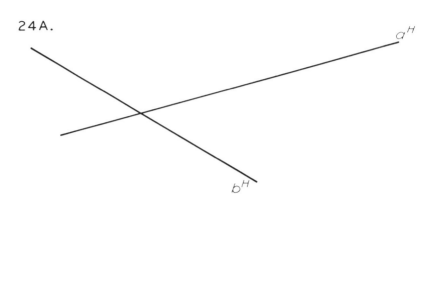

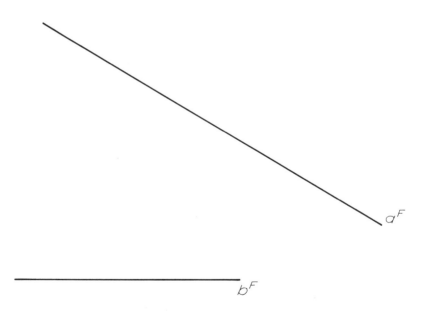

24B. In the isometric pictorial, establish the intersection of the cone and cylinder.

24 B.

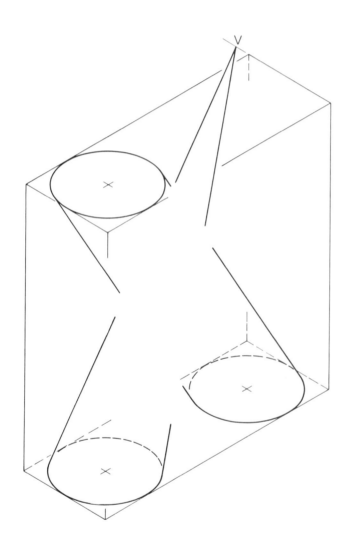

24C. and **24D.** Complete the perspective pictorial and obtain the shade and shadow using the light source at point S.

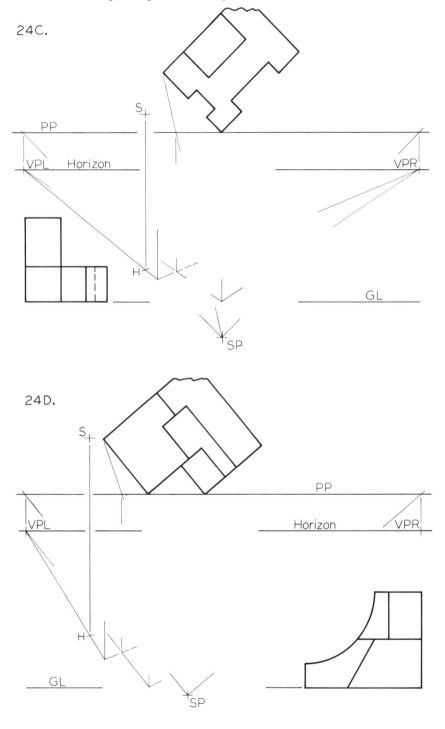

24C.

24D.

24E. Provide a line through point O parallel to each of the planes ABC and MNK.

24F. Obtain the intersection of AB and the conical surface. Then represent a plane tangent to the cone and parallel to AB.

24E.

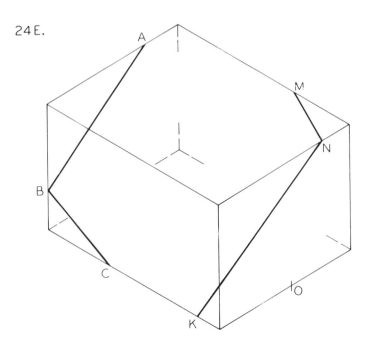

24F.

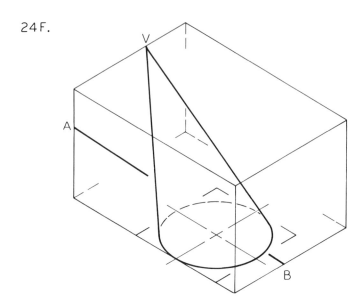

25 Computer Graphics Projects

The computer projects in this chapter have been designed to utilize FORTRAN–PLOT 10 subroutines acquired from the Tektronix Corporation and introduced into a Prime minicomputer. The graphics terminal most readily available at Washington State University is a Hewlett–Packard 2623 A. The digital plotter most frequently used is a Tektronix 4662, which has a plotting area of 11 × 17 inches. A FORTRAN compiler is available from Prime Computer and is based on ANSI X3.9–1978 standard FORTRAN.

Modification of these pre-prepared computer programs to accommodate other computer languages and equipment represents a preliminary obstacle that is inevitable until some uniformity of language and hardware is established. Such standardization is now being encouraged in Japanese industry.

However, of major import is the fact that the essential concepts of any computer program utilize the same geometric and mathematical analysis that is presented in these original programs.

376

THIS COMPUTER PROGRAM PROVIDES THE TL AND GRADE OF LINE AB. THE DESIRED TL IS THE HYPOTENUSE OF A RIGHT TRIANGLE THE RISE OF WHICH IS EQUAL TO THE HEIGHT Y. THE RUN FROM THE TOP VIEW IS THE HYPOTENUSE OF THE RIGHT TRIANGLE INVOLVING THE LEGS X AND Z. AN ARRAY OF VALUES IS SUGGESTED FOR COORDINATES X, Y, Z SO THAT A VARIETY OF SOLUTIONS MAY BE PRODUCED.

ARRAY OF VALUES FOR X, Y, Z IN CM

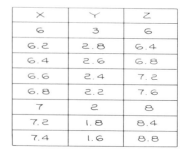

X	Y	Z
6	3	6
6.2	2.8	6.4
6.4	2.6	6.8
6.6	2.4	7.2
6.8	2.2	7.6
7	2	8
7.2	1.8	8.4
7.4	1.6	8.8

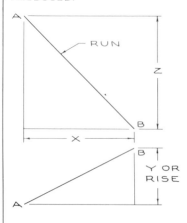

NOTE THAT ALL THESE TEXTBOOK COMPUTER GRAPHICS DRAWINGS ARE PROVIDED AT HALF SIZE.

GENERAL COMMENTS	RELATED COMPUTER INPUT
ACTIVATE COMPUTER, TYPE LOGIN	LOGIN
USER-ID, TYPE ME 102	USER ID
PASSWORD, TYPE ME 102	PASSWORD
AFTER OK, APPEARS, TYPE LD.	OK,
FROM PROJECT LIST, SELECT AND TYPE THE NAME, TLLINA	TLLINA
VARIABLES OUT (1) THRU OUT(10) STORES DATA UP TO 80 CHARACTERS EACH.	CHARACTER *80 OUT (10)
STORE ONE-LETTER RESPONSE. FOR EXAMPLE: Y FOR YES.	CHARACTER *1 ANS
STORE WHOLE NUMBER VARIABLES.	INTEGER *2 ID, I
FLOATING-POINT MODE VALUES.	REAL *4 X, Y, Z, RISE, RUN, TLAB, & GRADE
LOCATION OF DATA : ID = 0 FOR THE TERMINAL, ID = 22 FOR PLOTTER.	ID = 0
BRIEF PROGRAM EXPLANATION.	PRINT *, 'PROGRAM TLLINA'
	PRINT *, 'USER PROVIDES X, Y, Z'
	PRINT *, 'THE PROGRAM DISPLAYS'
	PRINT *, '1) TL OF LINE AB'
	PRINT *, '2) THE GRADE OF AB'
	PRINT *, 'PLOT OF THE TL'
REFERENCE LINE	100 CONTINUE
TYPE SELECTED X, Y, Z VALUES. FOR EXAMPLE : TYPE 6., 3., 6.	PRINT *, 'ENTER X, Y, Z VALUES'
STORE THESE VALUES, FOR AN ERROR, RETURN TO LINE 100.	READ(I, *, ERROR=100) X, Y, Z
STORE Y IN THE VARIABLE, RISE.	RISE = Y
RUN = $\sqrt{X^2 + Z^2}$	RUN = SQRT(X**2 + Z**2)
TLAB = $\sqrt{RISE^2 + RUN^2}$	TLAB = SQRT(RISE**2 + RUN**2)
GRADE = $\dfrac{RISE}{RUN}$ (100)	GRADE = RISE/RUN *100
PUT TERMINAL IN PLOT 10 MODE.	CALL HPPTON

COMPUTER GRAPHICS   TLLINA −1

```
REFERENCE LINE                          200 CONTINUE
ENTIRE PLOT AREA.                           CALL SWDV$W(ID,1024,0,1023,0,779)
WINDOW WITH ORIGIN AT (2,4).                CALL DWINDO(-2.,31.274,-4.,21.4)
MOVE ELEVATED PEN TO POINT A (0,0).         CALL MOVEA(0.,0.)
DRAW TL LINE FROM A TO B.                   CALL DRAWA(RUN,RISE)
MOVE PEN TO LETTERING LOCATION.             CALL MOVEA(2.,20.)
                                            CALL SEELOC(IX,IY)
SET MARGIN FOR LETTERING.                   CALL SETMRG(IX,KCM(33.274))
INTERNAL FILE-OUT DATA.                     WRITE(OUT,1000)X,Y,Z,TLAB,GRADE
THESE LINES PROVIDE FORMAT              1000 FORMAT('INPUT'/'X =',F5.2/'Y=',F5.2/
INSTRUCTIONS FOR LETTERING.                 & ',Z =',F5.2/'ANSWERS'/'TLAB=',F5.2,'CM'/
                                            & 'GRADE =',F5.2,'PERCENT')
SIZE 3 LETTERING.                           IF(ID.LT.0) CALL CHRSIZ(3)
DO LOOP FOR PLOT.                           DO 1010 I=1,7
                                            CALL AOUTST(20,OUT(I))
MOVE PEN TO NEXT LINE.                      CALL NEWLIN
                                       1010 CONTINUE
ALERT FOR PRINTING.                         CALL ANMODE
                                            CALL HPGTOF
CLEAR SCREEN WITH 23 BLANK LINES.           DO 1015 I=1,23
                                            PRINT *,' '
                                       1015 CONTINUE
ALPHA DISPLAY ON TERMINAL.                  CALL HPADON
PROVIDE YES OR NO RESPONSE.                 PRINT *,'PLOT ON PLOTTER?(YES OR NO)'
READ AND STORE ONE-LETTER ANS.              READ(1,1020) ANS
IF ANS. IS 'Y', STATEMENTS FROM 'IF'   1020 FORMAT(A1)
TO 'ELSE' ARE EXECUTED. FOR 'N' THE         IF(ANS.EQ.'Y') THEN ID=22
STATEMENTS FROM 'ELSE' TO 'END IF'
ARE PERFORMED.                         1050 CONTINUE
RESPONSE NEEDED.                            PRINT *,'IS PLOTTER READY TO PLOT?'
                                            PRINT *,'NEW PAPER,PEN UNCAPPED'
                                            READ(1,1020) ANS
IF THE PLOTTER IS PREPARED, RETURN          IF(ANS.EQ.'N') RETURN TO 1050
TO REFERENCE LINE 200.                      GO TO 200
                                            ELSE
                                            ID=0
RELEASE THE PLOTTER.                        CALL CHLINE(0)
                                            END IF
RESPONSE NEEDED.                            PRINT *,'ENTER NEW CASE?(YES OR NO)'
FOR YES, RETURN TO LINE 100.                READ(1,1020) ANS
                                            IF(ANS.EQ.'Y') GO TO 100
TERMINATE GRAPHICS DISPLAY.                 CALL HPGDOF
                                            PRINT *,'CAP PLOTTER PEN'
                                            CALL EXIT
                                            END
```

FOR A MORE EXTENSIVE COVERAGE OF COMPUTER
GRAPHICS PROJECTS, REFER TO THE MACMILLAN
PUBLICATION: COMPUTER GRAPHICS PROJECTS FOR
DESIGN AND DESCRIPTIVE GEOMETRY BY E. PARÉ
AND M. SHOOK.

COMPUTER GRAPHICS   TLLINA -2

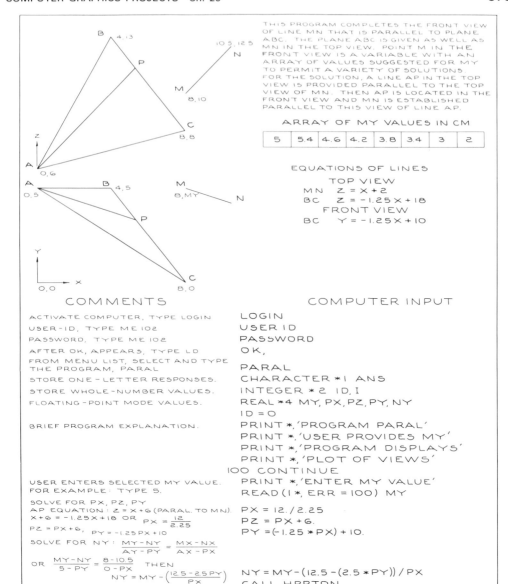

THIS PROGRAM COMPLETES THE FRONT VIEW
OF LINE MN THAT IS PARALLEL TO PLANE
ABC. THE PLANE ABC IS GIVEN AS WELL AS
MN IN THE TOP VIEW. POINT M IN THE
FRONT VIEW IS A VARIABLE WITH AN
ARRAY OF VALUES SUGGESTED FOR MY
TO PERMIT A VARIETY OF SOLUTIONS.
FOR THE SOLUTION, A LINE AP IN THE TOP
VIEW IS PROVIDED PARALLEL TO THE TOP
VIEW OF MN. THEN AP IS LOCATED IN THE
FRONT VIEW AND MN IS ESTABLISHED
PARALLEL TO THIS VIEW OF LINE AP.

ARRAY OF MY VALUES IN CM

5	5.4	4.6	4.2	3.8	3.4	3	2

EQUATIONS OF LINES
TOP VIEW
MN $Z = X + 2$
BC $Z = -1.25X + 18$
FRONT VIEW
BC $Y = -1.25X + 10$

COMMENTS

ACTIVATE COMPUTER, TYPE LOGIN
USER-ID, TYPE ME 102
PASSWORD, TYPE ME 102
AFTER OK, APPEARS, TYPE LD
FROM MENU LIST, SELECT AND TYPE
THE PROGRAM, PARAL
STORE ONE-LETTER RESPONSES.
STORE WHOLE-NUMBER VALUES.
FLOATING-POINT MODE VALUES.

BRIEF PROGRAM EXPLANATION.

USER ENTERS SELECTED MY VALUE.
FOR EXAMPLE: TYPE 5.
SOLVE FOR PX, PZ, PY
AP EQUATION: $Z = X + 6$ (PARAL. TO MN).
$X + 6 = -1.25X + 18$ OR $PX = \dfrac{12}{2.25}$
$PZ = PX + 6$, $PY = -1.25 PX + 10$
SOLVE FOR NY: $\dfrac{MY - NY}{AY - PY} = \dfrac{MX - NX}{AX - PX}$
OR $\dfrac{MY - NY}{5 - PY} = \dfrac{8 - 10.5}{0 - PX}$ THEN
$NY = MY - \left(\dfrac{12.5 - 2.5 PY}{PX}\right)$

WINDOW WITH ORIGIN AT (2,4).
MOVE PEN TO POINT A (TOP VIEW).
DRAW LINE FROM A TO B.
DRAW LINE FROM B TO C.

COMPUTER INPUT

LOGIN
USER ID
PASSWORD
OK,
PARAL
CHARACTER *1 ANS
INTEGER *2 ID, I
REAL *4 MY, PX, PZ, PY, NY
ID = 0
PRINT *, 'PROGRAM PARAL'
PRINT *, 'USER PROVIDES MY'
PRINT *, 'PROGRAM DISPLAYS'
PRINT *, 'PLOT OF VIEWS'
100 CONTINUE
PRINT *, 'ENTER MY VALUE'
READ (I*, ERR = 100) MY

PX = 12. / 2.25
PZ = PX + 6.
PY = (-1.25 * PX) + 10.

NY = MY - (12.5 - (2.5 * PY)) / PX
CALL HPPTON
200 CONTINUE
CALL SWDV $W (ID, 1024, 0, 1023, 0, 779)
CALL DWINDO (-2., 31.274, -4., 21.4)
CALL MOVEA (0., 6.)
CALL DRAWA (4., 13)
CALL DRAWA (8., 8.)

COMPUTER GRAPHICS   PARAL - 1

DRAW LINE FROM C TO A.	CALL DRAWA (O.,6.)
DRAW LINE FROM A TO P.	CALL DRAWA (PX, PZ)
MOVE PEN TO POINT M.	CALL MOVEA (8., IO.)
DRAW LINE FROM M TO N.	CALL DRAWA (IO.5, I2.5)
MOVE PEN TO POINT A (FRONT).	CALL MOVEA (O., 5.)
DRAW LINE FROM A TO B.	CALL DRAWA (4., 5.)
DRAW LINE FROM B TO C.	CALL DRAWA (8., O.)
DRAW LINE FROM C TO A.	CALL DRAWA (O., 5.)
DRAW LINE FROM A TO P.	CALL DRAWA (PX, PY)
MOVE PEN TO POINT M.	CALL MOVEA (8., MY)
DRAW LINE FROM M TO N.	CALL DRAWA (IO.5, NY)
	CALL ANMODE
	CALL HPGTOF
CLEAR SCREEN WITH 23 BLANK LINES.	DO IOI5 I = I,23
	PRINT *, ' '
	IOI5 CONTINUE
	CALL HPADON
PROVIDE YES OR NO RESPONSE.	PRINT *,'PLOT ON PLOTTER?(YES OR NO)'
	READ (I,IO2O) ANS
	IO2O FORMAT (AI)
FOR YES, ID = 22, PLOTTER USE IS REQUESTED.	IF (ANS. EQ.'Y') THEN
	ID = 22
	IO5O CONTINUE
RESPONSE NEEDED.	PRINT *,'IS PLOTTER PREPARED?'
	PRINT *,'NEW PAPER, PEN UNCAPPED'
	READ (I,IO2O) ANS
FOR YES, RETURN TO LINE 200 FOR PLOTTER DRAWING.	IF (ANS. EQ. 'N') GO TO IO5O
	GO TO 200
IF PLOTTER IS NOT REQUESTED, COME TO ELSE.	ELSE
	ID = O
	CALL CHLINE (O)
	END IF
RESPONSE NEEDED.	PRINT *,'ENTER NEW CASE?(YES OR NO)'
	READ (I,IO2O) ANS
	IF (ANS. EQ. 'Y') GO TO IOO
	CALL HPGDOF
PREPARE A PROGRAM THAT COMPLETES THE SIDE VIEW OF PLANE ABC THAT IS PARALLEL TO THE GIVEN LINE MN. PROVIDE ARRAY OF VALUES FOR LOCATION NZ SO THAT A DESIRABLE VARIETY OF SOLUTIONS IS AVAILABLE.	PRINT *, 'CAP PLOTTER PEN'
	CALL EXIT
	END

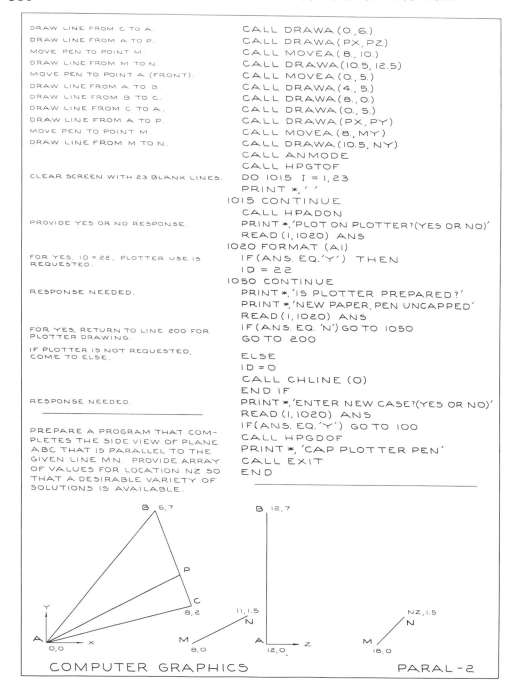

B 6,7

P

C
8,2

Y

A
0,0

x

M
8,0

B 12,7

11,1.5
N

A
12,0

z

NZ,1.5
N

M
18,0

COMPUTER GRAPHICS   PARAL-2

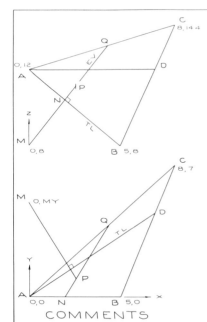

THIS PROGRAM SECURES THE LINE MP FROM POINT M
THAT IS PERPENDICULAR TO AND INTERSECTS PLANE
ABC AT THE FOOT OF THE PERPENDICULAR, POINT P.
IN BOTH VIEWS THE PERPENDICULAR IS INTRODUCED
NORMAL TO A TRUE LENGTH LINE OF THE PLANE
IN EACH VIEW RESPECTIVELY. THESE TL LINES ARE
AB IN THE TOP VIEW AND AD WHICH HAS BEEN
ADDED TO THE FRONT VIEW. THEN INTERSECTION
POINT P IS OBTAINED USING AN EV CUTTING
PLANE THRU MQ IN THE TOP VIEW.

AN ARRAY OF VALUES IS SUGGESTED FOR MY SO
THAT A VARIETY OF SOLUTIONS IS AVAILABLE.

ARRAY OF MY VALUES IN CM

5	5.2	5.4	5.6	5.8
6	6.25	6.5	6.75	7

EQUATIONS OF PERTINENT LINES

TOP VIEW	FRONT VIEW
AB $Z = -.8X + 12$	AB $Y = 0$
AC $Z = .3X + 12$	AC $Y = .875X$
BC $Z = 2.167X - 2.83$	BC $Y = 2.333X - 11.665$

MQ EQUATION, NORMAL TO LINE AB (TOP VIEW):
$Z = 1.25X + 8$ THEN SOLVE FOR NX, NY:
$1.25X + 8 = -.8X + 12$ OR $NX = 1.951$ AND $NY = 0$
SOLVE FOR QX, QY: $1.25X + 8 = .3X + 12$ OR
$QX = 4.21$ AND $QY = .875QX$ OR $QY = 3.684$

NQ EQUATION (FRONT) $Y = 1.631X - 3.182$

COMMENTS	COMPUTER INPUT
ACTIVATE COMPUTER, TYPE LOGIN	LOGIN
USER-ID, TYPE ME102	USER ID
PASSWORD, TYPE ME102	PASSWORD
AFTER OK, APPEARS, TYPE LD.	OK,
FROM MENU LIST, SELECT AND TYPE THE PROGRAM, PERPL	PERPL
STORE ONE-LETTER RESPONSES.	CHARACTER *1 ANS
STORE WHOLE-NUMBER VALUES.	INTEGER *2 ID, I
FLOATING-POINT MODE VALUES.	REAL *4 MY, DX, DY, PX, PY, PZ
	ID = 0
BRIEF PROGRAM EXPLANATION.	PRINT *, 'PROGRAM PERPL'
	PRINT *, 'USER PROVIDES MY'
	PRINT *, 'PROGRAM DISPLAYS'
	PRINT *, 'PLOT OF VIEWS'
USER ENTERS SELECTED MY VALUE. FOR EXAMPLE: TYPE 5.	100 CONTINUE
	PRINT *, 'USER ENTERS MY'
SOLVE FOR DX, DY : $12 = 2.167X - 2.83$	READ (1 *, ERR = 100) MY
OR DX = 6.844, $DY = 2.333(6.844) - 11.665$	DX = 6.844
THEN DY = 4.302 AD EQUATION :	DY = 4.302

SOLVE FOR DX, DY : $12 = 2.167X - 2.83$
OR DX = 6.844, $DY = 2.333(6.844) - 11.665$
THEN DY = 4.302   AD EQUATION :
$\dfrac{Y - 0}{X - 0} = \dfrac{0 - 4.302}{0 - 6.844}$ OR $Y = .629X$

MP EQUATION (NORMAL TO LINE AD):
$Y = -1.59X + MY$ SOLVE FOR PX, PY, PZ
$3.221 PX = 3.182 + MY$, $PX = \dfrac{3.182 + MY}{3.221}$,
$PY = 1.631 PX - 3.182$,
$PZ = 1.25 PX + 8$

PX = (3.182 + MY) / 3.221
PY = (1.631 * PX) - 3.182
PZ = (1.25 * PX) + 8.
CALL HPPTON
200 CONTINUE
CALL SWDV$W(ID, 1024, 0, 1023, 0, 779)

WINDOW WITH ORIGIN AT (2,4).
CALL DWINDO(-2., 31.274, -4., 21.4)

COMPUTER GRAPHICS   PERPL - 1

MOVE PEN TO POINT A (TOP VIEW).	CALL MOVEA (0., 12.)
DRAW LINE FROM A TO B.	CALL DRAWA (5., 8.)
DRAW LINE FROM B TO C.	CALL DRAWA (8., 14.5)
DRAW LINE FROM C TO A.	CALL DRAWA (0., 12.)
DRAW LINE FROM A TO D.	CALL DRAWA (DX, 12.)
MOVE PEN TO POINT M.	CALL MOVEA (0., 8.)
DRAW LINE FROM M TO P.	CALL DRAWA (PX, PZ)
MOVE PEN TO POINT A (FRONT).	CALL MOVEA (0., 0.)
DRAW LINE FROM A TO B.	CALL DRAWA (5., 0.)
DRAW LINE FROM B TO C.	CALL DRAWA (8., 7.)
DRAW LINE FROM C TO A.	CALL DRAWA (0., 0.)
DRAW LINE FROM A TO D.	CALL DRAWA (DX, DY)
MOVE PEN TO POINT M.	CALL MOVEA (0., MY)
DRAW LINE FROM M TO P.	CALL DRAWA (PX, PY)
	CALL ANMODE
	CALL HPGTOF
CLEAR SCREEN WITH 23 BLANK LINES.	DO 1015 I = 1, 23
	PRINT *, ' '
END OF THIS DO LOOP.	1015 CONTINUE
	CALL HPADON
PROVIDE YES OR NO RESPONSE.	PRINT *,'PLOT ON PLOTTER?(YES OR NO)'
	READ (1, 1020) ANS
	1020 FORMAT (A1)
FOR YES, ID = 22, PLOTTER USE IS REQUESTED.	IF (ANS. EQ. 'Y') THEN
	ID = 22
	1050 CONTINUE
RESPONSE NEEDED.	PRINT *, 'IS PLOTTER PREPARED?'
	PRINT *, 'NEW PAPER, PEN UNCAPPED'
	READ (1, 1020) ANS
FOR YES, RETURN TO THE LINE 200 FOR DRAWING ON THE PLOTTER.	IF (ANS. EQ. 'N') GO TO 1050
	GO TO 200
IF PLOTTER IS NOT REQUESTED, COME TO ELSE.	ELSE
	CALL CHLINE (0)
	END IF
RESPONSE NEEDED.	PRINT *,'ENTER NEW CASE?(YES OR NO)'
	READ (1, 1020) ANS
	IF (ANS. EQ. 'Y') GO TO 100
	CALL HPGDOF
	PRINT *, 'CAP PLOTTER PEN'
	CALL EXIT
	END

SUPPLEMENTARY PROJECT

EXPAND THIS COMPUTER PROGRAM TO OBTAIN
AND RECORD THE TRUE LENGTH OF PERPENDICULAR,
LINE MP.

COMPUTER GRAPHICS   PERPL - 2

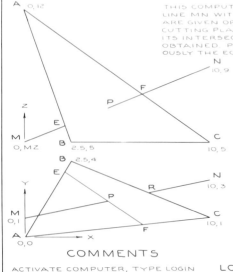

THIS COMPUTER PROGRAM PROVIDES THE INTERSECTION OF LINE MN WITH PLANE ABC. THE EQUATIONS OF THE LINES ARE GIVEN OR ESTABLISHED AS NEEDED. AN EDGE VIEW CUTTING PLANE THRU MN IN THE TOP VIEW IS EMPLOYED. ITS INTERSECTION, LINE EF, WITH PLANE ABC IS THEN OBTAINED. POINT E IS SECURED BY SOLVING SIMUTANE- OUSLY THE EQUATIONS FOR LINES MN AND AB. POINT F IS FOUND SIMILARLY USING THE EQUATIONS FOR MN AND AC. LINE EF IN THE FRONT VIEW IS OBTAINED BY INTRODUCING THE X COORDINATES OF E AND F INTO THE FRONT VIEW EQUATIONS OF LINES AB AND AC RESPECTIVELY. THE INTERSECTION OF LINES EF AND MN LOCATES THE DESIRED PIERCING POINT P OF MN AND PLANE ABC. AN ARRAY OF MZ VALUES IS SUGGESTED TO PERMIT A VARIETY OF SOLUTIONS.

ARRAY OF MZ VALUES IN CM

5	5.2	5.4	5.6	5.8	6	6.2	6.4
6.6	6.8	7	7.2	7.4	7.6	7.8	8

EQUATIONS OF LINES

FRONT VIEW	TOP VIEW
AB $Y = 1.6 X$	AB $Z = -2.8 X + 12$
AC $Y = .1 X$	AC $Z = -.7 X + 12$
BC $Y = -4X + 5$	BC $Z = 5$
MN $Y = .2 X + 1$	MN $Z = (9 - .1 MZ) X + MZ$

COMMENTS	COMPUTER INPUT
ACTIVATE COMPUTER, TYPE LOGIN	LOGIN
USER-ID, TYPE ME 102	USER ID
PASSWORD, TYPE ME 102	PASSWORD
AFTER OK, APPEARS, TYPE LD	OK,
FROM LISTED MENU, SELECT AND TYPE PROJECT, INLAP	INLAP
ONE-LETTER RESPONSE.	CHARACTER *.1 ANS
STORE WHOLE NUMBER VARIABLES.	INTEGER * 2 ID, I
FLOATING-POINT MODE VALUES.	REAL *.4 MZ, EX, EY, EZ, FX, FY, FZ, PX, & PY, PZ, RX, RY
	ID = O
BRIEF PROGRAM EXPLANATION.	PRINT *, 'PROGRAM INLAP'
	PRINT *, 'USER PROVIDES MZ'
	PRINT *, 'THE PROGRAM DISPLAYS'
	PRINT *, 'TWO-VIEW PLOT'
REFERENCE LINE.	100 CONTINUE
ENTER SELECTED MZ VALUE. FOR EXAMPLE: TYPE 5.	PRINT *, 'ENTER MZ VALUE'
SOLVE FOR E COORDINATES:	READ(1, *, ERROR = 100) MZ
$-2.8X + 12 = (.9 - .1 MZ) X + MZ$ THEN	EX = (12. - MZ)/(3.7 - .1 * MZ)
$EX = \frac{12 - MZ}{3.7 - .1 MZ}$, $EZ = \frac{2.8 MZ - 33.6}{3.7 - .1 MZ} + 12$	EZ = (2.8 * MZ - 33.6)/(3.7 - .1 * MZ) + 12.
$EY = 1.6 EX$ SOLVE FOR F COORDIN.	EY = 1.6 * EX
$-.7X + 12 = (.9 - .1 MZ) X + MZ$ OR	FX = (12. - MZ)/(1.6 - .1 * MZ)
$FX = \frac{12 - MZ}{1.6 - .1 MZ}$, $FZ = -.7 FX + 12$,	FZ = .7 * FX + 12.
$FY = .1 FX$	FY = .1 * FX
EF EQUATION SLOPE $= (EY - FY)/(EX - FX)$	S = (EY - FY)/(EX - FX)
THEN $Y = 5(X - FX) + FY$	
SOLVE FOR PX USING EQUATION FOR MN & EF.	PX = (1 - FY + FX * S)/(S - .2)
$PX = \frac{1 - FY + FX(5)}{5 - .2}$	
$PY = .2 PX + 1$, $PZ = (9 - .1 MZ) PX + MZ$	PY = .2 * PX + 1.
SOLVE FOR RX, RY: $2X + 1 = 4X + 5$ THEN	PZ = (.9 - (.1 * MZ)) * PX + MZ
$RX = 6.667$ AND $RY = 2RX + 1$, $RY = 2.333$	RX = 6.667
	RY = 2.333
TERMINAL IN PLOT 10 MODE.	CALL HPPTON
	200 CONTINUE
ENTIRE PLOT AREA.	CALL SWDV$W(ID, 1024, 0, 1023, 0, 779)

COMPUTER GRAPHICS   INLAP - 1

WINDOW WITH ORIGIN AT (2,4).	CALL DWINDO(-2.31,274,-4.,21.4)
MOVE PEN TO POINT A (TOP VIEW).	CALL MOVEA(0.,12.)
DRAW LINE FROM A TO B.	CALL DRAWA(2.5,5.)
DRAW LINE FROM B TO C.	CALL DRAWA(10.,5.)
DRAW LINE FROM C TO A.	CALL DRAWA(0.,12.)
MOVE PEN TO POINT M.	CALL MOVEA(0.,MZ)
DRAW LINE FROM M TO E.	CALL DRAWA(EX,EZ)
MOVE PEN TO POINT P.	CALL MOVEA(PX,PZ)
DRAW LINE FROM P TO N.	CALL DRAWA(10.,9.)
MOVE PEN TO POINT A (FRONT).	CALL MOVEA(0.,0.)
DRAW LINE FROM A TO B.	CALL DRAWA(2.5,4.)
DRAW LINE FROM B TO C.	CALL DRAWA(10.,1.)
DRAW LINE FROM C TO A.	CALL DRAWA(0.,0.)
MOVE PEN TO POINT M.	CALL MOVEA(0.,1.)
DRAW LINE FROM M TO P.	CALL DRAWA(PX,PY)
MOVE PEN TO POINT R.	CALL MOVEA(RX,RY)
DRAW LINE FROM R TO N.	CALL DRAWA(10.,3.)

```
                              1000 CONTINUE
                                   CALL ANMODE
                                   CALL HPGTOF
CLEAR THE SCREEN.                  DO 1015 I = 1, 23
                                   PRINT *,' '
                              1015 CONTINUE
                                   CALL HPADON
PROVIDE YES OR NO RESPONSE.        PRINT *,'PLOT ON PLOTTER?(YES OR NO)'
FOR YES, STATEMENTS FROM IF TO     PRINT *,'NEW PAPER, PEN UNCAPPED'
ELSE ARE EXECUTED.                 READ(1,1020) ANS
FOR NO, STATEMENTS FROM ELSE  1020 FORMAT (A1)
TO END IF ARE PERFORMED.           IF(ANS. EQ. 'Y') THEN ID = 22
                              1050 CONTINUE
PREPARE PLOTTER, PROVIDE YES       PRINT *,'IS PLOTTER READY?'
OR NO RESPONSE.                    PRINT *,'NEW PAPER, PEN UNCAPPED'
                                   READ(1,1020) ANS
IF PLOTTER IS PREPARED, RETURN     IF(ANS. EQ. 'N') GO TO 1050
TO REFERENCE LINE 200.             GO TO 200
                                   ELSE
                                   ID = 0
                                   CALL CHLINE(0)
                                   END IF
RESPONSE NEEDED.                   PRINT *,'ENTER NEW CASE?(YES OR NO)'
                                   READ(1,1020) ANS
FOR YES, RETURN TO LINE 100.       IF(ANS. EQ. 'Y') GO TO 100
TERMINATE GRAPHICS DISPLAY.        CALL HPGDOF
                                   PRINT *,'PLEASE CAP PLOTTER PEN'
                                   CALL EXIT
                                   END
```

ALTER THE FRONT VIEW OF THE LINE MN
AND CHANGE THE COMPUTER PROGRAM TO
COMPLY WITH THAT ALTERATION.

COMPUTER GRAPHICS   INLAP-2

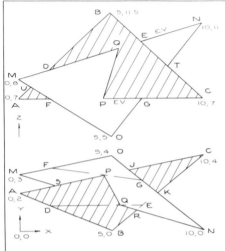

THIS PROGRAM PRODUCES THE INTERSECTION
OF THE TWO PLANES ABC AND MNO. NOTE THE
USE OF EV CUTTING PLANES IN THE TOP VIEW
THRU MN AND AC TO LOCATE THE CRITICAL
PIERCING POINTS Q AND P. OBSERVE THE
DESIRABLE ADDITION OF MECHANICAL SHAD-
ING TO ENHANCE THE VISIBILITY.

EQUATIONS OF LINES

TOP VIEW		FRONT VIEW	
AB	$Z = 9X + 7$	AB	$Y = -4X + 2$
BC	$Z = -9X + 16$	BC	$Y = 8X - 4$
AC	$Z = 7$	AC	$Y = 2X + 2$
MN	$Z = .3X + 8$	MN	$Y = -.3X + 3$
NO	$Z = 1.2X - 1$	NO	$Y = -.8X + 8$
MO	$Z = -.6X + 8$	MO	$Y = .2X + 3$

SOLVE FOR DX, DZ, DY:
$9X + 7 = .3X + 8$, $DX = .1667$, $DZ = .3DX + 8$, $DZ = 8.5$
$DY = -4DX + 2$, $DY = 1.333$

SOLVE FOR EX, EZ, EY:
$-.9X + 16 = .3X + 8$, $EX = 6.667$, $EZ = .3EX + 8$, $EZ = 10$
$EY = .8EX - 4$, $EY = 1.333$

SOLVE FOR FX, FZ, FY:
$7 = -.6X + 8$, $FX = 1.667$, $FZ = 7$, $FY = 2FX + 3$, $FY = 3.333$

SOLVE FOR GX, GZ, GY:
$7 = 1.2X - 1$, $GX = 6.667$, $GZ = 7$, $GY = -.8GX + 8$, $GY = 2.667$

DE EQUATION (FRONT VIEW): $Y = 1.333$

FG EQUATION: $Y = -.1333X + 3.555$

COMMENTS	COMPUTER INPUT

FROM MENU LIST, SELECT AND TYPE
THE PROGRAM, INPLA | INPLA
STORE ONE-LETTER RESPONSES. | CHARACTER *1 ANS
STORE WHOLE-NUMBER VALUES. | INTEGER *2 ID, I

FLOATING-POINT MODE VALUES. | REAL *4 DX, DZ, DY, EX, EZ, EY, FX, FZ, FY,
&GX, GZ, GY, PX, PY, PZ, QX, QY, QZ, SX, SY,
&JX, JY, KX, KY, RX, RY, UX, UZ, TX, TZ
| ID = 0

BRIEF PROGRAM EXPLANATION. | PRINT *, 'PROGRAM INPLA'
PRINT *, 'USER PROVIDES SCALE'
PRINT *, 'PROGRAM DISPLAYS'
PRINT *, 'PLOT OF VIEWS'
| 100 CONTINUE

USER ENTERS A SELECTED SCALE
FACTOR (SF). FOR EXAMPLE: TYPE
1., 1.2, 1.4, 1.6, 1.8, 2., 8, 6, 5, OR 4 | PRINT *, 'ENTER SCALE FACTOR'
READ(I*, ERR = 100) SF

SOLVE FOR QX, QY, QZ: $1.333 = .3X + 3$ OR | QX = 1.667/.3
$QX = \frac{1.667}{.3}$, $QY = 1.333$, $QZ = .3QX + 8$ | QY = 1.333
| QZ = (.3 * QX) + 8.

SOLVE FOR PX, PY, PZ: | PX = 1.555/.3333
$.2X + 2 = -.1333X + 3.555$, $PX = \frac{1.555}{.3333}$ | PY = (.2 * PX) + 2.
$PY = .2PX + 2$, $PZ = 7$ | PZ = 7.

SOLVE FOR SX, SY: | SX = 1./.5
$.2X + 2 = -.3X + 3$ OR $SX = \frac{1}{.5}$ | SY = (.2 * SX) + 2.
$SY = .2SX + 2$

SOLVE FOR JX, JY: | JX = 6.
$.2X + 2 = -.8X + 8$ OR $JX = 6$ | JY = (.2 * JX) + 2.
$JY = .2JX + 2$

SOLVE FOR KX, KY: $.8X - 4 = -.8X + 8$ | KX = 12./1.6
OR $KX = \frac{12}{1.6}$, $KY = .8KX - 4$ | KY = (.8 * KX) - 4.

SOLVE FOR RX, RY: $-.3X + 3 = .8X - 4$ | RX = 7./1.1
$RX = \frac{7}{1.1}$, $RY = .8RX - 4$ | RY = (.8 * RX) - 4.

COMPUTER GRAPHICS   INPLA-1

```
SOLVE FOR UX,UZ  9X+7=-6X+8  OR        UX = 1./1.5
UX = 1/1.5 ,  UZ = 9UX+7                UZ = (.9 * UX) + 7.
SOLVE FOR TX,TZ  12X-1=-9X+16  OR       TX = 17./2.1
TX = 17/2.1 ,  TZ = 12TX-1              TZ = (1.2 * TX)-1.
PUT TERMINAL IN PLOT 10 MODE            CALL HPPTON
                                        200 CONTINUE
CM WINDOW WITH ORIGIN AT (2,4).         CALL SWDV$W(ID,1024,0,1023,0,779)
NOTE 5F (SCALE FACTOR).                 CALL DWINDO(-2./5F, 31.274/5F,-4/5F, 214/5F)
MOVE PEN TO POINT A (TOP VIEW).         CALL MOVEA(0.,7.)
DRAW LINE FROM A TO U.                  CALL DRAWA(UX,UZ)
MOVE PEN TO POINT D.                    CALL MOVEA(1.667, 8.5)
DRAW LINE FROM D TO B.                  CALL DRAWA(5., 11.5)
DRAW LINE FROM B TO C.                  CALL DRAWA(10.,7.)
DRAW LINE FROM C TO P.                  CALL DRAWA(PX, PZ)
MOVE PEN TO POINT F.                    CALL MOVEA(1.667, 7.)
DRAW LINE FROM F TO A.                  CALL DRAWA(0., 7.)
MOVE PEN TO POINT M.                    CALL MOVEA(0., 8.)
DRAW LINE FROM M TO Q.                  CALL DRAWA(QX, QZ)
MOVE PEN TO POINT E.                    CALL MOVEA(6.667, 10.)
DRAW LINE FROM E TO N.                  CALL DRAWA(10., 11.)
DRAW LINE FROM N TO T.                  CALL DRAWA(TX, TZ)
MOVE PEN TO POINT G.                    CALL MOVEA(6.667, 7.)
DRAW LINE FROM G TO O.                  CALL DRAWA(5., 5.)
DRAW LINE FROM O TO M.                  CALL DRAWA(0., 8.)

MOVE PEN TO POINT A (FRONT VIEW).       CALL MOVEA(0., 2.)
DRAW LINE FROM A TO B.                  CALL DRAWA(5., 0.)
DRAW LINE FROM B TO R.                  CALL DRAWA(RX, RY)
MOVE PEN TO POINT K.                    CALL MOVEA(KX, KY)
DRAW LINE FROM K TO C.                  CALL DRAWA(10., 4.)
DRAW LINE FROM C TO J.                  CALL DRAWA(JX, JY)
MOVE PEN TO POINT P.                    CALL MOVEA(PX, PY)
DRAW LINE FROM P TO A.                  CALL DRAWA(0., 2.)
MOVE PEN TO POINT M.                    CALL MOVEA(0., 3.)
DRAW LINE FROM M TO S.                  CALL DRAWA(SX, SY)
MOVE PEN TO POINT Q.                    CALL MOVEA(QX, QY)
DRAW LINE FROM Q TO N.                  CALL DRAWA(10., 0.)
DRAW LINE FROM N TO O.                  CALL DRAWA(5., 4.)
DRAW LINE FROM O TO M.                  CALL DRAWA(0., 3.)
                                        1000 CONTINUE
                                        CALL ANMODE
                                        CALL HPGTOF
CLEAR SCREEN WITH 23 BLANK LINES.       DO 1015 I = 1, 23
                                        PRINT *, ' '
END OF THIS DO LOOP.                    1015 CONTINUE
                                        CALL HPADON
RESPONSE NEEDED.                        PRINT *,'PLOT ON PLOTTER?(YES OR NO)'
                                        READ(1,1020) ANS
                                        1020 FORMAT (A1)
FOR YES, ID = 22 AND PLOTTER USE        IF (ANS. EQ.'Y') THEN
IS IMPLEMENTED.                         ID = 22
```

COMPUTER GRAPHICS   INPLA - 2

RESPONSE NEEDED.

FOR YES RESPONSE, RETURN TO
THE REFERENCE LINE 200 FOR
DRAWING ON THE PLOTTER.
IF PLOTTER IS NOT REQUESTED,
COME TO ELSE.

RESPONSE NEEDED.

```
1050 CONTINUE
     PRINT *, 'IS PLOTTER PREPARED?'
     PRINT *, 'NEW PAPER, PEN UNCAPPED'
     READ (1,1020) ANS
     IF (ANS. EQ. 'N') GO TO 1050
     GO TO 200

     ELSE
     ID = 0
     CALL CHLINE (0)
     END IF
     PRINT *, 'ENTER NEW SF?(YES OR NO)'
     READ (1,1020) ANS
     IF (ANS. EQ. 'Y') GO TO 100
     CALL HPGDOF
     PRINT *, 'PLEASE CAP PLOTTER PEN'
     CALL EXIT
     END
```

PREPARE A SIMILAR COMPUTER PROGRAM
THAT WILL PROVIDE THE INTERSECTION OF
THE TWO PLANES ABC AND MNOR. NOTE
THE USE OF EV CUTTING PLANES THRU MN
IN THE FRONT VIEW AND AC IN THE SIDE
VIEW. PROVIDE A SUGGESTED ARRAY OF SCALE
FACTORS TO FIT THE PLOTTING AREA.

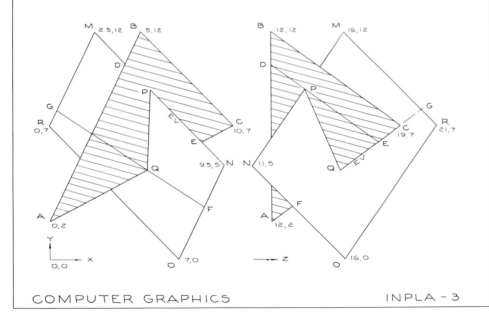

COMPUTER GRAPHICS   INPLA – 3

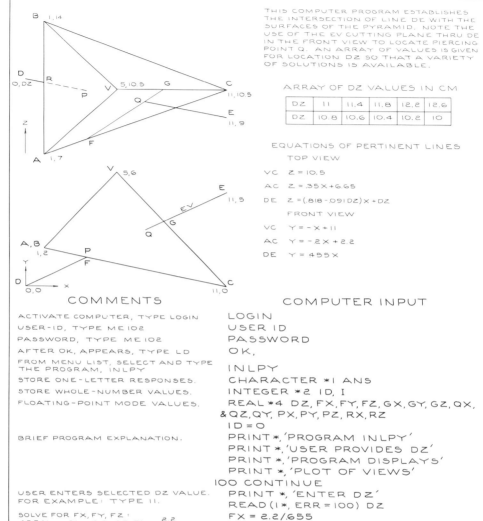

THIS COMPUTER PROGRAM ESTABLISHES
THE INTERSECTION OF LINE DE WITH THE
SURFACES OF THE PYRAMID. NOTE THE
USE OF THE EV CUTTING PLANE THRU DE
IN THE FRONT VIEW TO LOCATE PIERCING
POINT Q. AN ARRAY OF VALUES IS GIVEN
FOR LOCATION DZ SO THAT A VARIETY
OF SOLUTIONS IS AVAILABLE.

ARRAY OF DZ VALUES IN CM

DZ	11	11.4	11.8	12.2	12.6
DZ	10.8	10.6	10.4	10.2	10

EQUATIONS OF PERTINENT LINES
    TOP VIEW

VC $Z = 10.5$

AC $Z = .35X + 6.65$

DE $Z = (.818 - .091DZ)X + DZ$

    FRONT VIEW

VC $Y = -X + 11$

AC $Y = -2X + 2.2$

DE $Y = .455X$

COMMENTS

ACTIVATE COMPUTER, TYPE LOGIN	LOGIN
USER-ID, TYPE ME102	USER ID
PASSWORD, TYPE ME102	PASSWORD
AFTER OK, APPEARS, TYPE LD	OK,
FROM MENU LIST, SELECT AND TYPE THE PROGRAM, INLPY	INLPY
STORE ONE-LETTER RESPONSES.	CHARACTER *1 ANS
STORE WHOLE-NUMBER VALUES.	INTEGER *2 ID, I
FLOATING-POINT MODE VALUES.	REAL *4 DZ, FX, FY, FZ, GX, GY, GZ, QX,

COMPUTER INPUT

```
LOGIN
USER ID
PASSWORD
OK,

INLPY
CHARACTER *1 ANS
INTEGER *2 ID, I
REAL *4 DZ, FX, FY, FZ, GX, GY, GZ, QX,
&QZ, QY, PX, PY, PZ, RX, RZ
ID = O
```

BRIEF PROGRAM EXPLANATION.

```
PRINT *, 'PROGRAM INLPY'
PRINT *, 'USER PROVIDES DZ'
PRINT *, 'PROGRAM DISPLAYS'
PRINT *, 'PLOT OF VIEWS'
100 CONTINUE
```

USER ENTERS SELECTED DZ VALUE.
FOR EXAMPLE: TYPE 11.

```
PRINT *, 'ENTER DZ'
READ (1*, ERR = 100) DZ
```

SOLVE FOR FX, FY, FZ:
$455X = -2X + 2.2$ OR $FX = \dfrac{2.2}{655}$

$FY = 455 FX$ AND $FZ = .35 FX + 6.65$

```
FX = 2.2/655
FY = 455 * FX
FZ = (.35 * FX) + 6.65
```

SOLVE FOR GX, GY, GZ:
$.455X = -X + 11$ OR $GX = \dfrac{11}{1.455}$

$GY = 455 GX$ AND $GZ = 10.5$

```
GX = 11./1.455
GY = 455 * GX
GZ = 10.5
```

FG EQUATION (TOP VIEW)
$\dfrac{Z - Z1}{X - X1} = \dfrac{Z1 - Z2}{X1 - X2}$ OR $\dfrac{Z - 7.826}{X - 3.359} = \dfrac{7.826 - 10.5}{3.359 - 7.56}$

$Z = .637X + 5.688$

SOLVE FOR QX, QZ, QY:
$QX = \dfrac{5.688 - DZ}{.181 - .091DZ}$

$QZ = .637 QX + 5.688$

$QY = 455 QX$

```
QX = (5.688 - DZ)/(.181 - (.091 * DZ))
QZ = (.637 * QX) + 5.688
QY = 455 * QX
```

COMPUTER GRAPHICS   INLPY - 1

Comment	Code
PX = FX	PX = FX
PY = FY	PY = FY
PZ = (.818 - .091 DZ) PX + DZ	PZ = (.818 - (.091 * DZ)) * PX + DZ
RX = 1	RX = 1.
RZ = (.818 - .091 DZ) RX + DZ	RZ = (.818 - (.091 * DZ)) * RX + DZ
TERMINAL IN PLOT 10 MODE.	CALL HPPTON
	200 CONTINUE
	CALL SWDV$W(ID,1024,0,1023,0,779)
WINDOW WITH ORIGIN AT D (2,4).	CALL DWINDO(-2.,31.274,-4.,21.4)
MOVE PEN TO POINT A (FRONT).	CALL MOVEA(1.,2.)
DRAW LINE FROM A TO C.	CALL DRAWA(11.,0.)
DRAW LINE FROM C TO V.	CALL DRAWA(5.,6.)
DRAW LINE FROM V TO A.	CALL DRAWA(1.,2.)
MOVE PEN TO POINT E.	CALL MOVEA(11.,5.)
DRAW LINE FROM E TO Q.	CALL DRAWA(QX,QY)
MOVE PEN TO POINT P.	CALL MOVEA(PX,PY)
DRAW LINE FROM P TO D.	CALL DRAWA(0.,0.)
MOVE PEN TO POINT A (TOP).	CALL MOVEA(1.,7.)
DRAW LINE FROM A TO C.	CALL DRAWA(11.,10.5)
DRAW LINE FROM C TO B.	CALL DRAWA(1.,14.)
DRAW LINE FROM B TO A.	CALL DRAWA(1.,7.)
DRAW LINE FROM A TO V.	CALL DRAWA(5.,10.5)
DRAW LINE FROM V TO B.	CALL DRAWA(1.,14.)
MOVE PEN TO POINT V.	CALL MOVEA(5.,10.5)
DRAW LINE FROM V TO C.	CALL DRAWA(11.,10.5)
MOVE PEN TO POINT E.	CALL MOVEA(11.,9.)
DRAW LINE FROM E TO Q.	CALL DRAWA(QX,QZ)
MOVE PEN TO POINT P. (PEN NO. 3)	CALL MOVEA(PX,PZ)
DASH LINE FROM P TO R.	CALL DASHA(1.,RZ)
DRAW LINE R TO D. (RETURN TO PEN 2)	CALL DRAWA(0.,DZ)
	1000 CONTINUE
	CALL ANMODE
	CALL HPGTOF
CLEAR SCREEN WITH 23 BLANK LINES.	DO 1015 I=1,23
	PRINT *,' '
END OF THIS DO LOOP.	1015 CONTINUE
	CALL HPADON
PROVIDE YES OR NO RESPONSE.	PRINT *,'PLOT ON PLOTTER?(YES OR NO)'
	READ(1,1020) ANS
	1020 FORMAT (A1)
FOR YES, ID=22, PLOTTER IS USED.	IF(ANS.EQ.'Y') THEN ID=22
	1050 CONTINUE
	PRINT *,'IS PLOTTER PREPARED?'
	PRINT *,'NEW PAPER,PEN UNCAPPED'
	READ(1,1020) ANS
IF ANS IS YES, RETURN TO LINE 200	IF(ANS.EQ.'N') GO TO 1050
FOR DRAWING ON THE PLOTTER.	GO TO 200
IF PLOTTER IS NOT REQUESTED,	ELSE
COME TO ELSE.	ID=0
	CALL CHLINE(0)
	END IF
RESPONSE NEEDED.	PRINT *,'ENTER NEW CASE?(YES OR NO)'
	READ(1,1020) ANS
FOR AN ADDITIONAL PROJECT,	IF(ANS.EQ.'Y') GO TO 100
CHANGE LOCATION E (FRONT)	CALL HPGDOF
TO (11,4) AND ALTER PROGRAM	CALL EXIT
AS NEEDED.	END

COMPUTER GRAPHICS   INLPY-2

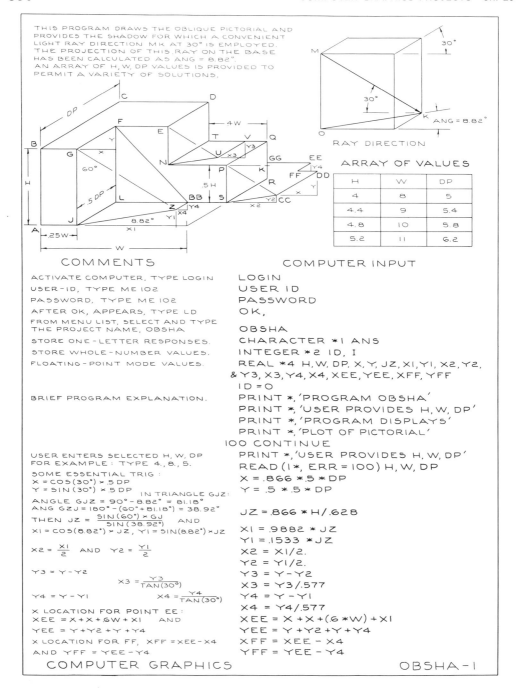

THIS PROGRAM DRAWS THE OBLIQUE PICTORIAL AND
PROVIDES THE SHADOW FOR WHICH A CONVENIENT
LIGHT RAY DIRECTION MK AT 30° IS EMPLOYED.
THE PROJECTION OF THIS RAY ON THE BASE
HAS BEEN CALCULATED AS ANG = 8.82°.
AN ARRAY OF H, W, DP VALUES IS PROVIDED TO
PERMIT A VARIETY OF SOLUTIONS.

RAY DIRECTION

ARRAY OF VALUES

H	W	DP
4	8	5
4.4	9	5.4
4.8	10	5.8
5.2	11	6.2

COMMENTS	COMPUTER INPUT
ACTIVATE COMPUTER, TYPE LOGIN	LOGIN
USER-ID, TYPE ME 102	USER ID
PASSWORD, TYPE ME 102	PASSWORD
AFTER OK, APPEARS, TYPE LD	OK,
FROM MENU LIST, SELECT AND TYPE THE PROJECT NAME, OBSHA	OBSHA
STORE ONE-LETTER RESPONSES.	CHARACTER *1 ANS
STORE WHOLE-NUMBER VALUES.	INTEGER *2 ID, I
FLOATING-POINT MODE VALUES.	REAL *4 H, W, DP, X, Y, JZ, X1, Y1, X2, Y2, & Y3, X3, Y4, X4, XEE, YEE, XFF, YFF
	ID = 0
BRIEF PROGRAM EXPLANATION.	PRINT *, 'PROGRAM OBSHA'
	PRINT *, 'USER PROVIDES H, W, DP'
	PRINT *, 'PROGRAM DISPLAYS'
	PRINT *, 'PLOT OF PICTORIAL'
	100 CONTINUE
USER ENTERS SELECTED H, W, DP	PRINT *, 'USER PROVIDES H, W, DP'
FOR EXAMPLE : TYPE 4., 8., 5.	READ (I*, ERR = 100) H, W, DP
SOME ESSENTIAL TRIG :	
$X = \cos(30°) \times .5\,DP$	X = .866 * .5 * DP
$Y = \sin(30°) \times .5\,DP$ IN TRIANGLE GJZ:	Y = .5 * .5 * DP
ANGLE GJZ = 90° - 8.82° = 81.18°	
ANG GZJ = 180° - (60° + 81.18°) = 38.92°	
THEN $JZ = \dfrac{\sin(60°) \times GJ}{\sin(38.92°)}$ AND	JZ = .866 * H / .628
$X1 = \cos(8.82°) \times JZ,\; Y1 = \sin(8.82°) \times JZ$	X1 = .9882 * JZ
	Y1 = .1533 * JZ
$X2 = \dfrac{X1}{2}$ AND $Y2 = \dfrac{Y1}{2}$	X2 = X1/2.
	Y2 = Y1/2.
$Y3 = Y - Y2$	Y3 = Y - Y2
$X3 = \dfrac{Y3}{\tan(30°)}$	X3 = Y3/.577
$Y4 = Y - Y1$ $X4 = \dfrac{Y4}{\tan(30°)}$	Y4 = Y - Y1
	X4 = Y4/.577
X LOCATION FOR POINT EE :	
XEE = X + X + 6W + X1 AND	XEE = X + X + (.6 *W) + X1
YEE = Y + Y2 + Y + Y4	YEE = Y + Y2 + Y + Y4
X LOCATION FOR FF, XFF = XEE - X4	XFF = XEE - X4
AND YFF = YEE - Y4	YFF = YEE - Y4

COMPUTER GRAPHICS   OBSHA-1

```
PUT TERMINAL IN PLOT IO MODE         CALL HPPTON
REFERENCE LINE.                      200 CONTINUE
                                     CALL SWDV$W(ID,1024,0,1023,0, 779)
CM WINDOW WITH ORIGIN AT (2,4).      CALL DWINDO(-2.,31.274,-4.,21.4)
MOVE PEN TO POINT B.                 CALL MOVEA(0.,H)
DRAW LINE FROM B TO G.               CALL DRAWA(.25*W, H)
DRAW LINE FROM G TO J.               CALL DRAWA(.25*W, 0.)
DRAW LINE FROM J TO A.               CALL DRAWA(0., 0.)
DRAW LINE FROM A TO B.               CALL DRAWA(0., H)
DRAW LINE FROM B TO C.               CALL DRAWA(X+X, Y+Y+H)
DRAW LINE FROM C TO D.               CALL DRAWA(X+X+(.6*W),Y+Y+H)
DRAW LINE FROM D TO E.               CALL DRAWA(X+(.6*W), Y+H)
DRAW LINE FROM E TO F.               CALL DRAWA(X+(.25*W), Y+H)
DRAW LINE FROM F TO G.               CALL DRAWA(.25*W, H)
MOVE PEN TO POINT J.                 CALL MOVEA(.25*W, 0.)
DRAW LINE FROM J TO L.               CALL DRAWA(X+(.25*W), Y)
DRAW LINE FROM L TO F.               CALL DRAWA(X+(.25*W), Y+H)
MOVE PEN TO POINT D.                 CALL MOVEA(X+X+(.6*W), Y+Y+H)
DRAW LINE FROM D TO T.               CALL DRAWA(X+X+(.6*W), Y+(.5*H)+Y)
DRAW LINE FROM T TO N.               CALL DRAWA(X+(.6*W), Y+(.5*H))
DRAW LINE FROM N TO E.               CALL DRAWA(X+(.6*W), Y+H)
MOVE PEN TO POINT L.                 CALL MOVEA(X+(.25*W), Y)
DRAW LINE FROM L TO S.               CALL DRAWA(W+X, Y)
DRAW LINE FROM S TO R.               CALL DRAWA(W+X+X, Y+Y)
DRAW LINE FROM R TO Q.               CALL DRAWA(W+X+X, Y+Y+(.5*H))
DRAW LINE FROM Q TO P.               CALL DRAWA(W+X, Y+(.5*H))
DRAW LINE FROM P TO N.               CALL DRAWA(X+(.6*W), Y+(.5*H))
MOVE PEN TO POINT P.                 CALL MOVEA(W+X, Y+(.5*H))
DRAW LINE FROM P TO S.               CALL DRAWA(W+X, Y)
MOVE PEN TO POINT T.                 CALL MOVEA(X+X+(.6*W),Y+(.5*H)+Y)
DRAW LINE FROM T TO Q.               CALL DRAWA(W+X+X, Y+Y+(.5*H))
MOVE PEN TO POINT S.                 CALL MOVEA(W+X, Y)
DRAW LINE FROM S TO CC.              CALL DRAWA(W+X+X2, Y+Y2)
DRAW LINE FROM CC TO DD.             CALL DRAWA(W+X+X2+X, Y+Y2+Y)
DRAW LINE FROM DD TO FF.             CALL DRAWA(XFF, YFF)
DRAW LINE FROM FF TO EE.             CALL DRAWA(XEE, YEE)
DRAW LINE FROM EE TO GG.             CALL DRAWA(W+X+X, YEE)
MOVE PEN TO POINT N.                 CALL MOVEA(X+(.6*W), Y+(.5*H))
DRAW LINE FROM N TO U.               CALL DRAWA(X+(.6*W)+X2,Y+(.5*H)+Y2)
DRAW LINE FROM U TO V.               CALL DRAWA(X+(.6*W)+X2+X3,Y+Y+(.5*H))
MOVE PEN TO POINT J.                 CALL MOVEA(.25*W, 0.)
DRAW LINE FROM J TO Z.               CALL DRAWA(X1+(.25*W), Y1)
DRAW LINE FROM Z TO BB.              CALL DRAWA(X1+(.25*W+X4, Y)
DRAW LINE FROM BB TO F.              CALL DRAWA(X+(.25*W), H+Y)
                                     CALL ANMODE
                                     CALL HPGTOF
CLEAR SCREEN WITH 23 BLANK LINES.    DO 1015 I=1, 23
                                     PRINT *,' '
                                     1015 CONTINUE
                                     CALL HPADON
RESPONSE NEEDED.                     PRINT *,'PLOT ON PLOTTER?(YES, NO)'
                                     READ(1,1020) ANS
                                     1020 FORMAT (A1)
FOR YES, PLOTTER USE IS NEEDED.      IF(ANS. EQ.'Y') THEN
                                     ID=22
```

COMPUTER GRAPHICS   OBSHA-2

```
                                  1050 CONTINUE
PROVIDE YES OR NO RESPONSE.           PRINT *,'IS PLOTTER PREPARED?'
                                      PRINT *,'NEW PAPER, PEN UNCAPPED'
                                      READ (1,1020) ANS
                                      IF(ANS. EQ. 'N') GO TO 1050
FOR YES, RETURN TO THE LINE 200       GO TO 200
FOR DRAWING ON THE PLOTTER.           ELSE
                                      CALL CHLINE (0)
                                      END IF
RESPONSE NEEDED.                      PRINT *,'ENTER NEW CASE? (YES OR NO)'
                                      READ (1,1020) ANS
                                      IF(ANS. EQ. 'Y') GO TO 100
                                      CALL HPGDOF
                                      PRINT *,'CAP PLOTTER PEN'
                                      CALL EXIT
                                      END
```

SUPPLEMENTARY PROJECT

PREPARE A PROGRAM THAT DRAWS THE OBLIQUE
PICTORIAL AND ITS SHADOW. AN ARRAY OF H, W,
AND DP VALUES IS LISTED.

H	8	8.5	9	9.5	10	7.5	7	6.5	6
W	13	13.4	13.8	14.2	14.6	15	12.6	12.2	11.8
DP	6	6.2	6.4	6.6	6.8	7	5.8	5.6	5.4

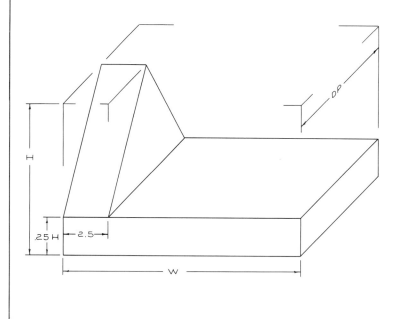

COMPUTER GRAPHICS   OBSHA - 3

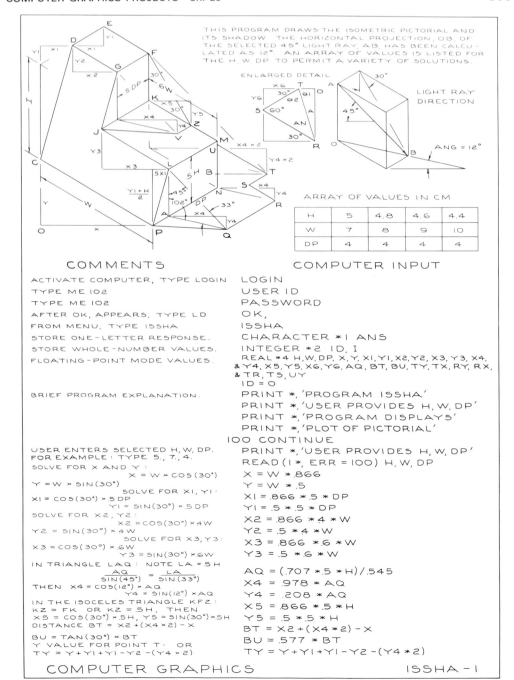

THIS PROGRAM DRAWS THE ISOMETRIC PICTORIAL AND ITS SHADOW. THE HORIZONTAL PROJECTION, OB, OF THE SELECTED 45° LIGHT RAY, AB, HAS BEEN CALCULATED AS 12°. AN ARRAY OF VALUES IS LISTED FOR THE H, W, DP TO PERMIT A VARIETY OF SOLUTIONS.

ENLARGED DETAIL

LIGHT RAY DIRECTION

ANG = 12°

ARRAY OF VALUES IN CM

H	5	4.8	4.6	4.4
W	7	8	9	10
DP	4	4	4	4

COMMENTS	COMPUTER INPUT
ACTIVATE COMPUTER, TYPE LOGIN	LOGIN
TYPE ME 102	USER ID
TYPE ME 102	PASSWORD
AFTER OK, APPEARS, TYPE LD	OK,
FROM MENU, TYPE ISSHA	ISSHA
STORE ONE-LETTER RESPONSE.	CHARACTER *1 ANS
STORE WHOLE-NUMBER VALUES.	INTEGER *2 ID, I
FLOATING-POINT MODE VALUES.	REAL *4 H, W, DP, X, Y, X1, Y1, X2, Y2, X3, Y3, X4,
	& Y4, X5, Y5, X6, Y6, AQ, BT, BU, TY, TX, RY, RX,
	& TR, TS, UY
	ID = 0
BRIEF PROGRAM EXPLANATION.	PRINT *, 'PROGRAM ISSHA'
	PRINT *, 'USER PROVIDES H, W, DP'
	PRINT *, 'PROGRAM DISPLAYS'
	PRINT *, 'PLOT OF PICTORIAL'
	100 CONTINUE

USER ENTERS SELECTED H, W, DP. FOR EXAMPLE: TYPE 5., 7., 4.

PRINT *, 'USER PROVIDES H, W, DP'
READ (I *, ERR = 100) H, W, DP

SOLVE FOR X AND Y:
$$X = W \times \cos(30°)$$
$$Y = W \times \sin(30°)$$

X = W * .866
Y = W * .5

SOLVE FOR X1, Y1:
$$X1 = \cos(30°) \times .5\,DP$$
$$Y1 = \sin(30°) \times .5\,DP$$

X1 = .866 * .5 * DP
Y1 = .5 * .5 * DP

SOLVE FOR X2, Y2:
$$X2 = \cos(30°) \times 4W$$
$$Y2 = \sin(30°) \times 4W$$

X2 = .866 * 4 * W
Y2 = .5 * 4 * W

SOLVE FOR X3, Y3:
$$X3 = \cos(30°) \times .6W$$
$$Y3 = \sin(30°) \times .6W$$

X3 = .866 * .6 * W
Y3 = .5 * .6 * W

IN TRIANGLE LAQ: NOTE LA = .5 H
$$\frac{AQ}{\sin(45°)} = \frac{LA}{\sin(33°)}$$
THEN $X4 = \cos(12°) \times AQ$
$$Y4 = \sin(12°) \times AQ$$

AQ = (.707 * .5 * H) / .545
X4 = .978 * AQ
Y4 = .208 * AQ

IN THE ISOCELES TRIANGLE KFZ:
KZ = FK OR KZ = .5H, THEN
$X5 = \cos(30°) \times .5H$, $Y5 = \sin(30°) \times .5H$
DISTANCE BT = X2 + (X4 × 2) − X

X5 = .866 * .5 * H
Y5 = .5 * .5 * H
BT = X2 + (X4 * 2) − X

BU = TAN(30°) × BT
Y VALUE FOR POINT T: OR
TY = Y + Y1 + Y1 − Y2 − (Y4 × 2)

BU = .577 * BT
TY = Y + Y1 + Y1 − Y2 − (Y4 * 2)

COMPUTER GRAPHICS   ISSHA − 1

LOCATION UY = TY + BU	UY = TY + BU
LOCATION TX	TX = X1 + X2 + X1 + (X4 * 2.)
LOCATION RY	RY = Y1 + Y1 - Y4
LOCATION RX , SOLVE FOR TR LENGTH :	RX = X + X1 + X1 + X4
$TR = \sqrt{(TY-RY)^2 + (RX-TX)^2}$	TR = SQRT((TY-RY)**2 + (RX-TX)**2)
ANGLE A IN DEGREES.	A = ATAN((RX-TX)/(TY-RY)) * 57.3
SOLVE FOR ST LENGTH : AN IN RADIANS	AN = (180. - (A + 30.))/57.3
$\frac{ST}{SIN(AN)} = \frac{TR}{SIN(60°)}$	ST = (TR * SIN(AN))/.866
SOLVE FOR XG, YG : XG = COS(30°) * ST	XG = .866 * ST
YG = SIN(30°) * ST	YG = .5 * ST
PUT TERMINAL IN PLOT IO MODE.	CALL HPPTON
	200 CONTINUE
AVAILABLE PLOTTING AREA.	CALL SWDV$W(ID,1024,0,1023,0,779)
CM WINDOW WITH ORIGIN AT (2,4).	CALL DWINDO(-2.,31.274,-4.,21.4)
MOVE PEN TO POINT C.	CALL MOVEA(0,Y)
DRAW LINE FROM C TO D.	CALL DRAWA(X1,Y+H+Y1)
DRAW LINE FROM D TO E.	CALL DRAWA(X1+X1,Y+H+Y1+Y1)
DRAW LINE FROM E TO F.	CALL DRAWA(X1+X1+X2,Y+H+Y1-Y2+Y1)
DRAW LINE FROM F TO K.	CALL DRAWA(X1+X1+X2,Y+Y1-Y2+(5*H)+Y1)
DRAW LINE FROM K TO M.	CALL DRAWA(X+X1+X1,Y1+Y1+(.5*H))
DRAW LINE FROM M TO N.	CALL DRAWA(X+X1+X1,Y1+Y1)
DRAW LINE FROM N TO P.	CALL DRAWA(X,0.)
DRAW LINE FROM P TO L.	CALL DRAWA(X+(.5*X1),(Y1+H)/2.)
DRAW LINE FROM L TO J.	CALL DRAWA(X+(.5*X1)-X3,Y3+(Y1+H)/2.)
DRAW LINE FROM J TO G.	CALL DRAWA(X1+X2,Y+H+Y1-Y2)
DRAW LINE FROM G TO D.	CALL DRAWA(X1,Y+H+Y1)
MOVE PEN TO POINT G.	CALL MOVEA(X1+X2,Y+H+Y1-Y2)
DRAW LINE FROM G TO F.	CALL DRAWA(X1+X1+X2,Y+H+Y1+Y2+Y1)
MOVE PEN TO POINT J.	CALL MOVEA(X+(.5*X1)-X3,Y3+(Y1+H)/2.)
DRAW LINE FROM J TO K.	CALL DRAWA(X1+X1+X2,Y+Y1-Y2+(5*H)+Y1)
MOVE PEN TO POINT L.	CALL MOVEA(X+(.5*X1),(Y1+H)/2.)
DRAW LINE FROM L TO M.	CALL DRAWA(X+X1+X1,Y1+Y1+(.5*H))
MOVE PEN TO POINT J.	CALL MOVEA(X+(.5*X1)-X3,Y3+(Y1+H)/2.)
DRAW LINE FROM J TO V.	CALL DRAWA(X1+X2+X4,Y+H+Y1-Y2-(.5*H)-Y4)
DRAW LINE FROM V TO Z.	CALL DRAWA(X1+X2+X1+X5,Y+(5*H)+Y1-Y2-Y5+Y1)
MOVE PEN TO POINT C.	CALL MOVEA(0.,Y)
DRAW LINE FROM C TO P.	CALL DRAWA(X,0.)
DRAW LINE FROM P TO Q.	CALL DRAWA(X+.5*X1+X4,.5*Y1-Y4)
DRAW LINE FROM Q TO R	CALL DRAWA(RX,RY)
DRAW LINE FROM R TO S.	CALL DRAWA(TX-XG,TY-YG)
DRAW LINE FROM S TO T.	CALL DRAWA(TX,TY)
DRAW LINE FROM T TO U.	CALL DRAWA(X+X1+X1,UY)
STORE MULTIPLOTTER PEN.	CALL PEN (0)
MOVE PEN AWAY FROM PICTORIAL.	CALL MOVEA(4096,4096)
	CALL ANMODE
	CALL HPGTOF
CLEAR SCREEN WITH 23 BLANK LINES.	DO 1015 I = 1,23
	PRINT *, ' '
END OF THIS DO LOOP.	1015 CONTINUE
	CALL HPADON
PROVIDE RESPONSE.	PRINT *,'PLOT ON PLOTTER? (YES OR NO)'
	READ(1,1020) ANS
	1020 FORMAT (A1)
FOR YES, PLOTTER 22 IS REQUESTED.	IF(ANS. EQ. 'Y') THEN
	ID = 22
GET NEW ID IF 22 IS OCCUPIED.	CALL GETP$W(ID)

COMPUTER GRAPHICS   ISSHA - 2

```
                          1050 CONTINUE
PROVIDE RESPONSE.             PRINT *, 'IS PLOTTER READY?
                              PRINT *, 'NEW PAPER, PEN UNCAPPED'
                              READ(1,1020) ANS
                              IF(ANS. EQ. 'N') GO TO 1050
IF PLOTTER IS PREPARED, GO TO 200.    GO TO 200
                              ELSE
                              ID = 0
                              CALL CHLINE (0)
                              END IF
RESPONSE NEEDED.              PRINT *, 'ENTER NEW CASE?(YES OR NO)'
                              READ(1,1020) ANS
                              IF(ANS. EQ. 'Y') GO TO 100
                              CALL HPGDOF
                              PRINT *, 'CAP PLOTTER PEN'
                              CALL EXIT
                              END
```

SUPPLEMENTARY PROJECT

PREPARE A SIMILAR PROGRAM THAT PRODUCES
THE PICTORIAL AND SHADOW OF THIS ISOMETRIC.
AN ARRAY OF H, W, AND DP VALUES IS LISTED.
EMPLOY THE SAME LIGHT RAY DIRECTION.

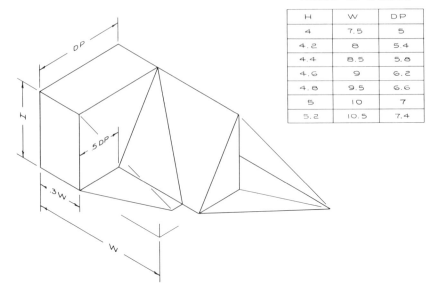

ARRAY OF VALUES

H	W	DP
4	7.5	5
4.2	8	5.4
4.4	8.5	5.8
4.6	9	6.2
4.8	9.5	6.6
5	10	7
5.2	10.5	7.4

COMPUTER GRAPHICS   ISSHA - 3

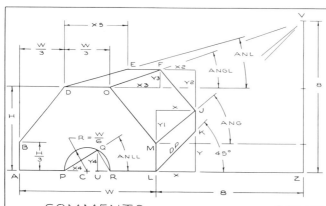

THIS COMPUTER PROGRAM DRAWS A ONE-POINT PERSPECTIVE. AN ARRAY OF VALUES IS SUGGESTED FOR THE H, W, AND DP TO PERMIT SEVERAL SOLUTIONS.

ARRAY OF VALUES

H	W	DP
4.5	7.5	3
4.8	8	3.4
5	8.5	3.8
5.2	9	4.2
5.5	9.5	4.6
6	10	5

COMMENTS	COMPUTER INPUT
ACTIVATE COMPUTER, TYPE LOGIN	LOGIN
USER-ID, TYPE ME 102	USER ID
PASSWORD, TYPE ME 102	PASSWORD
WHEN OK, APPEARS, TYPE LD	OK,
FROM MENU LIST, SELECT AND TYPE THE PROGRAM, OPPER.	OPPER
	CHARACTER *1 ANS
	INTEGER *2 ID, I
	REAL *4 H, W, DP, X, Y, Y1, X2, Y2, Y3, X3,
	& Y4, X4, X5, MV, MJ, OV, OF, PQ, ANGLES
	ID = O
BRIEF PROGRAM EXPLANATION.	PRINT *, 'PROGRAM OPPER'
	PRINT *, 'USER PROVIDES H, W, DP'
	PRINT *, 'PROGRAM DISPLAYS'
	PRINT *, 'PLOT OF PERSPECTIVE'
	100 CONTINUE
USER ENTERS SELECTED H, W, DP. FOR EXAMPLE: TYPE 4.5, 7.5, 3.	PRINT *, 'USER PROVIDES H, W, DP'
	READ (I *, ERR = 100) H, W, DP
SOME ESSENTIAL MATH: $X = \cos(45°) \times DP$	$X = .707 * DP$
$Y = \sin(45°) \times DP$ OR $Y = X$	$Y = X$
$\tan(ANG) = \dfrac{8 - H/3}{8}$ (IN RADIANS)	$ANG = ATAN((8. - H/3.)/8.)$
$Y1 = \tan(ANG) \times X$	$Y1 = TAN(ANG) * X$
$MV = \dfrac{8 - H/3}{\sin(ANG)}$ $MJ = \sqrt{X^2 + Y1^2}$	$MV = (8. - (H/3.))/SIN(ANG)$
	$MJ = SQRT(X**2 + Y1**2)$
$\tan(ANGL) = \dfrac{8-H}{(W+8)-.667W}$	$ANGL = ATAN((8.-H)/((W+8.)-(.667*W)))$
$OV = \dfrac{8-H}{\sin(ANGL)}$	$OV = (8.-H)/SIN(ANGL)$
SINCE OM AND FJ ARE PARALLEL: $\dfrac{OF}{OV} = \dfrac{MJ}{MV}$ THEN	$OF = MJ * OV/MV$
$X3 = OF \times \cos(ANGL)$	$X3 = OF * COS(ANGL)$
$Y3 = OF \times \sin(ANGL)$	$Y3 = OF * SIN(ANGL)$
$\tan(ANL) = \dfrac{8-H}{8+.667W}$	$ANL = ATAN((8.-H)/(8.+(.667*W)))$
$X5 = Y3 / \tan(ANL)$	$X5 = Y3 / TAN(ANL)$
$X2 = (W+X) - (.667W + X3)$	$X2 = (W+X) - ((.667*W) + X3)$
$Y2 = (H+Y3) - (H/3 + Y1)$	$Y2 = (H+Y3) - ((H/3.) + Y1)$
$\tan(ANLL) = 8/(.667W + 8)$	$ANLL = ATAN(8./((.667*W) + 8.))$
IN RIGHT TRIANGLE PQR: $\cos(ANLL) = \dfrac{PQ}{PR}$	$PQ = (W/3.) * COS(ANLL)$
$Y4 = PQ \times \sin(ANLL)$	$Y4 = PQ * SIN(ANLL)$
$X4 = PQ \times \cos(ANLL)$	$X4 = PQ * COS(ANLL)$
	CALL HPPTON
	200 CONTINUE

COMPUTER GRAPHICS

OPPER - 1

```
                                              CALL SWDV$W(ID,1024,0,1023,0,779)
WINDOW WITH ORIGIN AT A (2,4).                CALL DWINDO(-2.,31.274,-4.,21.4)
MOVE PEN TO POINT P.                          CALL MOVEA(W/3.,O.)
DRAW LINE FROM P TO A.                        CALL DRAWA(O.,O.)
DRAW LINE FROM A TO B.                        CALL DRAWA(O.,H/3.)
DRAW LINE FROM B TO D.                        CALL DRAWA(W/3.,H)
DRAW LINE FROM D TO E.                        CALL DRAWA(W/3.+X5,H+Y3)
DRAW LINE FROM E TO F.                        CALL DRAWA(.667*W+X3,H+Y3)
DRAW LINE FROM F TO J.                        CALL DRAWA(W+X,H/3.+Y1)
DRAW LINE FROM J TO K.                        CALL DRAWA(W+X,Y)
DRAW LINE FROM K TO L.                        CALL DRAWA(W,O.)
DRAW LINE FROM L TO R.                        CALL DRAWA(667*W,O.)
DRAW ARC FROM R TO P.                         CALL ARC(W/2.,O.,W/6.,O.,180.)
DRAW LINE FROM P TO Q.                        CALL DRAWA(W/3.+X4,Y4)
MOVE PEN TO POINT D.                          CALL MOVEA(W/3.,H)
DRAW LINE FROM D TO O.                        CALL DRAWA(667*W,H)
DRAW LINE FROM O TO F.                        CALL DRAWA(667*W+X3,H+Y3)
MOVE PEN TO POINT O.                          CALL MOVEA(.667*W,H)
DRAW LINE FROM O TO M.                        CALL DRAWA(W,H/3.)
DRAW LINE FROM M TO L.                        CALL DRAWA(W,O.)
MOVE PEN TO POINT M.                          CALL MOVEA(W,H/3.)
DRAW LINE FROM M TO J.                        CALL DRAWA(W+X,H/3.+Y1)
                                              CALL ANMODE
                                              CALL HPGTOF
CLEAR SCREEN WITH 23 BLANK LINES.             DO 1015 I = 1,23
                                              PRINT *,' '
END OF THIS DO LOOP.                     1015 CONTINUE
                                              CALL HPADON
PROVIDE YES OR NO RESPONSE.                   PRINT *,'PLOT ON PLOTTER?(YES OR NO)'
                                              READ(1,1020) ANS
                                         1020 FORMAT (A1)
FOR YES, ID = 22 AND PLOTTER                  IF(ANS. EQ.'Y') THEN
USE IS REQUESTED.                             ID = 22
                                         1050 CONTINUE
RESPONSE NEEDED.                              PRINT *,'IS PLOTTER PREPARED?'
                                              PRINT *,'NEW PAPER, PEN UNCAPPED'
                                              READ(1,1020) ANS
                                              IF(ANS. EQ.'N') GO TO 1050
IF PLOTTER IS PREPARED, RETURN                GO TO 200
TO REFERENCE LINE 200.                        ELSE
                                              ID = O
                                              CALL CHLINE(0)
                                              END IF
RESPONSE NEEDED.                              PRINT *,'ENTER NEW CASE?(YES OR NO)'
                                              READ(1,1020) ANS
                                              IF(ANS. EQ.'Y') GO TO 100
                                              CALL HPGDOF
                                              PRINT *,'CAP PLOTTER PEN'
                                              CALL EXIT
                                              END
```

FOR A SUPPLEMENTAY PROJECT, ALTER
THE LOCATION OF VANISHING POINT V
AND REVISE THE PROGRAM AS NEEDED.

COMPUTER GRAPHICS   OPPER-2

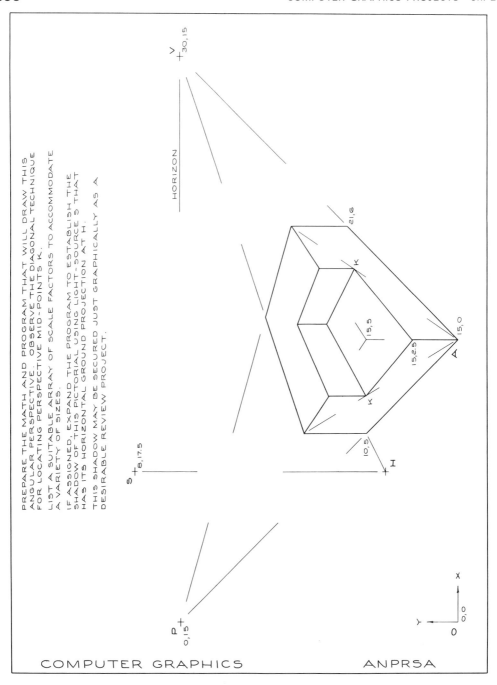

PREPARE THE MATH AND PROGRAM THAT WILL DRAW THIS ANGULAR PERSPECTIVE. OBSERVE THE DIAGONAL TECHNIQUE FOR LOCATING PERSPECTIVE MID-POINTS K.

LIST A SUITABLE ARRAY OF SCALE FACTORS TO ACCOMMODATE A VARIETY OF SIZES.

IF ASSIGNED, EXPAND THE PROGRAM TO ESTABLISH THE SHADOW OF THIS PICTORIAL USING LIGHT-SOURCE S THAT HAS ITS HORIZONTAL GROUND PROJECTION AT H.

THIS SHADOW MAY BE SECURED JUST GRAPHICALLY AS A DESIRABLE REVIEW PROJECT.

COMPUTER GRAPHICS   ANPRSA

1,2 PREPARE THE ANALYSIS AND PROGRAM THAT WILL DRAW THE
OBLIQUE PICTORIAL AND WILL LOCATE THE PIERCING POINTS OF
LINE MN WITH THE PICTORIAL SURFACES.

PROVIDE AN ARRAY OF VALUES FOR THE VARIABLE MY TO
ACCOMMODATE A VARIETY OF PIERCING POINT SOLUTIONS.

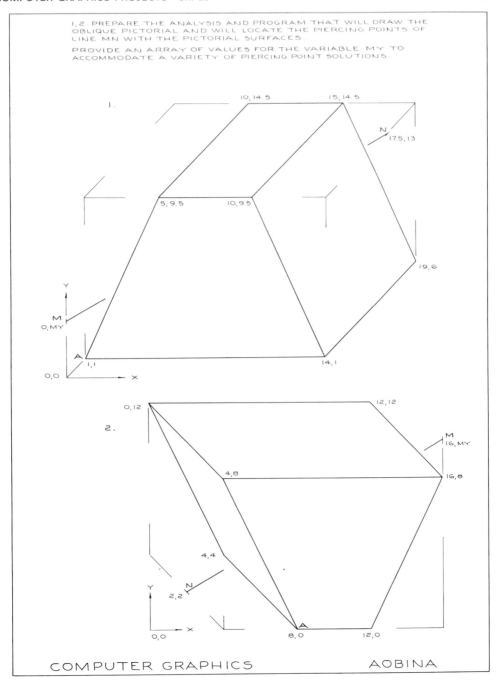

COMPUTER GRAPHICS   AOBINA

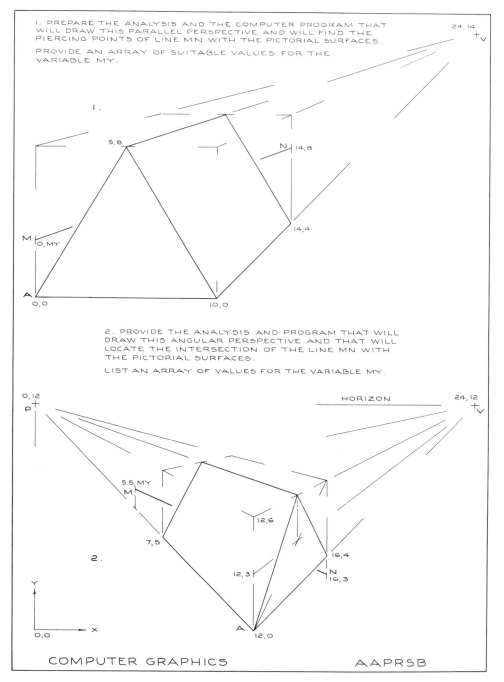

1. PREPARE THE ANALYSIS AND THE COMPUTER PROGRAM THAT WILL DRAW THIS PARALLEL PERSPECTIVE AND WILL FIND THE PIERCING POINTS OF LINE MN WITH THE PICTORIAL SURFACES.

PROVIDE AN ARRAY OF SUITABLE VALUES FOR THE VARIABLE MY.

24,14
+V

1.

5,8

N 14,8

14,4

M
0,MY

A
0,0

10,0

2. PROVIDE THE ANALYSIS AND PROGRAM THAT WILL DRAW THIS ANGULAR PERSPECTIVE AND THAT WILL LOCATE THE INTERSECTION OF THE LINE MN WITH THE PICTORIAL SURFACES.

LIST AN ARRAY OF VALUES FOR THE VARIABLE MY.

0,12
P +

HORIZON

24,12
+
V

5.5,MY
M

12,6

7,5

2.

16,4

12,3

N
16,3

Y

X

0,0

A 12,0

COMPUTER GRAPHICS

AAPRSB

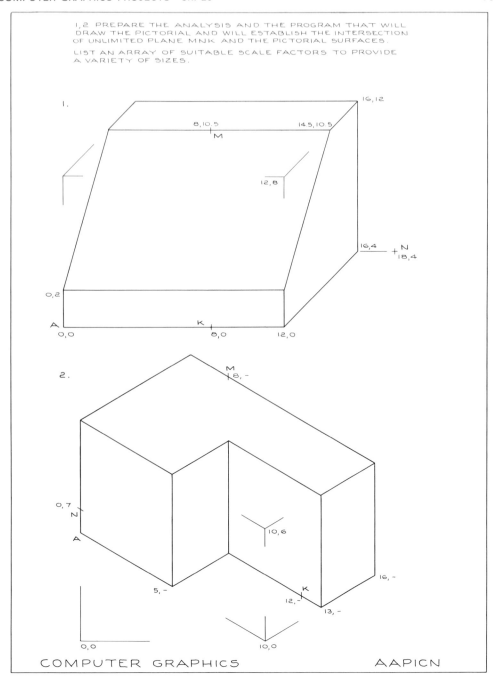

1,2 PREPARE THE ANALYSIS AND THE PROGRAM THAT WILL
DRAW THE PICTORIAL AND WILL ESTABLISH THE INTERSECTION
OF UNLIMITED PLANE MNK AND THE PICTORIAL SURFACES.

LIST AN ARRAY OF SUITABLE SCALE FACTORS TO PROVIDE
A VARIETY OF SIZES.

COMPUTER GRAPHICS   AAPICN

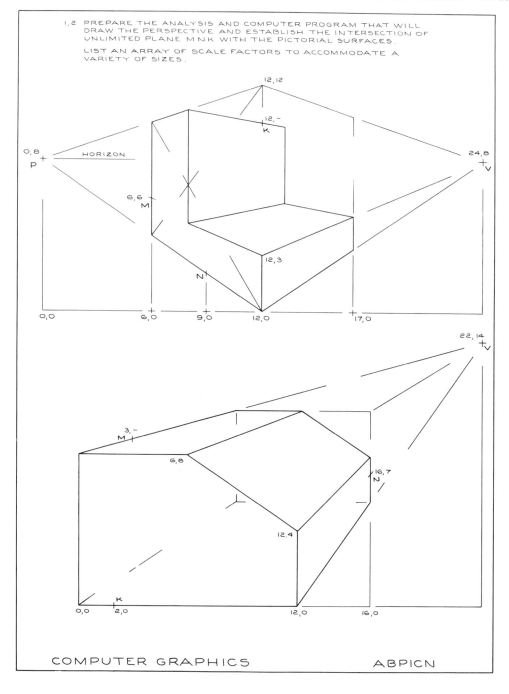

1,2 PREPARE THE ANALYSIS AND COMPUTER PROGRAM THAT WILL
     DRAW THE PERSPECTIVE AND ESTABLISH THE INTERSECTION OF
     UNLIMITED PLANE MNK WITH THE PICTORIAL SURFACES.

     LIST AN ARRAY OF SCALE FACTORS TO ACCOMMODATE A
     VARIETY OF SIZES.

COMPUTER GRAPHICS   ABPICN

Appendices

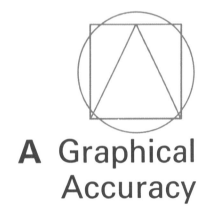

A Graphical Accuracy

The accuracy of a graphic solution may be affected by many factors, several of the more important of which are discussed in the following material. In the interest of economy of time the engineer should realize that graphical results can be no more significant than the original data. Then, too, if a problem is solved for which a large *safety factor* is introduced, the engineer, for economy of time and effort, needs only to maintain reasonable accuracy. With these considerations in mind, the engineer can select a scale for the work that will help produce the desired accuracy, for it is the scale of the drawing that is perhaps the one factor that most drastically affects the accuracy. For instance, an error of 0.5 mm in 200 mm is an error of only 0.25%, while a similar error in 25 mm represents an error of 2%. Thus it may be stated that the scale should be as large as can be conveniently handled with available drafting equipment.

A.1   Drawing Paper, Equipment, and Working Conditions

To assure best graphical accuracy, materials and working conditions should be carefully selected. A sharp, hard pencil should be used. A top-quality paper, film, or cloth that will retain its shape is essential; a metal surface and a scriber may be used for best results. The drawing paper should be firmly attached to the board and left in the original position until the job is completed. Ideally, the temperature and humidity should be kept constant, since changes in these conditions may stretch or shrink the working surface. In lieu of this often impractical requirement, it is best to complete a particular construction in as short and continuous a working time as possible.

Top-quality drafting instruments and other tools are needed to maintain accuracy. A small protractor should be used for only the roughest of work. If accurate angular measurements must be made, drafting machines or large vernier protractors are accurate tools for this purpose.

In the absence of such equipment an angular dimension may be laid off or measured by means of coordinates. For example, an angle of 38.5° can be established by using the tangent of the angle. The tangent of this angle is 0.7954, as found in a table of natural tangents. As large a triangle as practicable is used to lay off this angle by means of its tangent, Figure A.1. The adjacent side AC of the triangle is 100 units—any

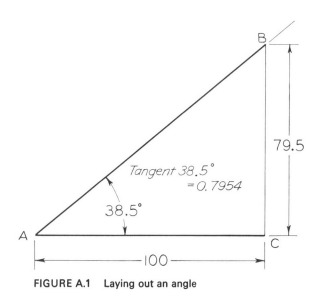

FIGURE A.1   Laying out an angle

convenient unit such as 100 mm may be used. The length of side CB is then 100 times the tangent of the desired angle, or 79.5 mm. Actually, with the unaided eye a scale may be read only to an accuracy of approximately 0.25 mm. Hence the side CB is set off with the scale to the nearest 0.25 mm, as indicated.

A.2   Drafting Practices

Although the student of descriptive geometry can be expected to exercise good drafting practices, it may be well to point out certain practices somewhat peculiar to this field. If, as in Figure A.2, only the top view of point C on line AB is given, the front view of this point cannot be precisely located by direct projection because the projection line from the top view of point C is too nearly parallel to a^Fb. Point C can be located more accurately by the use of an additional view, as shown in Figure A.2(a), or by means of revolution as illustrated in Figure A.2(b).

In Figure A.3 point K is in plane ABC. If either view of point K is given and it is required to locate the other view, the construction line

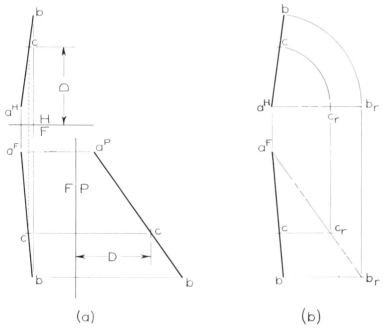

(a)   (b)

FIGURE A.2   Accurate location of point on line

through point K and in the plane should be carefully chosen. For
example, use of line 1-2 would probably introduce an error in a manner
similar to that of Figure A.2. It is better to draw the construction line
through an established point of the plane and at a fairly large angle with
the projection lines between the views. Thus use of construction line A-3
gives a more dependable result than use of line 1-2. In Figure A.4 the

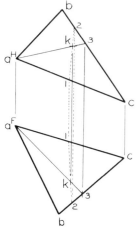

FIGURE A.3   Accurate location of point in plane

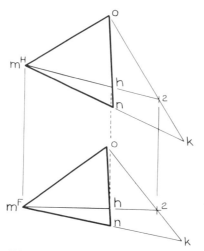

FIGURE A.4   Accurate location of line in plane

top view of the horizontal line MH in plane MNO can be more accurately established by extending the plane to a point K, as shown, than by using the plane as given.

Figure A.5 illustrates another case in which location of a view of a

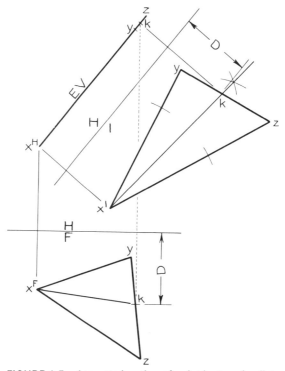

FIGURE A.5   Accurate location of point by transfer distance

point by projection to a given line is likely to be inaccurate. In this problem it is required to find the front and top views of the bisector of angle YXZ. The actual bisection is performed in the true-size auxiliary view, and the top view of point K on the bisector is readily found by projection. The front view of given line YZ is, however, nearly parallel to the vertical projection line from the top view of point K. Thus it is preferable, for accuracy, to locate the front view of point K by the transfer distance D available in the auxiliary view, rather than with the projection line.

When a line is established by locating two points of the line, the two points should not be too close together. It is physically impossible to draw a line precisely through two points, especially since *points* and *lines* in pencil or ink actually have diameters and widths in contrast to their theoretical definitions. Consequently, it must be assumed that when an attempt is made to draw a line through a point, a certain amount of error will always be present. As an illustration, let it be required to draw a line through points A and B of Figure A.6(a) and another line through points X and Y of Figure A.6(b). For simplicity

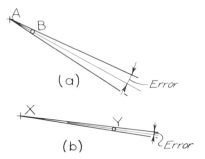

FIGURE A.6   Line located by two points

let it be assumed that the lines are actually drawn accurately through points A and X and that inaccuracy is confined to the drawing of the lines through points B and Y, the errors being represented by the radii of the small circles. The angular error in the direction of a line thus drawn is obviously much greater when the points are relatively close together, as in Figure A.6(a). When this situation occurs in the solution of a problem, a third point on the required line should be secured if possible.

In the measurement of grade, Figure A.7, the construction triangle should be as large as practicable. A scale error in measuring the distance labeled "rise" would be considerably less, percentage-wise, than a similar error in the measurement of distance Y.

Accumulative error is a frequent source of inaccuracy in drafting. If, for instance, dividers are used to step off a series of equal distances along a line, a seemingly insignificant error in the setting of the dividers may accumulate to an objectionable degree if the number of successive

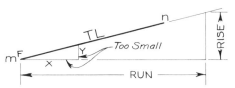

FIGURE A.7   Measurement of grade

distances is large. This may be avoided if the overall length of the line
is known or can be calculated. This length may then be set off to scale
and the divisions determined by the parallel-line method (Appendix B.1)
or by trial with dividers. A similar situation arises when a series of
dimensions are set off with a scale along a line. The scale should not be
shifted for each dimension but should be left in one position for all
settings, even though this may involve some mental arithmetic.

B Geometric Constructions

For quick reference a number of constructions frequently needed in the solution of descriptive geometry problems are given here. The methods shown are those which draftsmen prefer because of the rapidity and ease with which they are performed.

B.1 Parallel-Line Method of Division of Line into Segments

Given line AB is to be divided into three equal parts, Figure B.1. A scale is placed at any convenient angle with line AB and with its zero index at point A. Three equal divisions are marked and the third division point, point K, is connected by a construction line to point B. Lines are drawn through the intermediate points parallel to line BK. Their intersections with line AB establish the required divisions of line AB.

The construction is made somewhat more readily by first drawing a vertical (or horizontal) construction line through point B. The scale is

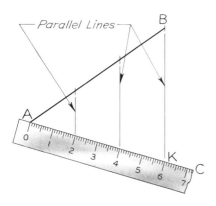

FIGURE B.1   Parallel-line method of division of line into segments

then placed at such an angle that the third division point K falls on this construction line. This makes the drawing of the intermediate construction lines more convenient.

B.2 Construction of a Square, Given the Center and One Corner

Point O is the center and point C the given corner of a square, Figure B.2. Line OC is one-half of a diagonal. The complete diagonal

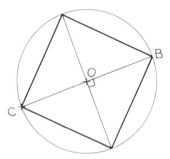

FIGURE B.2   Construction of a square, given the center and one corner

is distance CB, with distance OB made equal to distance OC. The other diagonal is perpendicular to diagonal OB, and its endpoints are located by divider distances equal to distance OC or by drawing a circle with center at point O and line OC as a radius.

B.3   Drawing a Circle Through Three Points

Points A, B, and C are given points on the required circle, Figure B.3. Lines AB and AC are chords of this circle. The perpendicular bisectors of

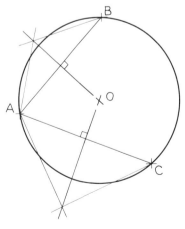

FIGURE B.3   Drawing a circle through three points

these chords are constructed by the drafter's method of equiangular lines, and they intersect at the center O of the required circle. The circle is then drawn with distance OA, OB, or OC as the radius.

B.4   Transferring a Polygon

While several methods, including triangulation, may be used to transfer a polygon from one position to another, the method shown in Figure B.4 is probably the most accurate. Let it be required to transfer the polygon from the given position at (a) to the position at (b) with edge 1-2 given at (b).

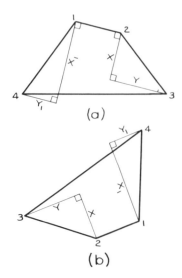

FIGURE B.4   Transferring a polygon

Construction lines X_1 and X are drawn perpendicular to lines 1-2 in (a) and (b) at points 1 and 2, respectively. In (a), construction lines Y and Y_1 are drawn parallel to edge 1-2 from the remaining corner points 3 and 4. The lengths of lines X and X_1 are then transferred from (a) to (b), and lines Y and Y_1 are drawn parallel to line 1-2 in (b). The lengths of lines Y and Y_1 are transferred to locate corner points 3 and 4 and the polygon is completed in (b).

B.5   Construction of Ellipse by Concentric Circle Method

The major axis XX_1 and the minor axis YY_1 are given, Figure B.5. Points on the ellipse may be located as follows: Two concentric circles are drawn with the given axes as diameters. Lines such as diameter 2-2

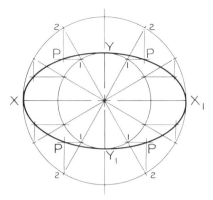

FIGURE B.5   Construction of ellipse by concentric circle method

are then drawn, intersecting the smaller circle at points 1. Lines 2-P are drawn from points 2 parallel to minor axis YY_1. Other lines 1-P are drawn from points 1 parallel to major axis XX_1. These lines intersect the first lines at points P, which are points on the required ellipse. This process is repeated for other diameters until a sufficient number of points is secured to locate a smooth, accurate curve.

B.6   Construction of Ellipse by Trammel Method

The methods illustrated in Figure B.6 are perhaps the most popular among draftsmen for the construction of an ellipse. Given the major and minor axes XX_1 and YY_1, a *short trammel*, Figure B.6(a), is constructed by marking along a straight edge of a strip of stiff paper the division points A, B, and C spaced and arranged as shown. If the trammel is placed at any angle with points C and B in contact with the major and minor axes, respectively, point A is a point on the ellipse. By repeated shifting of the trammel, the desired number of points may be found

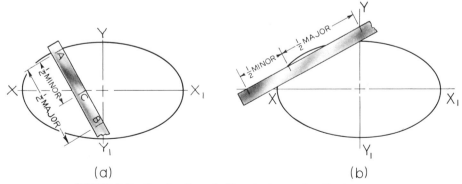

(a)   (b)

FIGURE B.6   Construction of ellipse by trammel method

rapidly and easily. For the *long trammel* construction, Figure B.6(b), the distances equal to the semimajor and semiminor axes do not overlap but are set off end to end along the straight edge of the strip of stiff paper. Place the trammel on the respective axes as illustrated. Move the trammel as necessary for a sufficient number of points to ensure a smooth and symmetrical ellipse. This version is recommended for small ellipses.

B.7 Drawing Parallel and Perpendicular Lines

To draw lines parallel and/or perpendicular to the given line AB, a triangle is placed as shown in combination with the T-square or another triangle, Figure B.7. Note that the side of the triangle *opposite* the

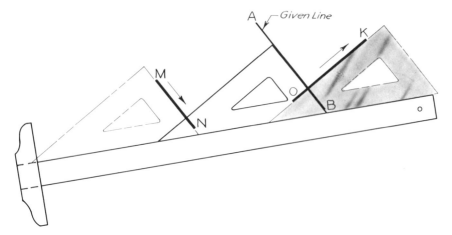

FIGURE B.7   Parallel and perpendicular lines

90° angle is placed in contact with the T-square. The two instruments are then moved as a unit until one leg of the triangle lies along the given line AB. With the T-square held firmly, the triangle may now be shifted along the T-square until the other leg is aligned with a given point such as point O. Line OK drawn along this leg is perpendicular to line AB.

If instead, the triangle is moved along the T-square until the leg which coincided with line AB is aligned with a given point such as point M, a line MN drawn along this leg is parallel to line AB.

B.8 Drawing a Line Tangent to a Circle from a Point Outside the Circle

The perpendicularity construction discussed in Appendix B.7 is frequently employed in the construction of lines tangent to circles. In the

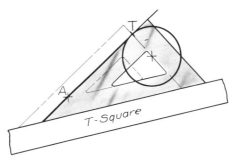

FIGURE B.8   Drawing a line tangent to a circle from a point outside the circle

illustration shown, Figure B.8, the given point is point A. *A line tangent to a circle is perpendicular to the radius drawn to the point of tangency.*

The triangle and T-square are arranged as shown, with one leg of the triangle aligned with point A and tangent to the circle. The triangle is then moved along the T-square until its other leg passes through the center of the given circle. This leg then represents the radius perpendicular to the required tangent line, and the tangency point T is thus located and is marked. With the triangle returned to its original position, the required tangent is drawn through point A to point T.

B.9 Approximate Location of a Tangent to a Noncircular Curve

A tangent at given point T, Figure B.9, may be located with sufficient accuracy for many practical purposes by assuming that a small portion

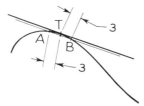

FIGURE B.9   Approximate location of a tangent to a noncircular curve

of the curve in the vicinity of point T is circular. Setting off small equal distances such as 3 mm on both sides of point T locates a chord AB to which the required tangent line is drawn parallel.

B.10 Drawing an Arc Tangent to Two Straight Lines

The given lines are AB and CD, Figure B.10, and the required tangent arc has a radius of 16 mm. Centered at any convenient points such as

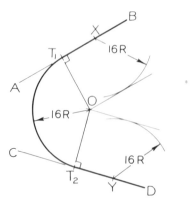

FIGURE B.10   Drawing an arc tangent to two straight lines

X and Y, one on each line, construction arcs are drawn with the required radius of 16 mm. By the method of Appendix B.7, construction lines are drawn parallel respectively to the given lines and tangent to the arcs. The construction lines intersect at point O, the required center. Again, as in Appendix B.7, perpendiculars are extended from point O to line AB, and to line CD to locate points of tangency T_1 and T_2. The required arc is then drawn between these points.

B.11   Approximate Rectification of an Arc

The length of arc XY, Figure B.11, may be approximated by drawing line XQ tangent to arc XY at point X and stepping off suitable small distances Y-1, 1-2, . . ., 6-7, with dividers along the arc from point Y

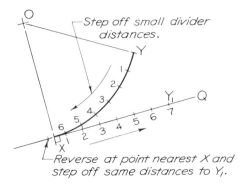

FIGURE B.11   Approximate rectification of an arc

toward point X. When near point X and without lifting the nearer divider point from the paper, the same number of distances is stepped off along line XQ to establish point Y_1. Length of line XY_1 will be very near the length of arc XY, particularly if angle XOY is small (less than 60°). Larger angles may be divided into equal parts for improved accuracy of this method.

The procedure may be reversed for establishing an arc approximately equal in length to the given line segment. The same general idea may be adapted to set off approximately equal lengths along arcs of different radii.

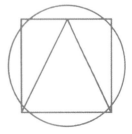

C Classification of Geometric Forms

Forms from plane and solid geometry are reviewed here under three classifications: plane figures and forms composed of plane figures, single-curved surfaces, and double curved and warped surfaces.

C.1   Plane Figures and Polyhedra

Polygons are plane figures bounded by straight sides, Figure C.1. A *triangle* is a three-sided polygon and a *quadrilateral* is a four-sided polygon. *Squares* and *rectangles* are included among the quadrilaterals. The square is also one of the regular polygons, a regular polygon being any polygon with equal sides and equal interior angles.

Polyhedra are three-dimensional forms whose *faces* or surfaces are polygons, the most common examples being *pyramids* and *prisms*. A pyramid is a polyhedron with a *base* having three or more sides, and a corresponding number of *lateral faces* which are triangular and have a common point, the *vertex* or *apex* of the pyramid. A pyramid is commonly classified by the position of its principal center line or *axis* with respect to its base, plus the shape of its base, as "right square pyramid" or "oblique rectangular pyramid." The term *tetrahedron* means "four-faced," and this form could be described as an "oblique (or right) triangular pyramid."

Prisms are composed of two parallel and congruent polygons as bases, joined by lateral faces that are parallelograms. If the lateral faces (and edges) are perpendicular to the planes of the bases, the prism is a *right* prism. A *cube* is a special example of a right square prism.

Parallelepiped is a term assigned to prisms having parallelograms for bases and thus having three sets of parallel edges.

C.2   Single-Curved Surfaces

Single-curved surfaces, Figure C.2, are surfaces that may be *generated* by a straight line, the *generatrix*, moving in contact with a curve, the

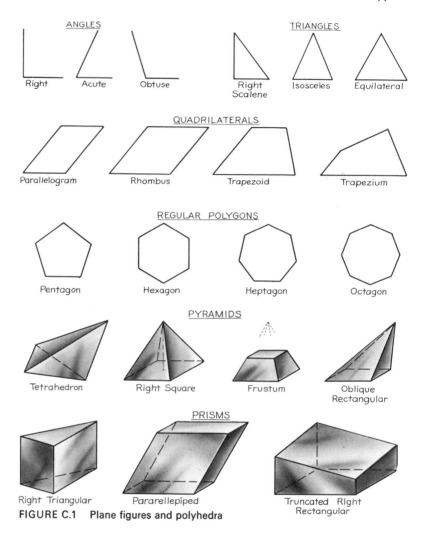

FIGURE C.1   **Plane figures and polyhedra**

directrix, in such manner that any two successive positions of the generatrix either intersect or are parallel. The various positions of the generatrix are called *elements* of the surface. A single-curved surface is developable. It is a *ruled* surface, a term that includes any surface which can be generated by a moving straight line. Other ruled surfaces are *warped surfaces* (Appendix C.3) and *planes*.

Cylinders. If the generatrix of a single-curved surface moves in such a manner that it always remains parallel to its original position, the surface is called *cylindrical*. A cylinder is formed when a cylindrical surface is intersected by two cutting planes forming bases. Cylinders are classified according to the position of their axes and the shape of their *right sections* as "right circular," "oblique elliptical," "oblique parabolic."

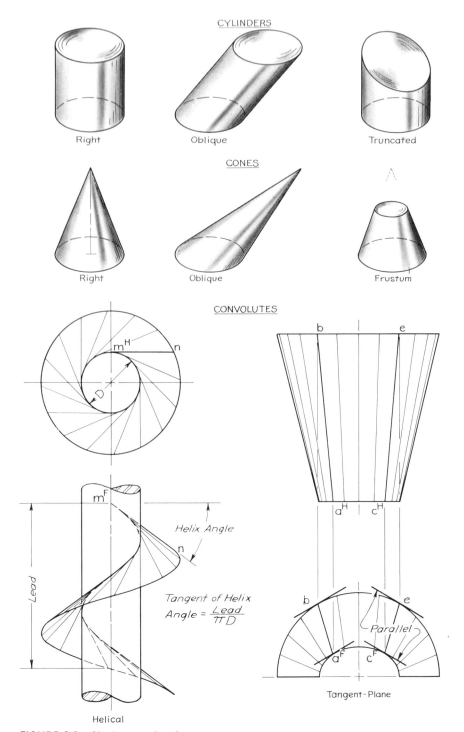

FIGURE C.2   Single-curved surfaces

Cones. If the generatrix of a single-curved surface moves so that it always passes through a fixed point, the vertex, the surface generated is *conical.* If the generatrix extends through the vertex, the conical surface consists of two portions called *nappes.* A cone consists of one nappe intersected by a cutting plane forming a base. Cones are classified in the same manner as cylinders.

Convolutes. If the directrix is a double-curved line (a curve whose points do not all lie in the same plane) and the generatrix is always tangent to the directrix, the surface generated is a *convolute.* It is a single-curved surface because any two consecutive elements (if reasonably close together) intersect and thus lie in the same plane.

The most common practical form of the convolute is the *helical convolute,* in which the directrix is a helix. In order to fulfill the requirement that the generatrix be tangent to the directrix, the elements of the helical convolute in the illustration shown are drawn at a true slope equal to the *helix angle.*

A convolute may also be generated as the *envelope* of a series of planes tangent to two curved lines that do not lie in the same plane. The line connecting the two points of tangency of any plane is an element of the surface. The *tangent-plane convolute* is represented by drawing a number of elements spaced around the two curves. A common form of this convolute occurs when two dissimilar curves, such as the circle and ellipse of the illustration in Figure C.2 are connected with a smooth surface. Elements are located as follows: Any point such as point B is selected on the circle. The tangent line at point B is a frontal line, and the line tangent to the ellipse at the other end of the element through point B is also a frontal line. Since these two frontal lines are in the same tangent plane, they are parallel. Hence the other end, point A, of the element through point B is located by drawing a line tangent to the ellipse (by eye) and parallel to the line tangent at point B. The point of tangency of this second line is the desired point A, and line AB is one element of the convolute.

C.3   Double-Curved and Warped Surfaces

A *double-curved surface* is a surface generated by a moving curved line and containing no straight-line elements, Figure C.3. Double-curved surfaces are not developable but, in practice, are approximately developed by substituting small segments of developable surfaces for the double-curved surface.

The most common form of the double-curved surface is the *sphere,* generated by revolving a circle about one of its diameters.

The *torus* is formed when a circle is revolved in a circular path about an axis outside the circle but in the plane of the circle.

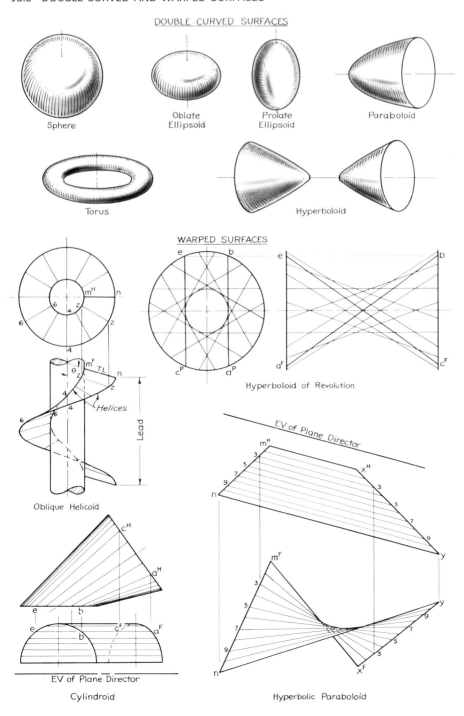

DOUBLE CURVED SURFACES

Sphere

Oblate
Ellipsoid

Prolate
Ellipsoid

Paraboloid

Torus

Hyperboloid

WARPED SURFACES

Hyperboloid of Revolution

Oblique Helicoid

Helices

TL

Lead

EV of Plane Director

Cylindroid

EV of Plane Director

Hyperbolic Paraboloid

FIGURE C.3   Double-curved and warped surfaces

Oblate and *prolate* ellipsoids are generated by revolving ellipses about their minor and major axes, respectively.

A parabola revolved about its axis generates a *paraboloid*, and a hyperbola revolved about its axis generates a *hyperboloid*.

Warped surfaces are ruled surfaces in which no two successive positions of the straight-line generatrix either intersect or are parallel. Like double-curved surfaces they are not developable but are sometimes approximated in practice by substituting for them small portions of developable surfaces.

A common form of warped surface is the *helicoid*, in which the generatrix moves along two concentric helices and remains at a constant angle with the axis of the helices. Theoretically the generatrix need not necessarily intersect the axis, but in practice it normally does. If the angle between the generatrix and the axis, angle θ in the illustration, is other than 90°, the helicoid is called an *oblique helicoid*. The sloping sides of V or Acme threads are examples of oblique helicoids. The right helicoid, used frequently in screw conveyors, is formed when the generatrix is perpendicular to the axis. An example is shown in Figure 17.17.

An *hyperboloid of revolution* is generated by revolving a straight-line generatrix about an axis that is neither parallel to nor intersects the generatrix. In the illustration the generatrix could be either line AB or line CE, since these lines are in the same relative positions with respect to the axis except that they slope in opposite directions. The hyperboloid thus has two sets of straight-line elements and is a *double-ruled* surface.

The *cylindroid* is a warped surface generated by a straight line moving parallel to a plane director and in contact with two curves. In practice the curves are usually similar but lie in nonparallel planes.

The *hyperbolic paraboloid* is commonly used as a smooth transition between walls of different slopes. It is generated by a straight line moving parallel to a plane director and in contact with two straight-line directors that are skew lines. In the illustration the straight-line directors are lines MN and XY.

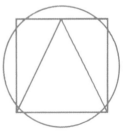

D Solutions of Self-Testing Problems

Solutions of the self-testing problems at the ends of chapters are reproduced on the following pages. The scale of the solutions is one-half that of the problem layouts.

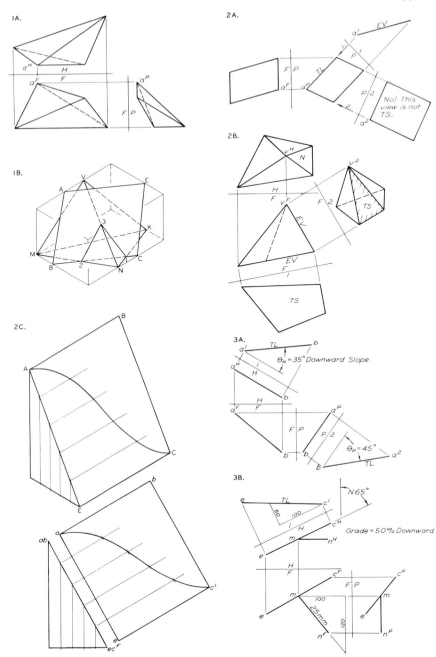

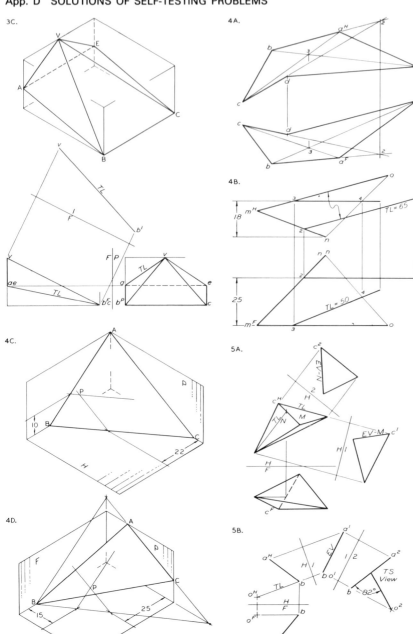

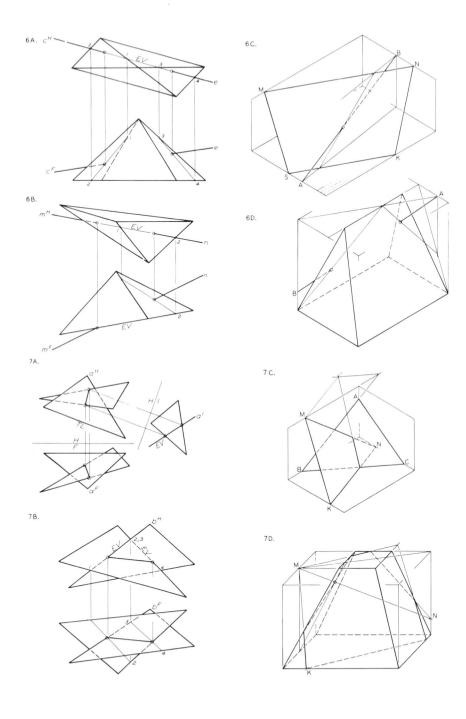

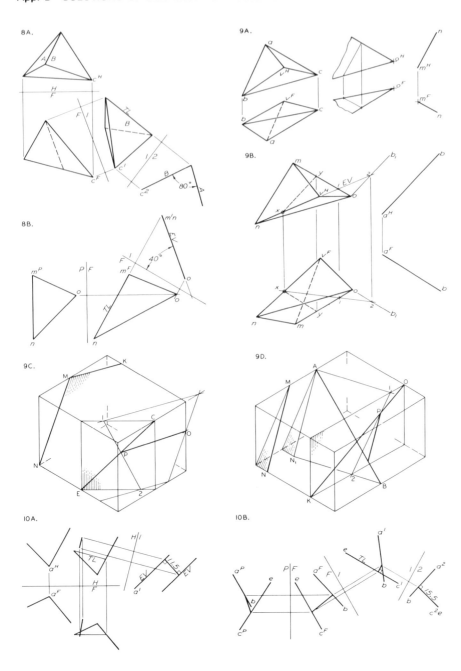

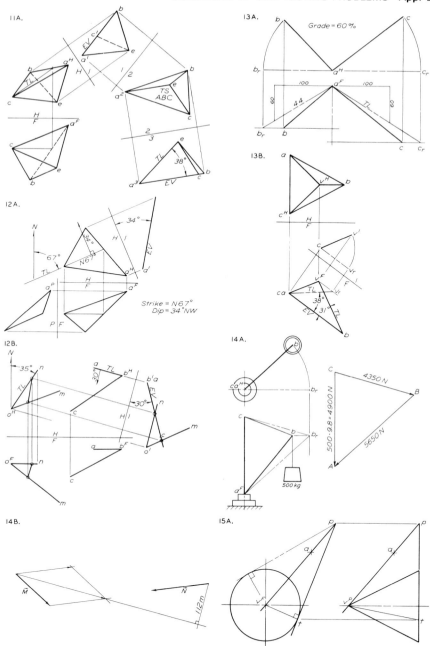

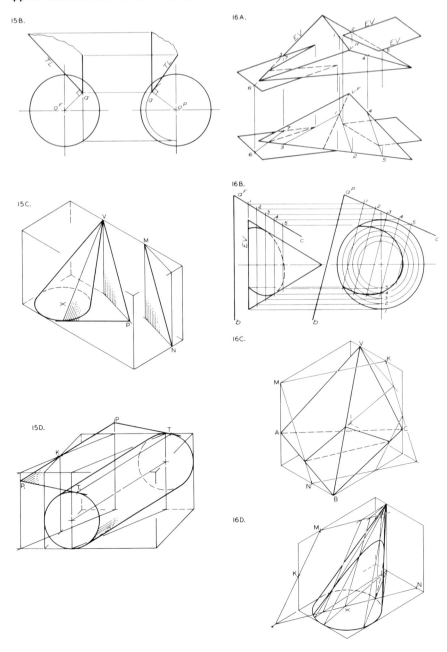

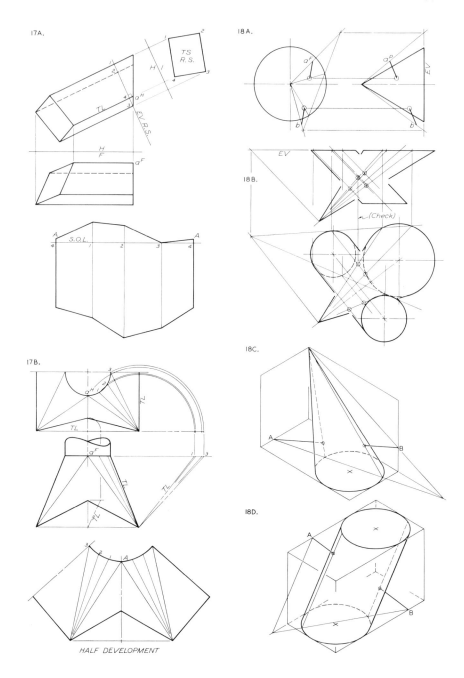

HALF DEVELOPMENT

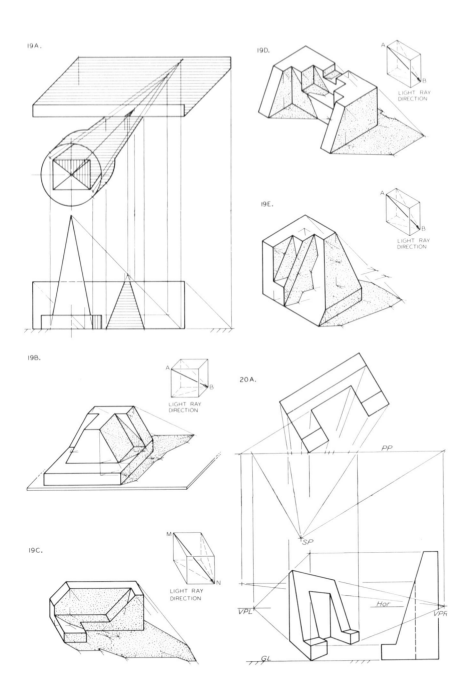

19A.

19D.

LIGHT RAY
DIRECTION

19E.

LIGHT RAY
DIRECTION

19B.

LIGHT RAY
DIRECTION

20A.

19C.

LIGHT RAY
DIRECTION

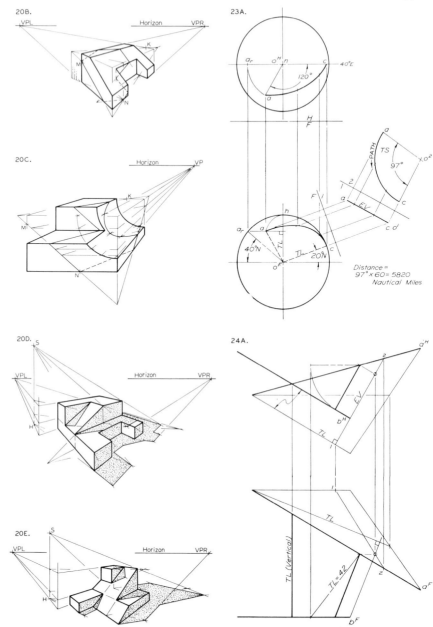

20B.

20C.

20D.

20E.

23A.

Distance =
97° × 60 = 5820
Nautical Miles

24A.

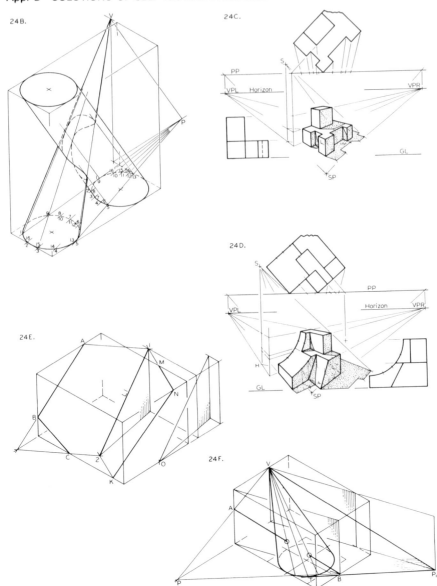

24B.

24C.

24D.

24E.

24F.

Index